A History of Platinum
and its
Allied Metals

A History of Platinum
and its
Allied Metals

Donald McDonald
and
Leslie B. Hunt

JOHNSON MATTHEY
Hatton Garden, London, EC1

FIRST PUBLISHED IN GREAT BRITAIN IN 1982

COPYRIGHT © JOHNSON MATTHEY

ISBN 0 905118 83 9

Distributed by

EUROPA PUBLICATIONS LIMITED
18 Bedford Square, London WC1B 3JN, England

by arrangement with

JOHNSON MATTHEY

This book has been typeset in Photon Bodoni and Baskerville
and printed and bound in England by
STAPLES PRINTERS ST ALBANS LIMITED
at The Priory Press, Hertfordshire

Foreword

BY THE RT. HON. LORD ROBENS OF WOLDINGHAM

Chairman of Johnson Matthey

Since man first discovered fire and invented the wheel there has been a continuous effort through the centuries to push back the frontiers of knowledge and this still goes on with marked success. From the very beginnings of modern industry the inherent characteristics of platinum – its high melting point combined with its exceptional resistance to corrosion – opened up fields for its application and gave it, in fact, a unique place in the history of research and invention.

The researches of Faraday, Davy, and of those who followed them in developing our knowledge of electricity show how greatly they relied upon platinum to provide a means of carrying, making and breaking a current, while the development of the electric telegraph, of the incandescent lamp and later of the thermionic valve all involved the use of platinum, as did the early internal combustion engines, first for igniter tubes and later for magneto contacts. Some of these pioneering applications have been superseded by other materials, but further and broader demands have arisen and become widely established. Today very large amounts of platinum are employed, for example, as a catalyst in the production of nitric acid and thence in the manufacture of huge quantities of fertilisers to increase the growing of food for an ever-increasing and hungry world population, while the reforming of crude petroleum with a platinum catalyst yields not only our high octane petrol but also a range of chemical intermediates required for the production of plastics, synthetic fibres, dyestuffs and pharmaceuticals. The manufacture of optical glass and of fibre-glass would not be possible without the use of platinum, while the invention of the fuel cell enabling space to be conquered and man's journey to the moon to be accomplished and now being used to generate electricity also depends upon the metal.

The usefulness of platinum's five allied metals – palladium, rhodium, iridium, ruthenium and osmium – was later in development than with platinum itself, but over the past fifty years or so they too have found growing applications in chemical, electrical and electronic engineering.

In all these and in many other developments the platinum metals have played a very special part in the lives of mankind, and in this book the authors, whose combined length of service with Johnson Matthey exceeds ninety years, have traced their history from the day the qualities of platinum were first discovered in 1750. It is a work of historical importance and will undoubtedly encourage the scientific and technical researchers to seek pastures new with this remarkable noble metal which has improved the quality of life in many diverse ways.

One of the latest achievements has been the use of a complex platinum compound to produce a drug known as Cisplatin which, used either as a single agent or more usually in combination therapy, is active against several types of human tumours. Since this discovery further research is leading to a range of anti-tumour drugs for which many cancer sufferers will be grateful.

But the end of the product line by the use of platinum has not yet been reached. Research goes on ceaselessly and new benefits to man will flow from the efforts of the scientists who already see so much yet to conquer in the development of platinum usage.

Yes, platinum is a noble metal and a noble task has been accomplished by the authors who have been associated with it for most of their working lives and have seen so much that is good for mankind emerge from its use.

Most unfortunately Mr Donald McDonald, who first compiled a history of platinum in 1960, died at the age of 92 while this book, much enlarged and brought up-to-date by Dr Leslie Hunt, was in the press.

Robens of Woldingham

Preface

Of all the chemical elements platinum has attracted the active interest of more distinguished scientists than any other since it was first brought to their attention in 1750. Its high melting point and the great difficulties encountered in rendering its invaluable properties available for practical use frustrated many able men over a long period of years and the story of their struggles is one of perseverance and ingenuity. Only after the discovery in 1803 and 1804 that native platinum did not consist of just one element but contained at least four others – palladium, rhodium, iridium and osmium – and their identification and separation was it possible to obtain platinum in a state of relatively high purity.

Then the well-known work of William Hyde Wollaston in devising a powder metallurgy process to bring it into malleable form, published only after his death in 1828, led to its first commercial applications.

From then onwards the refining and fabrication of platinum, as well as the study of its chemistry and metallurgy, were taken up more widely and a number of industrial concerns commenced operations in England, France and Germany, this leading on to the search for further sources of mineral. This slow but continuous development over some two hundred years was the subject of "A History of Platinum from the Earliest Times to the 1880's" written by my senior colleague Donald McDonald and published by Johnson Matthey in 1960. This was based not only upon his long experience in working with platinum but also on a great deal of painstaking research in the early scientific literature in five languages as well as in the company's archives. During the intervening twenty-two years a considerable amount of research has been carried out by historians of science on the lives and work of a number of the men associated with platinum, and with the enthusiastic agreement of Mr. McDonald, who sadly died while this book was in the press, I have undertaken the preparation of a completely revised and enlarged volume to record in greater detail the history of the platinum group of metals. Also, the scale of production and use of these metals has increased almost five-fold since 1960 so that they now play an even more important part in many phases of industry and in daily life and the period covered has therefore been extended by some seventy-five years or so to include the major discoveries of new sources of supply as well as their more recent applications.

The present volume retains the general structure of Donald McDonald's book but includes considerably more biographical material on those who feature in the story, with many of their portraits, while I have attempted to bring out both their personal motivation in taking up the work and their influence one upon another. In this I have also tried to follow the precept laid down by Dr. Johnson in another context. In introducing his "Life of Addison" he wrote:

> Not to name the school or the masters of men illustrious for literature is a kind of historical fraud by which honest fame is injuriously diminished.

My thanks are due to a great many people for making this book possible. First, to my former colleagues on the Board of Johnson Matthey for sponsoring its publication and then to Dr. W. A. Smeaton, a tower of strength and advice to me on the history of science over the past twenty years and more. To Professor Melvyn Usselman of the University of Western Ontario, and to Dr. John Chaldecott, formerly of the Science Museum, I owe a great debt for their continuing researches on the life and work of Wollaston, as I do to the late Dr. A. E. Wales for similar researches on Smithson Tennant, to Dr. Peter Collins for his study of Döbereiner and to Professor George Kauffman for his researches on Russian scientists. To Mr. Max Wood and Air Commodore F. J. P. Wood I am most grateful for information on their ancestor Charles Wood who first brought specimens of platinum to England, while to Dr. J. R. Fisher of the University of Liverpool I owe guidance on the early history of platinum mining in the Spanish colonies of South America. To M. Roget Christophe I am indebted for his research on the early history of platinum fabrication in France and to Dr. Hermann Renner for providing full details of the early history of DEGUSSA. To Dr. Peta Buchanan I am indebted for painstaking research on a number of genealogical problems and to Mrs. V. E. Harding, until recently librarian of Johnson Matthey, for procuring copies of publications difficult of access. I have also had the benefit of access to the archives of Johnson Matthey, including the long correspondence of George Matthey, and their early Board minutes.

For facilitating my own researches I have pleasure in acknowledging the helpfulness of the librarians and their staffs of The British Library, The Royal Society, the Royal Society of Chemistry, The Royal Institution, The Science Museum, The University of Cambridge, The University of Edinburgh, The Institution of Mining and Metallurgy, the Geological Society, The Victoria and Albert Museum, Guy's Hospital, The Wellcome Institute, The Society of Friends, the Bibliotèque Nationale, The Académie Royal des Sciences, The Bundesarchiv, The Württemburgische Landesbibliothek, the Pfälzische Landesbibliothek, The Bayerischen Staatsbibliothek and the Royal Swedish Academy of Sciences.

For answering my many questions on specific points or about individual scientists or for providing valuable information I am also most grateful to Dr. Robert Anderson, Dr. Robert Bud, Dr. Brian Bowers, Dr. Jill Austin and Mr. Peter Mann, all of the Science Museum, to Dr. Warwick Bray of the Institute of Archaeology, University of London, to Mrs. Shirley Bury of the Victoria and Albert Museum, Mr. Henry Wollaston, Professor Maurice Crosland, Professor Cyril Stanley Smith, Mr. Peter Embrey of the Natural History Museum, Dr. Lindsey Hughes of the University of Reading, Mrs. Una des Fontaines, Mr. Robert Copeland, Dr. Ian Fraser, Miss Ann Petrie, Mrs. Clare Le Corbeiller of the Metropolitan Museum of Art, New York, Mr. James V. Crawford, President of U.O.P., M. Oliver Soulet of Paris, Mr. K. W. Maxwell of Rustenburg Platinum Mines, Professor Arne Fredga of the Swedish Academy of Sciences, Mr. Niels Gram and Dr. Frode Galsbøl of Copenhagen, Herr U. Kunz of DEGUSSA, M. Jean Pierre Savard of Paris, Dr. Christoph Raub of Schwäbisch Gmünd, Dr. F. W. J. McCosh, Mr. Robert Barker, Dr. Janet Cutler, Mr. Paul Weindling, Mr. P. A. Lovett and Mr. S. T. Payne of Inco Europe, Mr. Graham Dyer of the Royal Mint, Mr. Vincent Newman, Miss Barbara Pyrah, Herr Willy Fuchs of Frankfurt, Professor Gian Maria Gros-Pietro and Dr. Donna d'Oldenico of Turin, Dr. Hans Prescher of Dresden, Madame Tamara Preaud of the Manufacture nationale de Sèves, Dr. J. G. Bruijn and Dr. E. Bock of the Netherlands and Professor Francisico Aragon de la Cruz of Madrid.

Lastly it is a pleasure to acknowledge the generous help received from my colleagues on *Platinum Metals Review,* my secretary Mrs. Julie Adams who nobly struggled with the typing of the manuscript and its numerous revisions and additions, Mr. I. E. Cottington for reading the proofs, suggesting many improvements and preparing the name index, Mrs. Susan Ashton for much library work involving wrestling with innumerable incorrect references in the literature and for preparing the subject index, Miss Pavla Knopova for searching for patents and for translation from Russian. Finally my thanks also go to Mr. H. D. Smith, Managing Director of Staples Printers St. Albans Limited, for his lively personal interest in the typography, design and production of the book.

Hatton Garden   LESLIE B. HUNT
London
June 1982

Contents

Iulius Cæsar **SCALIGER** . *a great Reſtorer of Learninge. He died at Agen in France . Anº. Dñi. 1558. aged 75 yeares.* W.M.ſculp:

Julius Caesar Scaliger
1484–1558

The first known literary reference to platinum is to be found in the *Exotericarum Exercitationum* compiled by Scaliger and published in Paris in 1557. This was largely a polemical work criticising the great work of another Italian scholar, Hieronimo Cardan, who had defined a metal as "a substance that can be melted and which hardens on cooling"

1

The Beginning of the Story

"A substance which it has not hitherto been possible to melt by fire or by any of the Spanish arts."

JULIUS CAESAR SCALIGER

Until the present century the only form in which naturally occurring platinum was available to commerce was that of the water-borne grains found in alluvial gravels, usually in conjunction with gold. These grains were metallic and contained roughly 50 to 80 per cent of platinum, the remainder consisting of the other members of the platinum group, ruthenium, rhodium, palladium, osmium and iridium in varying amounts, together with a certain amount of iron, copper and other base metals. In both cases samples were contaminated with heavy black mineral sands such as chromite and magnetite, from which it was difficult to effect complete separation. There were also some spangles of gold and in the early specimens always some globules of mercury. This was the material which confronted the first craftsmen and scientists who came into contact with it and it presented them with many difficulties. But it soon became evident that the presence of the mercury was due to the attempts made at the place of origin to remove the gold by means of amalgamation. The earliest attempts to work the metal were confined to the selection of single grains and these were then hammered out or soldered together. All the early attempts to melt the material in bulk failed, and much patient research over many years at the hands of a succession of brilliant scientists was necessary before a reasonably malleable metal was produced.

The Egyptian Find

It is very doubtful whether platinum was recognised as a separate body in the early civilisations. Occasionally traces of it have been found among artifacts from ancient Egypt, the best known example being a small strip of native platinum set on the surface of a box among many hieroglyphic inscriptions, made of gold on one side and of silver on the other. This had originally come from Thebes and is dated to the seventh century BC. In 1900 it was submitted by the Keeper of Egyptian Antiquities in the Louvre for examination by the French scientist Marcelin Berthelot, who found that one of the characters on the side

1

having hieroglyphics in silver differed considerably from the others. His careful series of tests showed that it was almost insoluble in aqua regia – "a very singular resistance to attack" as he described it in the course of a paper to the Académie des Sciences (1), "surpassing that of gold or of pure platinum". He considered that it must be a complex alloy containing several of the metals of the platinum group. Berthelot also found no evidence that the Egyptian craftsman had noticed any distinction between this particular piece of metal and the silver he had used for the other characters. It had been hammered out in the same way, and its occurrence among the other was probably quite fortuitous – it had simply been mistaken for silver.

Since Berthelot's time many examples of platinum metal inclusions in gold artifacts from ancient times have been recorded, and it has been established that they occur quite commonly in alluvial gold, most often as complex osmium-iridium-ruthenium-platinum alloys or compounds. Their presence and composition have been well reviewed by J. M. Ogden (2).

The Classical Writers

Several attempts have been made over the years to show that references to platinum can be found in ancient Greek and Roman literary works. The earliest of these was a dissertation by an Italian scholar, Don Angelo Maria Cortinovis, with the title "Della Platina Conosciuta degli Antichi" written in 1778 and published in 1790 (3). In this he sought to prove that the name "electrum", the natural alloy of gold and silver much used in antiquity, also referred to platinum.

In 1845 Professor J. S. C. Schweigger (1779–1857) of Halle, the inventor of the galvanometer, wrote a long paper after his retirement "On Platinum, Old and New" (4) in which he maintained that the suggestion of Cortinovis could not be supported, but that Pliny's description of "plumbum album", white lead, of the same weight as gold, in the "Natural History" Book XXXIV, chapter 47, generally accepted as referring to tin, should be interpreted as platinum. He went on further to suggest at some length that a reference in the Guide to Greece written by the traveller and geographer Pausanius in the second century AD to a kind of electrum "found as a natural product in the sands of the Eridamos (the River Po) which is extremely rare and valuable" must necessarily be construed as meaning platinum. Schweigger also drew upon Homer, who described the costly armour of Agamemnon in the Iliad (Book XI) being made of stripes, "twelve of shining gold and twenty of tin". He considered the latter as a poor safeguard against the thrust of a spear and that they could well have been platinum.

Five years later a French chemist named Paravey, in a letter to the Académie des Sciences (5), again argued for Pliny's "plumbum album", found in the gold mines of Lusitania and Galicia in Spain, being platinum. Following up this theme many years afterwards a Spanish physicist, Piña de Rubies (6) again

2

advanced the view that platinum was in fact the "plumbum album" of Pliny and that it had been discovered in Spain before the first century AD.

A more recent suggestion that Pliny's "adamas", described in his Book XXXVII, Chapter 15, was in fact platinum has been made by Ogden (2) and has a little more feasibility. The quotation reads in translation:

> "Adamas was the name given to the knot of gold found very occasionally in mines in association with gold. . . . The hardness of adamas is indescribable, and so too that property whereby it conquers fire." (7)

But the last word on this rather dubious argument about the knowledge of platinum in antiquity is probably that of Professor J. F. Healy who writes in the chapter on native metals in his recent study of the subject:

> "The second group, platinum, iridium and osmium, plays no part in Greek or Roman mining or metallurgy." (8)

Early European Literature

It is significant that there is no reference whatever in the numerous books on metallurgy, assaying and chemistry written by knowledgable and experienced observers that began to appear in Europe in the sixteenth century when the invention of printing had made these possible. The well known works of Biringuccio, Agricola, Ercker, Glauber, Kunckel, and Libavius, for example, show no trace of anything that might be identified as platinum. The only possible reference is contained in the voluminous works of a Bohemian priest and historian, Bohuslav Balbinus (1621–1688), who referred briefly in his "Miscellanea Historica Regni Bohemia," published in Prague in 1679 (9), to

> "Aurum album (argentum esse jurares, nisi pondus et quaedam tamen fulvedo per metallum fusa aliud svaderent) album aurum, inquam, in montibus effosum, vidi non semel." ("White gold (which one would swear to be silver except that its weight and a certain yellowish tinge pervading the metal persuaded otherwise), white gold, I say, which is dug out of the mountains, I have seen more than once.")

This passage was referred to by Michael Bernhard Valentini (1657–1729), a physician and scientist of Giessen in Germany, who was elected to the Royal Society in 1717. In his "Historia Literaria Academie Naturae Curiosorum" published in 1708 he considerably amplified the brief statement made by Balbinus:

> "Who would have believed that white gold deprived of all its colour had been discovered, but the authority for this is the most honourable in reputation the Jesuit Father Balbinus who states that one would swear it was silver if it was not for the familiar properties found in gold such as its weight, ductility, resistance to fire and to nitric acid and its solubility in aqua regia." (10)

A letter about a specimen of white gold from the Arch Duke Ferdinand of Bohemia, written in 1560 to the Master of the Mint in Prague, is quoted by Count Kaspar Maria Sternberg (1761–1838) in his two-volume work on the

3

EXERCITATIO LXXXVIII.

Quæ ad Metalla.

METALLVM, inquis, est quod liquescere potest : & cũ redit, durum manet. Oblito tibi magni illius uicarij metallorum hæc exciderunt. Aurum metallorum rex est : Argentum uiuum tyrannus. Quod ex tua definitione metallum non erit. Est autem tyrannus, quia cætera omnia absumit. Plinius in uicesimo nono ita scribit: Auro liquescenti si Gallinæ carnes admisceantur, ab illis rapi. Itaque auri uenenum esse. Quòd si uerum est : sanè sic præsentius atque commodius adipiscemur aurum esculentum, quàm ex tua inani indicatione, aurum potabile. Præterea scito, in Funduribus, qui tractus est inter Mexicum, & Dariem, fodinas esse orichalci: quòd nullo igni, nullis Hispanicis artibus hactenus liquescere potuit. Adhæc non omnibus metallis uerbum, liquescere, uidemus conuenire.

The passage from Scaligers' commentary dealing with the nature of metals and rebutting Cardan's assertion, concluding that there are mines in South America that contain a "substance that it has not hitherto been possible to melt by fire or by any of the Spanish arts"

history of mining in Bohemia, published in the last years of his life (11). He suggests that the remarks of Balbinus might have referred to this specimen, but says that gold was never found in the mountains of Bohemia.

Rumours from the New World

While therefore very little can really be said of platinum in the Old World it seems that rumours from the New began to reach Europe during the sixteenth century. The earliest known reference having some solid foundation occurs in the writings of Julius Caesar Scaliger or Della Scala, a well known Italian scholar and poet. In 1551 the Italian mathematician and philosopher Hieronimo Cardan had published his great work "De Subtilitate Rerum", combining the soundest physical knowledge of his time with some advanced speculation. His contemporary Scaliger then had the highest scientific and literary reputation of any man in Europe and had written a number of commentaries, including one on Cardan's work. This, his "Exercitationes" on the "De Subtilitate", was published in Paris in 1557 and shows both an encyclopedic knowledge and a vigorous polemical style (12).

It is in Chapter 88 that the historic reference to platinum is to be found. Cardan had defined a metal as "a substance that can be melted and which hardens on cooling", and Scaliger was at pains to fault him as far as he could, first by reference to mercury and then by citing the unmeltable metal – undoubtedly platinum – of which he had heard from South America. The passage is illustrated here and the essential parts, the first and last sentences, read in translation:

"Metal you say is something that can be melted but when it cools remains hard. . . . Moreover, I know that in Honduras, a district between Mexico and Darien, there are mines containing a substance which it has not hitherto been possible to melt by fire or by any of the Spanish arts. Thus we see that the word melt cannot be applied to all metals."

This quotation can reasonably be held to apply to platinum, although the metal has never been found occurring naturally in the areas mentioned by Scaliger. It is, however, quite likely that platinum objects were transported about Spanish America and might well have been encountered in the hands of people living in the country between Mexico and the Isthmus of Darien. On the other hand, there was at the time of Scaliger's writing a great deal of uncertainty in Europe about the geography of these regions.

The next example of this sort of reference from South America occurs in a chapter on gold and silver in "Historia Natural y Moral De Las Indias" published in Seville in 1590 by the Jesuit priest José de Acosta (1539–1600), one of a number of Spanish missionaries sent to the colonies in 1571. This reads in a contemporary translation:

"Yea there is another kinde which the Indians call *papas de plata* and sometimes they find pieces very fine and pure, like to small round rootes, the which is rare in silver but usual in gold."

There is a tantalising reference to this in Robert Boyle's "The Sceptical Chymist" published in London in 1661. After quoting Acosta's paragraph he adds:

"I myself have seen a lump of Oar not long since digged up, in whose stony part there grew, almost like trees, divers parcels though not of gold, yet of (what perhaps Mineralists will more wonder at) another metal which seemed to be very pure or unmixt with any heterogeneous substances, and were some of them as big as my finger, if not bigger. But upon observations of this kind, though perhaps I could, yet I must not at present dwell any longer." (16)

A further and equally vague mention is made by Alonso Barba, a Spanish priest who, like Acosta, spent many years in the Spanish Colonies. He lived in the famous silver mining district of Potosi and wrote his "Arte de los Metales", published in Madrid in 1640, from long observation and experience. He refers to:

"Chumpi, called thus on account of its grey colour is a stone of the nature of Emery with some Iron. It shines rather darkly; and the treatment thereof is difficult, because

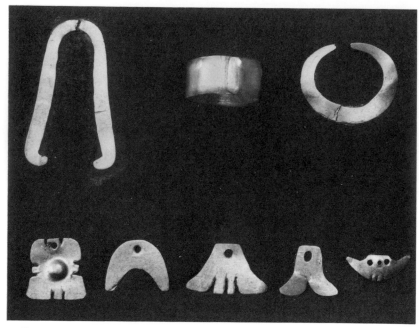

Specimens of platinum alloy jewellery found in the Esmeraldas region of Ecuador and now in the National Museum in Copenhagen. These date from several centuries before the Spanish conquest of South America

it offers resistance to heat. It is found with Black Sulphide and Ruby-Silver in Potosi, Chocaya and other places." (15)

The alchemist Johann Joachim Becher (1635–1682) in his "Physica Subterranea" published in Frankfurt in 1669 referred to the adulteration of gold with "smiridis hispanica", and this was later commented upon by William Lewis in his translation of "The Chemical Works of Caspar Neumann" published in London in 1759. In the course of considerable additions to the original work Lewis wrote in a footnote about Platina:

"These properties, together with the place where it is found, and the prohibition said to be laid upon its exploitation by the King of Spain, afford sufficient grounds to presume that the Smiris Hispanica of the alchemists, employed for augmenting gold, was no other than this Platina or some mineral containing it; more especially as Becher expressly declares that this augmentation was really an abuse; that the Gold so augmented was pale and brittle; and that though it stood all the established tests of perfect Gold, yet it would not bear amalgamation with quicksilver, the Mercury retaining the Gold, and throwing out the Smiris in form of a reddish powder. Platina mixed with Gold is thrown out in the same manner; though it is not easy by this method to obtain a perfect separation." (16)

This smiris may well have been a mixture of crude platinum and magnetic sand, but by the time that Lewis made these observations more was becoming

6

known of the occurrence of a somewhat unusual metal in the Spanish colonies. Long before this, however, the native Indians of the north-western part of South America had interested themselves in the new mineral and taken advantage of its properties.

The Pre-Columbian Indians of Ecuador and Colombia

This remarkable case of the successful exploitation of platinum in an isolated area in the New World of the Americas many hundreds of years before their discovery came to light in the nineteenth century after Ecuador had gained its independence from Spanish colonial rule. Following some years of internal dissension a more able and energetic president, Gabriel Moreno, came to power in 1865 and sought the help of a number of Jesuit scientists and teachers to aid in the development of his country. Among them was Theodor Wolf, (1841–1924) a German geologist engaged to report on the mineral resources of the country who remained to become Professor of Geology at Quito. While exploring the coastal region in north-western Ecuador, in the department of Esmeraldas and at a place called Lagarto, Wolf unearthed a number of small worked trinkets of gold and of platinum that had apparently been washed out of native burial mounds by tidal fluctuations. Among these was a tiny ingot of platinum which he analysed; it contained 84.95 per cent platinum, 4.64 per cent palladium, rhodium and iridium, 6.94 per cent of iron and a little over one per cent of copper. Other specimens were of gold alloys containing only modest amounts of platinum.

Wolf, whose findings were published in 1879 in the last of a three-volume work describing his journeys (17), realised that the territory of Ecuador had been in the possession of the Incas for only half a century before the arrival of the Spaniards and he concluded:

"A race that knew how to produce the alloys that I have just enumerated can certainly not be called uncultured and in metallurgy at least it was not inferior to the Incas as long as we can assume that the old Lagarto Indians carried out this industry for themselves and did not acquire these objects by trade. The presence of the platinum, pure and alloyed with gold, is a strong argument in support of the first supposition, that is to say of a native industry."

An expedition to the province of Esmeraldas in Ecuador in 1907, led by Professor Marshall Saville of New York, explored a number of large burial mounds on the small island of La Tolita at the mouth of the Santiago River. Here they excavated a large quantity of very small pieces of jewellery, rings, pendants, minute masks of filigree work, nose, ear and lip ornaments and so on, some of these being made in platinum or of platinum and gold combined together. Professor Saville reported that:

"The use of platinum is a unique feature of the section of South America extending from this province northward into the region of the Choco River. ... In this particular phase of ancient art the Esmeraldas people seem to stand alone." (18)

7

Paul Bergsøe
1872–1963

After retirement from directing his secondary tin smelter in Copenhagen Dr Paul Bergsøe embarked on a study of the many small platinum objects from Ecuador that had been found during panning for alluvial gold. He not only carried out analyses of these ornaments but by careful examination of pieces of metal in various stages successfully established their method of working as an ingenious process of sintering in the presence of a liquid phase. Bergsøe was also an active lecturer and broadcaster on scientific subjects and in 1959 was presented with the Oersted Gold Medal of the Society for Natural Sciences by the King of Denmark

Further examples of the natives' skill in fabricating both gold and platinum were found on the edge of an artificial mound on the island of La Tolita in 1912 and were the subject of a later report by William Curtis Farabee (1865–1925), the distinguished anthropologist of the University of Pennsylvania Museum where the collection is preserved. (19)

He wrote, after describing the enormous number of ornaments of an immense variety of forms and designs found there:

"The native Indian workers of Esmeraldas were metallurgists of marked ability; they were the only people who manufactured platinum jewellery. In our collection will be seen objects of pure platinum, objects with a platinum background set with tiny balls of gold used to form a border, and objects with one side platinum and the other gold."

An Early Use of Powder Metallurgy

Similar objects found on the coast of Esmeraldas, mostly at La Tolita were actively studied by Dr. Paul Bergsøe, the founder of a secondary tin smelter in Copenhagen, after he had retired from the control of his works in the early 1930s. Bergsøe not only carried out a number of analyses of many small objects, finding platinum contents ranging from 26 to 72 per cent with small amounts of iridium and the other platinum metals, some gold and a little silver, but gave a great deal of attention to the likely method of their manufacture.

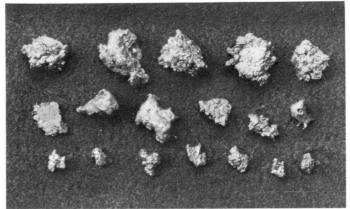

Some of the tiny fragments of platinum which gave Bergsøe the clue to the method employed by the native Indians. They were first coated with gold dust and then heated by means of a blowpipe on pieces of charcoal; the molten gold then served to sinter them so that they could be forged

Photograph by courtesy of the Danish National Museum, Copenhagen

The clue to his discovery he found in a number of very small grains of platinum that were mixed with gold dust. These, illustrated here, consisted of flat pieces weighing from 1 to 20 grams and were clearly in a half-finished stage of working. Bergsøe came to the conclusion that the native Indians had used a quite sophisticated technique of powder metallurgy – sintering in the presence of a liquid phase – a technique which he admits greatly astonished him. He wrote a preliminary account of his findings in 1935 in a letter to *Nature* (20), and later published a monograph in Copenhagen giving full details and many illustrations (21). His collection of specimens, including the part-finished grains, is now in the Danish National Museum. He described this technique as he envisaged it:

> "The small grains of platinum were mixed with a little gold dust and small portions placed upon a piece of wood charcoal; when the gold runs it will coat the grains of platinum with gold . . . the grains are simply 'soldered together'. If the piece is now further heated by means of the blowpipe, . . . a portion of the fused gold permeates the platinum and simultaneously a little of the latter is dissolved in the molten gold.
>
> This mixture of gold and platinum can now withstand a light blow of the hammer, especially when hot. By alternately forging and heating it is possible gradually to build up a homogeneous mixture."

Some of the pieces of jewellery examined by Bergsøe showed evidence of platinum cladding over gold, either on one side or on both. Again, he considered that a small piece of the sintered platinum alloy would have been placed on top of a bead of gold, heated and then hammered out together with occasional annealings. Useful reviews of this work on the Pre-Columbian metallurgy of platinum were published by Rivet and one of his colleagues in the 1940s. (22)

More recent studies have confirmed Bergsøe's concepts and the high degree of craftsmanship of the natives of this small area in South America in the working of crude platinum. In a paper by D. A. Scott and W. Bray recently published (23) the authors report on the use of modern metallographic and analytical techniques to examine specimens from both the National Museum of

9

Two platinum nose ornaments made by the Indians of Colombia. That on the left is in a native iron–platinum alloy with small platinum inclusions visible on the surface; on the right is a nose ring made in a natural copper–platinum alloy with small inclusions of osmiridium

Photograph by courtesy of
The Museo del Oro, Bogotá

Denmark and the Museo del Oro in Bogotá, Colombia. From the latter source two nose rings, illustrated here, were shown to consist of native iron-platinum alloy (left) and of a natural copper-iron-platinum alloy with small inclusions of osmiridium. Photomicrographs of sections of some of Bergsøe's specimens, both starting materials and finished objects, clearly showed the presence of sintering and the dispersion of platinum particles in a gold matrix.

Unfortunately few of the platinum finds from Ecuador or neighbouring Colombia have an established archaeological context, most of them having been unearthed by treasure hunters and tomb robbers. There is also evidence that a few objects of platinum were exported from the manufacturing region and made their way either southwards along the coast or inland to the Andean highlands. Their dating is therefore difficult to establish, but the site at La Tolita from which most of the platinum items in museums have come has been the subject of radio-carbon determinations ranging from the first to the fourth centuries AD while it had probably been abandoned early in the ninth century at latest although jewellery continued to be made in this area up to the time of the conquest, presumably including objects of platinum (23). But it was many hundreds of years before the Spanish settlers in South America re-discovered the source of platinum and even longer before European scientists succeeded in rendering it malleable and useful.

10

References for Chapter 1

1　M. Berthelot, *Compt. rend.*, 1901, **132**, 729–734; *Ann. Chim.*, 1901, **23**, 5–32

2　J. M. Ogden, *J. Hist. Met. Soc.*, 1977, **11**, (2), 53–71

3　A. M. Cortinovis, Opuscoli Scelti sulle Scienza e sulle Arte, 1790, Tom XIII, (4), 217–242

4　J. S. C. Schweigger, *J. Prakt. Chem.*, 1845, **34**, 385–420

5　C. de Paravey, *Compt. rend.*, 1850, **3 1**, 179

6　P. de Rubies, *An. Soc. Espanola Fisica y Quimica*, 1915, 420–433

7　Pliny, Natural History, Book XXXVII, Chap. XV, 55–60; Loeb edition, 207–209

8　J. F. Healy, Mining and Metallurgy in the Greek and Roman World, London, 1978, 35

9　Bohuslav Balbinus, Miscellanea Historica Regni Bohemiae, Prague, 1679, Book I, Chap. XIV, 40

10　M. B. Valentini, Historia Literarca S.R.I. Academiae Naturae Curiosorum, Giessen, 1708, 12

11　Graf K. Sternberg, Umrisse einer Geschichte der Böhmischem Bergwerke, Prague, 1837, Vol. I, Part 2, 38–41

12　J. C. Scaliger, Exotericarum exercitationum liber quintus decimus de Subtilitate ad Hieronymum Cardanum, Paris, 1557, 134–135

13　J. de Acosta, Historia Natural y Moral De Las Indias, Seville, 1590, Book 4, Chap. 4, 201; The Naturall and Morall Historie of the East and West Indies, Trans. E. Grimston, London 1604, 212

14　Robert Boyle, The Sceptical Chymist, London, 1661, 371–373; Everyman edition, London, 1964, 198

15　Alonso Barba, El Arte de los Metales, Madrid, 1640, Chap. XIII, 12

16　W. Lewis, The Chemical Works of Caspar Neumann, London, 1759, 43

17　T. Wolf, Viajes Cientificos por la Republica del Ecuador, Vol III, Memoria Sobre la Geografia y Geologia de la Provincia de Esmeraldas, econ una Carta Geografica, Guayaquil, 1879

18　M. H. Saville, Verhandlungen XVI Internat Amerikanisten Kongres, Vienna, 1908, **2**, 331–345

19　W. C. Farabee, *Museum Journal, Univ. Pennsylvania*, 1921, **12**, 43–52

20　P. Bergsøe, *Nature*, 1936, **137**, (1), 29

21　P. Bergsøe, The Metallurgy and Technology of Gold and Platinum among the Pre-Columbian Indians', Ingenioervidensk. Skr. (A44), Copenhagen, 1937

22　P. Rivet, *Revista Inst. Etnologico Nacional*, 1943–4, 1, 39–45; P. Rivet and H. Arsandaux, Trav. et. Mem. Inst. d'Ethnologie, 1946, 39, 113–115

23　D. A. Scott and W. Bray, *Platinum Metals Rev.*, 1980, **24**, (4), 147–157

Don Antonio de Ulloa
1716–1795

The Spanish naval officer, astronomer and mathematician who was a
member of a French expedition to Ecuador in the years 1736 to 1743.
The journal of his voyage, published in 1748 in Madrid and soon
translated into other languages, contained a reference to the new metal
platinum which quickly aroused the interest of scientists in Europe

2

The Platinum of New Granada

*"This platina is a Stone of such Resistance that
it is not easily broken by a blow upon an Anvil.
It is not subdued by Calcination, and it is very
difficult to extract the Metal it contains even
with much Labour and Expence."*

ANTONIO DE ULLOA

From the preceding chapter there seems to be no doubt that the platinum that
began to reach Europe from the Spanish colonies in South America in the
middle years of the eighteenth century was the first to be recovered in any
quantity and the first to attract the curiosity of scientists. The source was in the
western part of what is now the Republic of Colombia, but before embarking
upon a study of its history it is necessary for the reader to have a little background knowledge of the political circumstances of the time. Following upon the
discovery of the Americas by Columbus, the Spaniards in the early years of the
sixteenth century overran in a comparatively short time a vast area of both
northern and southern continents, while at the same time the Portuguese were
colonising in Brazil. After some bickering the possessions of the two countries
were defined by a line drawn by the Borgia Pope Alexander VI in 1494, running
north and south more or less along the forty-seventh degree of longitude from the
mouth of the Amazon to southern Brazil. The Spaniards proceeded to develop
their possessions to the west of this line and to exclude all other peoples from
settlement in them or trade with them. It is important to understand this overall
prohibition of trade and export because later on in our story we shall find some
authorities saying that the mining and export of platinum was forbidden by
definite Government action and other authorities of apparently equal standing
denying that any special regulations to this end were issued. These two points of
view are reconcilable if in fact the prohibition was real with regard to export to
non-Spanish destinations but was merely one aspect of a general commercial
exclusiveness.

The Spanish territories north of the Isthmus of Panama constituted the vice-
royalty of New Spain, and those in and to the south of the Isthmus the vice-
royalty of Peru, the capital of the former being Mexico and of the latter Lima. In

13

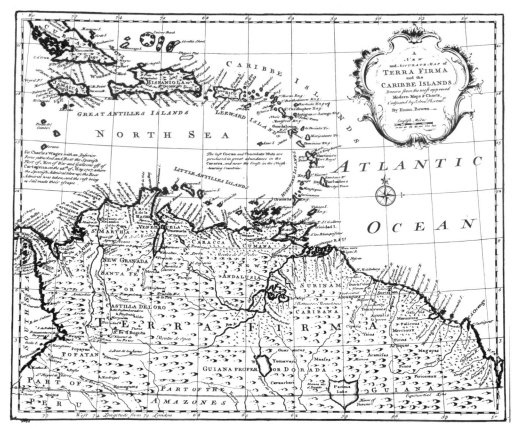

A map dating from about 1730 showing the territory of New Granada in the north-western part of the Spanish colonies of South America. Platinum from the Chocó district near the Pacific coast was smuggled from the port of Cartagena, mainly to British controlled Jamaica

1718, for strategic reasons arising out of attacks by other countries in the Caribbean, a viceroyalty of New Granada was carved out of that of Peru, having its seat at the ancient Indian city of Santa Fé de Bogotá, now the capital of the Republic of Colombia under the last part of this name alone. This vice-royalty was suppressed in 1722 but was revived in 1740. It included the Isthmus and the north-west of South America, and in it were all the areas concerned with the occurrence of platinum. The Empire was administered at home in Spain by a Council of the Indies, operating through a Secretary of State. This body acquired voluminous archives still in existence at Seville and among them many references to platinum can be found.

Apart from various small and strictly local occurrences, the main deposits of the metal are in the south central part of the Chocó region, a long narrow strip of country between the main Cordillera of the Andes and the Pacific. Quite early

14

in their explorations of the coastal areas the Spaniards learnt that there was gold in the Chocó, but the inland region was most difficult of access and subject to high temperatures and extremely heavy rainfall in a setting of dense jungle, swamps and meandering rivers as well as being occupied by hostile natives. By the 1560s several small expeditions into the basins of the Rivers Atrato and San Juan, where the richest platinum finds were eventually made, had also reported deposits of gold, and no doubt the rumours reaching Julius Caesar Scaliger, mentioned in the previous chapter, owed their origin to these forays.

It was not, however, until about 1690 that the Chocó was really settled and pacified, nearly two centuries after the beginning of the conquest, and almost immediately organised exploitation of the mineral resources began. Over the next twenty years Spanish immigration increased markedly, as did the importation of slaves from West Africa to work the alluvial deposits. (1)

These sources of gold – the great objective of the conquistadores – were among the richest then known in the world and were eagerly worked, but the nuisance caused by the presence of platinum must quickly have become apparent. It concentrated with the gold in washing in the form of white grains like small shot accompanied by heavy black magnetic sands and it demanded a good deal of labour to remove it, either by extended amalgamation – an expensive and unpopular method as the miners found mercury difficult to obtain and not too effective – or by careful and laborious sorting by hand. It occurred in the placer deposits in varying amounts, and the greater the proportion of platinum to gold the less inclination there was to work the deposit and in some cases it would be abandoned for this reason.

The Spaniards called this white metal Platina, a derogatory diminutive of plata, their word for silver, and it became known as Platina del Pinto, a small river near Popayan in New Granada no longer known by that name. It was also known as "oro blanco", white gold, or "juan blanco". The local archives contain the first reference to "platina" in 1707, by which time it had been used to adulterate the gold and a decree had been issued forbidding this practice (1). Despite this restriction, and other regulations imposing confiscation and fines on anyone caught so doing, adulteration continued for many years, but platinum itself was regarded as worthless right up to the 1780s, and was discarded and either thrown back into the rivers or scattered on the ground. It was thought by some that this annoying metal was a kind of unripe gold that had not been long enough in the ground to mature and to turn yellow.

However, it seems that at least some platinum was put to good use, most probably by the method originally devised by the native Indians that was described in Chapter 1. It is likely that the Incas who overran what is now Colombia well before the Spanish conquest adopted these techniques to make ornaments of various kinds and that a few articles were similarly fabricated by Spanish craftsmen. According to the chemist Juan Fages y Virgili some memoranda left by the Marques de los Castillejos contain a reference to a gift made in about 1730 to Don Jorge de Villalonga, the Viceroy of New Granada, of

The almost inaccessible region of the Chocó in New Granada in which deposits of platinum were discovered by the Spanish Conquistadores late in the seventeenth century. This shows the confluence of the Rivers San Juan and Condoto where the platinum occurred to the richest extent among the alluvial gold.

"a rapier guard and a set of buckles in platinum, although it is asserted that it had not sufficient coherence and was a brittle metal, although heavier than gold, with which it was mixed as a dross in the mines of the province of Citaro in the district of Chocó." (2)

Some further evidence of this is to be found in one of the letters to the Royal Society written in 1750 by Dr. William Brownrigg that will be dealt with more fully in Chapter 3. Here he mentions:

"The Spaniards have a Way of melting it down, either alone, or by means of some Flux, and cast it into Sword-hilts, Buckles, Snuff-boxes, and other Utensils." (3)

Antonio de Ulloa's Voyage of Discovery

The first person to make Europe familiar with the name "platina", however, was a young Spanish naval officer named Antonio de Ulloa, and this came about in a most indirect manner. In the 1730s one of the questions being discussed among astronomers was whether the earth was an oblate spheroid, as Newton had pre-

16

dicted, or a prolate spheroid; that is to say whether it was flattened or sharpened at the poles. To resolve this it was necessary to measure the length of a degree of longitude at the equator and again at somewhere as near as possible to one of the poles. The Académie des Sciences in Paris took the matter in hand and arranged for one expedition to go to Lapland, in the far north of Sweden, led by the famous mathematician Maupertuis, while on its behalf Louis XV sought the permission of his uncle King Philip V of Spain for another mission to visit Quito. The exclusion of foreigners from the Spanish colonies had been much relaxed under the Bourbon Kings of Spain, and permission was given with the proviso that two Spanish scientifically trained naval officers should be included in the party. The expedition, led by the French geographers Charles Marie de la Condamine (1701–1774) and Pierre Bouguer (1698–1758), was therefore to be accompanied by Don Jorge Juan and Don Antonio de Ulloa, then respectively only twenty-one and nineteen years old, and both promoted to the rank of frigate lieutenants. They left Spain in May 1735 and awaited their French colleagues in Cartagena, the complete party arriving at Quito in May of the following year. After making their many observations under great difficulties Ulloa had ample time to explore the territory before the party set out on their return journey, sailing round Cape Horn. North of the Azores their ship was captured by an English privateer but managed to escape, only to be seized by a British naval vessel when they reached Louisbourg in Nova Scotia in August 1745. Ulloa and his companions were imprisoned and conveyed to London, where all his papers were confiscated by the Admiralty. However, he was befriended by Martin Folkes, then President of the Royal Society, and by William Watson, who was later to play a major part in the discovery of platinum. Ulloa was elected a Fellow of the Royal Society in 1746, had his papers restored to him, and in the

The passage in the book by Don Antonio de Ulloa describing the occurrence of platinum among the alluvial gold workings in the district of Chocó

Lib. VI. **606** RELACION DE VIAGE
Cap. X. regularmente mientras los unos eſtàn lavando ſe emplean los otros en cortar Material; y aſsi no tienen lugar de parar los *Lavaderos*. La *Ley* de eſte *Oro* es por lo regular de 22. *Quilates*; alguno paſſa de ella, y llega haſta 23; y por el contrario baxa tambien, aunque no es comun que ſea menos de 21. *Quilates*. En el Partido del *Chocò*, haviendo muchas Minas de *Lavadero*, como las que ſe acaban de explicar, ſe encuentran tambien algunas, donde por eſtàr disfrazado, y envuelto el *Oro* con otros Cuerpos Metalicos, Jugos, y Piedras, neceſsita para ſu beneficio del auxilio del *Azogue*; y tal vez ſe hallan Minerales; donde la *Platina* (Piedra de tanta reſiſtencia, que no es facil romperla, ni deſmenuzarla con la fuerza del golpe ſobre el Yunque de Acero) es cauſa de que ſe abandonen; porque ni la calcinacion la vence, ni hay arbitrio para extraer el Metal, que encierra, ſino à expenſas de mucho trabajo, y coſto. Tam-

17

same year was allowed to return to Spain. Here he compiled an account of the expedition, published in Spanish in 1748 (4) and soon translated into several other languages. A dozen copies were sent to members of the Royal Society in London, and undoubtedly copies were sent to scientists in other European countries.

It was in this book that a reference was made to platinum and a translation of the relevant passage, given at the head of this chapter, was read to the Royal Society by Watson in 1750. A more modern version reads:

"In the district of Chocó are many mines of lavadero or wash gold, like those we have just described. There are also some where mercury must be used, the gold being enveloped in other metallic bodies, stones and bitumens. Several of the mines have been abandoned on account of the platina, a substance of such resistance, that, when struck on a anvil of steel, it is not easy to be separated; nor is it calcinable; so that the metal enclosed within this obdurate body could only be extracted with infinite labour and charge."

The joint leader of the expedition, Pierre Bouguer, also made a brief mention of platinum in his own account, "La Figure de la Terre" published in Paris in 1749. After describing the separation of the gold from the sandy material by the use of certain plant extracts he wrote:

"Sometimes also they have recourse to a quite different expedient: they make use of mercury, and are often obliged to do so in the Chocó where the metal is mixed with platina, a kind of pyrites peculiar to the region." (5)

The Irish Naturalist William Bowles

It has often been stated that Ulloa brought home to Madrid a sample of platinum, but there is no real evidence for this. However, his book and his later travels were to have a considerable effect upon the early scientific work on platinum. After he returned to Madrid Ulloa, now promoted to Lieutenant-General in the Spanish Fleet, was commissioned by the new King of Spain, Ferdinand VI, to undertake an extensive journey throughout Europe to study scientific progress and in the course of the years 1750 to 1752 he visited France, Holland, Denmark and Sweden. It was on his return through Paris that he met an Irish naturalist, William Bowles (1705–1780) and proposed to him that he enter the service of the Spanish Government. Bowles had left his native country in 1740, spending the intervening years in investigating the mineral and vegetable resources of France, and Ulloa realised the potential value of his experience in superintending the mines of Spain. Bowles accepted the offer and spent the remainder of his life in his newly adopted country.

In 1753, early in his work in Spain, he received from the Minister for the Indies in Madrid a small bag of platinum together with a note reading:

"In the Bishopric of Popayan, Suffragan of Lima, there are several gold mines among which there is one called Choco. In a part of the mountains which contains it there is a large quantity of a sort of sand which the people of the country call platina and white gold."

18

The opening page of the dissertation on platinum by the Irish naturalist William Bowles, recruited by Don Antonio de Ulloa to superintend the Spanish mines. Before describing the experiments he carried out in 1753 he gives a warning against the fraudulent use of platinum and of "letting it loose in commerce." Later he was more enthusiastic about the potential value and applications of platinum. The book was dedicated to King Carlos III

DISERTACION SOBRE LA PLATINA Y LOS
ANTIGUOS VOLCANES DE ESPAÑA.

En 1753 el Ministerio me hizo entregar una porcion suficiente de Platina con órden de hacer mis experiencias, y decir mi parecer acerca del uso bueno ó malo que podía tener. El saquillo de Platina venía acompañado de la nota siguiente. *En el Obispado de Popayan, sufraganeo de Lima, hai muchas minas de Oro, y entre ellas úna que se llama* Chocó. *En una porcion de la montaña en que está hai gran cantidad de una especie de arena que los del pais llaman* Platina, *y* Oro blanco.

En mi vida había oido hablar de tal arena, y comenzando á exâminarla hallé que era una materia mui pesada, y que tenía mezclados varios granos de oro de color de hollin. Separados éstos quedaban los granos de la Platina como municion menuda ó perdigones de plomo, y con mas propiedad se parecía en el color á aquel semimetal que los Alemanes llaman *Speis,* el qual es un régulo de Cobalto que se halla muchas veces enclavado en el *Safre* [1]. El peso de la Platina me sorprehendió, por que efectivamente es mas pesada que el oro de veinte quilates. Puse algunos granos sobre un yunque, y batiéndolos con un mar-
V 2   ti-

(1) Quando se trate del Cobalto de Aragon se verá lo que es *Safre.*

The Minister asked Bowles to advise him on the use – good or bad – to which this could be put, and after making a number of experiments on this quite new mineral, including determining its high density, its insolubility in simple acids and its miscibility with gold, he reported:

"Platina is a metallic sand that is *sui generis* which can be very pernicious in the world because it mixes easily with gold and because, although by chemistry it is easy to find the means of recognising the fraud and of separating the two metals, since this means would be available only in the hands of a few people and as cupidity is a general malady, temptation seductive, the means of deceiving easy and in everybody's reach, there can only be great danger in letting platina loose in commerce." (6)

Counterfeiting and Smuggling

The government paid attention to this report and confirmed the prohibition on the export of platinum from New Granada to Europe, but it was too late either to prevent its fraudulent use or to halt the vigorous smuggling trade that went on

19

The warning given by William Bowles against the fraudulent use of platinum was not without significance. From as early as 1763 dishonest employees of the mints in Bogota and Popayan began to forge the gold coinage by using platinum and then gilding the pieces. This specimen of a forged 8-escudo coin dated 1778, shows the bust of Carlos III

Photograph by courtesy of
Herr Willy Fuchs

from Cartegena to Jamaica, chiefly with British merchants there. One example of this was given in a letter from the distinguished London mineralogist Emanuel Mendes da Costa, read to the Royal Society in December 1750:

"In January 1742–3 there were brought from Jamaica in a Man of War, several Bars (as thought) of Gold, consigned from different Merchants of that Island to their different Correspondents here, as Bars of Gold. These Bars had the same specific Gravity, or rather more than Gold, and were equally like that Metal in Colour, Grain, etc. A Piece of one of these counterfeit Bars was sent to the Mint to be tested, and it was found to be twenty one Carats three Grains worse than Standard." (7)

In other words these bars contained something like 10 per cent only of gold, the balance presumably being platinum.

A similar case was reported by William Brownrigg, of whom much more will be said in the next chapter, in one of his letters to the Royal Society in 1750:

"I am told that one Mr. Ord, formerly a Factor to the South Sea Company, took in payment from some Spaniards Gold to the value of 500 1 Sterling which being mix'd with Platina was so brittle that he could not dispose of it, neither could he get it refined in London, so that it was quite useless to him." (3)

Another form of fraudulent practice involved the use of platinum in the forgery of coinage. Specimens of colonial escudos and of the famous doubloons have been found, made of platinum and then gilded to give the true appearance of gold, bearing dates from 1763 onwards, while the practice was also followed in Spain beginning in the late 1770s. (8) These counterfeits were made by

20

dishonest employees of the mints at Popayan and Santa Fé de Bogotá and later of course in Madrid and Seville, using the genuine dies employed for the gold coinage.

Leblond's Memoir

The best description of where the platinum was found and how it was recovered was not given until some years later when a paper was read to the Académie des Sciences in Paris. This was in June 1785, and when the memoir was printed a few months later in Rozier's *Observations sur la Physique* (9) the author was given only as "M.L." and for very many years his anonymity was maintained. In 1969, however, Dr. W. A. Smeaton successfully identified him as Monsieur Jean Baptiste Leblond (1747–1815) who had spent some years in South America and who had brought home to Paris no less than 200 pounds of platinum, almost certainly illegally (10), which he then attempted to sell in England through the good offices of Sir Joseph Banks, the President of the Royal Society.

Leblond records in his paper that the only neighbourhood where there was any considerable occurrence of platinum was in the two provinces of Novita and Citara in the district of the Chocó. The gold and platinum occurred mixed together in alluvial deposits of which the grains of both were approximately the same size. The proportion of the two varied in different localities and according to Leblond it might be one, two, three, four or even more ounces per pound of gold. The greater the proportion of platinum to gold the less inclination there was to work the deposit and it might even be abandoned, since the more there was of platinum the less the amount of gold obtained from the same amount of labour and costs. The gold and platinum were recovered together by washing

A detailed account of where the platinum was found in the Chocó and how it was recovered was read to the Académie des Sciences in 1785 by one M.L. who preferred to remain anonymous. His identity has only recently been established as Jean Baptiste Leblond

MÉMOIRE

SUR LA PLATINE OU OR BLANC;

Lu à l'Académie Royale des Sciences en Juin 1785;

Par M. L.

LE point d'où se développe l'Amérique méridionale, la Cordillère, est le théâtre à la fois grand & terrible, où l'œil surpris voit avec admiration ces abîmes profonds que creusent les torrens qui se précipitent des montagnes; ces énormes rochers qui menacent ruine, se détachent & entraînent dans leur chûte épouvantable, les arbres, les plantes, les terres & les minéraux; enfin, ces monts superbes dont la blancheur éblouit & la hauteur étonne, la plupart couronnés d'affreux volcans, dont l'explosion subite & terrible bouleverse & menace le monde d'une destruction prochaine; la terre tremble; des cendres, des rochers calcinés sont lancés dans les airs; d'immenses amas de neige sont fondus, un déluge en est formé: les hommes & les animaux que surprend ce désastre, fuyent saisis d'horreur, leurs habitations sont détruites & les campagnes dévastées par ces impétueux courans-d'eau, dont la violence entraîne tout; ces débris emportés par les torrens, forment

The conclusion of the letter from Jean Baptiste Leblond, dated July 30, 1780, to Sir Joseph Banks, shortly to become President of the Royal Society. In this he seeks to dispose of 200 pounds of platinum he had brought to Paris, and asks Banks "to accept his memoir on the natural history of platinum that I read last year at the Académie des Sciences"

and were then separated grain by grain with a knife blade or other similar means on a very smooth board, the last stages being achieved by the assistance of amalgamation with mercury. All the Chocó gold output, Leblond states, went to the two Mints at Santa Fé de Bogotá and Popayan, and here a second very careful separation was effected of any platinum still remaining present:

"The King's Officers keep this and when they have a certain quantity of it they go, accompanied by witnesses, to throw it into the Bogotá River two leagues from Santa Fé or into the Cauca which is one league from Popayan; this was a prudent expedient thought out by the Government, which in addition forbids its export in order to prevent fraud arising from melting it with gold; it appears that today it is sent to Spain. Formerly an alloy was made of it with different metals like copper, antimony, etc., but such work has ceased because of the labour, always expensive in America, which considerably increases the value without real improvement in the utility."

Leblond believed that the platinum had an origin in the mountains, since he says:

"The larger size the gold and platinum have, the more they seem to be nearer to the place of their origin; on the contrary, the smaller they are the further they appear to be from it; the aspect of the country confirms this for large grains of platinum or gold are rarely found in the plains at some distance from the mountains."

22

**Jean Baptiste Boussingault
1802–1887**

Following the formation of the Republic of Colombia from the former Spanish colony of New Granada, a small team of scientists was recruited to investigate the economic potential of the country. Boussingault, a Frenchman who had trained in mining and metallurgy and was later to become a pioneer in agricultural chemistry, was one of this group and was responsible for locating the reef which was the source of the alluvial platinum deposits in the Chocó district

The Source of the Alluvial Platinum

The mineralised source was not, in fact, discovered until forty years after Leblond read his memoir. It was located by Jean Baptiste Boussingault, a French metallurgist and mining engineer recruited by Alexander von Humboldt in 1822 at the request of Simón Bolivar as one of a team of scientists called upon to investigate the mineral and agricultural potential of the newly independent Republic of Colombia. He was first of all appointed a professor at the Escuela Nacional de Mineros in Bogotá, but as with the other members of the group he was given assignments that took him to various parts of the country as inspector of mines, prospector, assayer and surveyor (11). On one of these missions in 1826 he ascended a 9000 feet high plateau to the village of Santa Rosa de Osos near Medellin in the province of Antioquia where gold mining was active. There he found rounded grains of platinum, mixed with oxides or iron, in the syenite veins that were being mined for gold. These grains were entirely similar in form and appearance to those extracted in the valleys of the Chocó. Boussingault reported his discovery of the origin of the platinum in a letter to his friend and patron Humboldt, who published it immediately in the *Annales de Chimie et de Physique.* (12)

Persistent statements will be found in the literature on platinum that the Spanish Government prohibited its mining and recovery, but Leblond denies this categorically:

"Those who believe that the Spanish Minister has caused the platina mines to be shut up have certainly been misinformed, since there are no such mines for platinum

23

alone. All he has done is to prohibit its introduction into Europe, because of the inconvenience which might have resulted from its being alloyed with gold, which was not easy to recognise at that time; a wise precaution which has assuredly forestalled a large number of frauds in the gold trade."

He goes on to indicate that the question carried its own cure in that it was not profitable at that time to collect the platinum, the only uses suggested for it round about 1740 being as small shot or in small bags for clock weights. It was thrown away anywhere and everywhere, and the story of the rivers no doubt arose solely from the practice of the two Mints described above. Casual throwing away led to curious scenes years afterwards when platinum had become a valuable commodity, and there is a story of how the mining village of Quibdo was wrecked completely in an enthusiastic search for discarded metal among its foundations (13).

Samples of Platinum sent to Europe

Small samples had reached London, Paris and Stockholm during the 1740s, as will be seen in the next chapter, and rumours began to reach the Spanish authorities of the intense scientific interest being aroused by this newly discovered metal. In 1759 the Viceroy of New Granada was ordered to collect a large quantity of platinum from the dumps around the mints at Popoyan and Bogotá and to send it to Spain (14), and from this shipment, over the next thirty years or so, quite substantial amounts were willingly forwarded on request and without charge to chemists and scientific institutions throughout Europe, often in exchange for mineralogical specimens for the Royal Cabinet of Natural History established in Madrid.

These samples were sent among others, to Macquer, the Comte de Milly, the Comte de Buffon, Guyton de Morveau and Jean Darcet in France and to William Lewis in England, all of whose work on platinum will be reviewed in Chapters 3 and 4, but as interest increased and practical applications began to appear the demand grew for both legitimate supplies and through the extensive smuggling operations from Cartagena to Jamaica.

A French party led by the celebrated naturalist Joseph Dombey (1742–1794) arrived in 1777, nominally with botanical objectives, but before they left on their voyage the great statesman Anne Robert Jacques Turgot, at that time Comptroller-General to Louis XVI, instructed Dombey

"to spare no pains to procure for men of science such a quantity of platina as might be useful in their researches." (15)

Dombey was successful in this part of his mission, for he was able to write to Condorcet at the Académie des Sciences:

"I have put on board the vessel Bueno Consefo which leaves Callao on the 3rd April for Cadiz, seven cases for the King's Cabinet of Natural History. M. le Comte de Buffon, to whom I have addressed all of them, will send to you 11 livres of platina which I earmark for the Royal Academy of Sciences so that it can make some experiments on a metal which has become the object of the curiosity of savants. I send a like

24

**Don José Celestino Mutis
1732–1808**

Going to New Granada in 1760 as physician to the Viceroy, Mutis made a number of scientific journeys to study the flora and the mineral deposits of the area. After his death his records were brought back to Madrid and provide a valuable source of information on the early exploitation of platinum

amount of it to M. Turgot and another eleven livres to M. le Comte de Buffon to be placed in the King's Cabinet. However, if the 33 livres are required for a large-scale experiment, I have no doubt about the zeal that animates these two great men." (16)

The Spaniards also sent out their own scientists to study the natural history of their colonies. Of these the first and most famous was Don José Celestino Mutis (1732–1808) who originally went out as physician to the Viceroy Pedro Messia de la Cerdia in about 1760 but later made many scientific journeys to study both the flora and the mineral deposits. Mutis included in his long and intensive investigations the history and possible applications of platinum and in a manuscript letter in 1774 he referred to two portrait medallions of King Charles III that had been made by Francisco Benito, a Spaniard employed as an engraver at the mint in Bogotá (17). More will be heard of both Mutis and Benito later in this work when discussing the platinum industry in Spain.

In 1777, in an attempt to reduce or halt the vigorous smuggling trade in both gold and platinum, the Spanish authorities ordered the erection of a refinery at Novita in the Chocó, stipulating that only gold cast into bars would be allowed

25

to be exported once the plant was in operation (18). This move was strongly resented and resisted by the mine owners when the refinery opened in 1782, particularly on account of the high charges made for melting losses and for "displatinar", the removal of the platinum from the gold sent in. They also complained of the inefficiency of the operation. In fact the illegal traffic in both metals was stimulated rather than curtailed.

Platinum after Independence from Spain

In 1810 a revolutionary movement began in the colonies against the continuing Spanish domination, and a period of insurrection continued for many years, culminating in the founding of the Republic of Gran Colombia by Simón Bolívar in 1820 but with the obvious ill effects on the economy and particularly on the mineral industry.

A scathing commentary on the administration of the new republic in relation to its mineral wealth was made by an English naval officer, Captain Charles Stuart Cochrane, who sought two years' leave from the Admiralty to visit Colombia in the years 1823 and 1824. After describing the methods of separating platinum from gold, and saying that the mines considered worth working for platinum "give two pounds of platinum to six of gold", he goes on:

"The government are now endeavouring to buy up all the platina and having it sent to Bogotá in order, as report says, to make a coinage of it. But as British merchants here offer eight or ten dollars a pound for it, about five sixths are obtained by them and smuggled to Jamaica. It is great impolicy that the Congress does not entirely do away with the old Spanish system of monopoly; if they would put on a moderate duty and allow the exportation of gold and platina, they would secure a handsome revenue from it; but as they entirely prohibit the exportation of these metals, the whole is smuggled to Jamaica and at a moderate rate. The consequence is that scarcely a pound of gold dust remains in Colombia and hardly a shilling is drawn from the mines of Chocó towards the exigencies of the state. What blind policy." (19)

On several occasions, beginning in 1821, the new government nurtured ideas of a platinum coinage, and in 1825 a London merchanting house, Thomson Bonar and Company, was asked to arrange for specimen coins to be struck by the Royal Mint in London (20). This unusual request was agreed to and a few coins were made, but nothing more was heard of the matter.

There is one other story to relate concerning platinum in the newly independent republic. In 1825 the Congress voted to erect an equestrian statue, to be cast in platinum, of their national hero Bolivar, this to be erected in the centre of Bogotá. Boussingault, who had just discovered the plantiniferous reef, was appointed to superintend its casting and installation. The embarrassed scientist informed the Minister of Finance that he could not possibly undertake this work as the necessary quantity of platinum was so great that all the mines in Colombia could not produce it in the course of a century, and that since platinum was infusible by the usual processes it would not in any case be possible to cast such a statue. Boussingault records that the Minister said to him

26

that while all this was correct, it showed that Congress was ignorant, which would be inexcusable, and that he should merely write to the effect that he would spare no effort to assure the success of this important work; in this way nobody would be offended and the enterprise would soon be forgotten. This duly happened, and Boussingault received only two kilograms of platinum which he used to make several pieces of laboratory apparatus. (11)

In the meantime the platinum placers were being worked only in an intermittent manner by casual prospectors. Labour was scarce as the new regime had emancipated the negro slaves, while the more accessible deposits were becoming exhausted. But by this time other sources of platinum were becoming available to European refiners from Russia and it was not until very much later that Colombia again became a major factor.

References for Chapter 2

1   R. C. West, Colonial Place Mining in Colombia, Baton Rouge, 1952, 63–64

2   J. Fages y Virgili, Los Quimicos de Vergara, Discursos leidos ante la Real Academia de Ciencias exactas, fisicas y naturales, 27 Junio 1909, 41

3   W. Brownrigg, *Phil. Trans.*, 1749/50, **46**, 587

4   J. Juan and A. de Ulloa, Relacion historica del viaje a la America Meridional hecho de order de S. Mag, Madrid, 1748, **2**, Book 6, Chap. 10, 606

5   P. Bouguer, La Figure de la Terre, Paris, 1749, lxii

6   W. Bowles, Disertacion sobre la platina, Introduccion a la Historia Natural y de La Geografia fisica de Espana, Madrid, 1775, 155–167

7   E. M. da Costa, *Phil. Trans.*, 1749/50, **46**, 589

8   W. Fuchs, Platinmünzen und Medaillen, Walldorf, Hessen, 1975, 26–27; 38–39

9   M. L. (J. B. Leblond), *Obsns. Physique (Rozier)*, 1785, **27**, 362–373

10 W. A. Smeaton, *Platinum Metals Rev.*, 1969, **13**, 111–113

11 F. W. J. McCosh, *Platinum Metals Rev.*, 1977, **2 1**, 97–99

12 J. B. Boussingault, *Ann. Chim.*, 1826, **32**, 204–212

13 Enciclopedia Universal Ilustrada Europeo-Americana, Barcelona, 1907, **45**, 559

14 Archivo General de Indias, Santa Fé, 835

15 A. M. Rochon, *Phil. Mag.*, 1798/9, **2**, 21

16 A. Lacroix, Figures des Savants, Paris, 1938, **3**, 131

17 A. F. Gredilla, Biografia de José Celestino Mutis, Madrid, 1911, 157–163

18 W. F. Sharp, Slavery on the Spanish Frontier, Norman, Oklahoma, 1976, 54–60

19 C. S. Cochrane, Journal of a Residence and Travels in Colombia, London, 1825, **2**, 421–422

20 Public Record Office, Mint 1/24

Charles Wood

1702–1774

Born in Wolverhampton, the sixth son of the famous William Wood, ironmaster and the producer of the copper coinage known as Wood's Halfpence, Charles spent some time in Jamaica after the failure of a scheme promoted by his father for smelting iron with coal in the north-west of England. In 1741 he returned to England for a time, bringing with him the first samples of native platinum to reach Europe and to be submitted to scientific examination

3

Early Scientific Enquiries into the Properties and Nature of Platinum

"Upon the whole this Semi-metal seems a very singular Body that merits an exacter Inquiry into its Nature than hath hitherto been made."

WILLIAM BROWNRIGG

The first platinum to be subjected to scientific investigation by European scientists came to England, brought home in 1741 by Charles Wood from Jamaica. Earlier samples had reached Europe in one way or another, particularly through Spain, but beyond being regarded as a curious substance with somewhat remarkable properties, no published work was done upon them. Thus Charles Wood must be given priority since his samples led to full scientific examination, identification and publication.

When this book was first published very little was known about Wood, but as a result of the exhaustive genealogical researches of his great-great-grandson Mr. M. H. Wood some details of his life and activities became available to the original author in 1965 and formed the subject of a published paper (1). Charles was born at Wolverhampton, the sixth son of the famous William Wood (1671–1730), an ironmaster and a man of great enterprise among whose undertakings was the making of copper coinage for both Ireland and the American Colonies – a work that brought upon him a vitriolic attack by Dean Swift.

In 1729 William Wood, who enjoyed the patronage of the Prime Minister Robert Walpole, together with two of his sons, Francis and Charles, promoted a company financed by public subscription to establish an iron works at Frisington near Whitehaven in Cumberland and there to produce malleable iron by smelting with coal. William died in the following year and by 1733 the enterprise had collapsed, leaving the two sons bankrupt. A ruined and disappointed man, Charles Wood went off to Carolina for a time, returned to Cumberland where he married in 1735, and then settled in Jamaica where the first child of the marriage was born in 1739. Here he seems to have engaged in metal mining of some kind, and he certainly practised as an assayer. He has been described by a number of

William Brownrigg
1711–1800

A native of Cumberland, Brownrigg combined a distinguished scientific career with a modest and retiring nature. After taking his M.D. in Leyden he settled down to practice medicine in Whitehaven and was elected a Fellow of the Royal Society in 1742. He became friendly with Charles Wood, who passed to him the specimens of platinum, but although Brownrigg later carried out one or two experiments on them he passed them on to the Royal Society for further examination

From a portrait painted in about 1790 formerly in the Board Room of the Whitehaven Hospital

writers as the Assay Master to the Jamaican Government at this time, but the post was not in fact created until 1747. More recently, however, research by Mr. Robert Barker in Jamaica (2) has established that Wood again spent some time there and that he was appointed the first Assay Master in that year, several pieces of Jamaican silverware made during the period 1747 to 1749 showing his assayer's mark, C.W.

During his first period there some samples of native platinum had reached Wood from Cartagena, no doubt from a smuggler, and on his return to England in 1741 he passed them on to William Brownrigg, a doctor practising in Whitehaven who had published the results of a number of researches on subjects ranging from the gases found in mineral waters and the problem of "fire damp" in coal mines to the purification of common salt, his papers being presented to the Royal Society by his friends in London. Nothing was apparently done with these specimens until Wood's final return to England on giving up his appointment in 1749, when he set up with others an iron forge at Low Mill, near Whitehaven. Renewing his friendship with Brownrigg, whose brothers-in-law were among the investors in the Low Mill enterprise, Wood now persuaded him

30

to examine the samples of platinum and to undertake a few experiments upon them.

In the meantime, presumably in his laboratory in Kingston, Jamaica, Wood had carried out some examinations of his own, finding that the metal was present in small white shot-like grains among a black magnetic sand and that it could be melted only after mixing with more fusible metals such as copper, silver and tin. He submitted the native metal to cupellation with lead and found no alteration in either its behaviour or its weight, while it also withstood a twelve-hour digestion with nitric acid.

The specimens that Wood brought home with him were well chosen to show the essential facts about the metal. They consisted of:

The covering letter from Dr. William Brownrigg, read to the Royal Society by William Watson on December 13, 1750. With this he enclosed a number of samples of native platinum "first presented to me about nine years ago" and an account of the preliminary experiments carried out by Charles Wood and himself. In his second letter Brownrigg described Wood as "a skilful and inquisitive metallurgist who is not ambitious to appear in print"

XII. *Several* Papers *concerning a new* Semi-Metal, *called* Platina; *communicated to the* Royal Society *by Mr.* Wm. Watfon *F. R. S.*

I.

Extract of a Letter from William Brownrigg *M. D. F. R. S. to* Wm. Watfon *F. R. S.*

Dear Sir,   *Whitehaven, Dec. 5, 1750.*

Read Dec. 13. I TAKE the Freedom to inclofe to you
1750.   an Account of a Semi-metal call'd
Platina di Pinto; which, fo far as I know, hath not been taken notice of by any Writer on Minerals. Mr. *Hill,* who is one of the moft modern, makes no mention of it. Prefuming therefore that the Subject is new, I requeft the Favour of you to lay this Account before the *Royal Society,* to be by them read and publifhed, if they think it deferving thofe Honours. I fhould fooner have publifhed this Account, but waited, in hopes of finding Leifure to make further Experiments on this Body with fulphureous and other Cements; alfo with Mercury, and feveral corrofive *Menftrua.* But thefe Experiments I fhall now defer, until I learn how the above is receiv'd. The Experiments which I have related were feveral of them made by a Friend, whofe Exactnefs in performing them, and Veracity in relating them, I can rely on: However, for greater Certainty, I fhall myfelf repeat them I am, dear Sir,

Your moft obedient Servant,

W. Brownrigg.

William Watson
1715–1787

The English physician and scientist well known for his researches on electrical phenomena. Elected to the Royal Society in 1741, Watson was a most active Fellow as well as a founder member of the Royal Society Club, and enjoyed a wide circle of friends among contemporary scientists. It was to him that Brownrigg passed the samples of platinum, with an accompanying letter on his preliminary findings, for presentation to the Royal Society. In the last year of his life he received a knighthood for his services to science

1 Platinum grains mixed with black sand (magnetite)
2 Native platinum grains separated from the sand
3 Platinum that had been fused (after alloying)
4 A piece of such fused metal fashioned into part of the pommel of a sword

Brownrigg carried out a few preliminary experiments on the native platinum, including unsuccessful attempts to bring about either its fusion or its cupellation with lead, and then decided to pass the specimens on to the Royal Society together with an account of Wood's experiments and of his own views on the material and its occurrence. His friend William Watson, a distinguished physicist and a member of the Royal Society, was asked to make the presentation and this he did on December 13th, 1750, adding some comments of his own to Brownrigg's contribution (2).

In the first letter Brownrigg described how he had come into possession of the samples of platinum and went on:

"It is found in considerable quantities in the Spanish West Indies (in what part I could not learn) and is there known by the name of Platina di Pinto. The Spaniards probably call it Platina from the resemblance in colour that it bears to silver. It is bright and shining, and of a uniform texture; it takes a fine polish, and is not subject to tarnish or rust."

32

After describing his few experiments he concluded:

"It appears that no known body approaches nearer to the nature of gold, in its most essential properties of fixedness and solidity, than the semi-metal here treated of; and that it also bears a great resemblance to gold in other particulars."

Wood's first wife had died and in February 1756 he married Brownrigg's widowed sister Jemima and then in 1765 moved from Cumberland to South Wales where he built an iron works at Carfarfa near Merthyr Tydfil. This grew in importance to become the largest of its time in Wales and here Wood spent his last years and died in 1774.

Before leaving Charles Wood, however, there must be reported a curious isolated incident that has come to light from an entirely unexpected source, namely The Memoirs of Casanova. The reference is to a visit paid by Casanova in 1757 in Paris to a wealthy woman, the Marquise d'Urfé, who was interested in alchemy and the occult and had expressed a wish to meet him. In the course of the visit she took him to her alchemical laboratory and showed him a vessel containing some platine del Pinto which she was about to convert into gold . . .

"C'était M. Vood en personne qui lui en avait fait présent l'année 1742."

He was shown the platinum resisting the action of sulphuric, nitric and hydrochloric acids separately but yielding to aqua regia. She was melting it by means of a burning-mirror, saying that alone it could not be melted otherwise, which showed it to be superior to gold. She also showed him how it was precipitated by sal-ammoniac, "which has never been able to precipitate gold". No further explanation of these statements has so far emerged but it must be borne in mind that Casanova did not actually write his memoirs until 1792 and by that time most of the facts about platinum were known and had been widely published. (4)

William Watson's Contributions

After reading Brownrigg's letter to the Royal Society, giving all the details of Wood's experiments, Watson read a further communication of his own, quoting as mentioned in the preceding chapter, the important passage from Ulloa's journal. The Presidential Chair was still occupied by Martin Folkes, who together with Watson had befriended Ulloa during his confinement in London in 1746. Then in the following February he was asked to read to the Royal Society a further letter from Brownrigg (3) in which the writer disclosed for the first time the name of Charles Wood and went on to describe his experiment to show that platinum did not resist the action of lead in cupellation as he had previously thought.

As one of his biographers has written (5), while Brownrigg's actual contribution to our knowledge of the properties of platinum was not perhaps of major importance, he none the less played a vital role in that the publication of his findings brought them to the notice of scientists throughout England and continental Europe. Possibly his most significant comment was that quoted at the head of this chapter.

William Watson played a large part in this dissemination of knowledge. In January 1751 he wrote an informative letter to his friend Georg Matthias Bose (1710–1761), Professor of Natural Philosophy at the University of Wittenberg, with whom he had been in constant correspondence on the new subject of static electricity. Bose immediately published the letter in German in the first number of a journal on popular science, *Physikalische Belustigungen*, just established by Christopher Mylius in Berlin (6). A second letter from Watson followed in May of the same year in which Watson apologised for the delay in answering Bose's letter of acknowledgement of February 20th, this being occasioned by the death of the Prince of Wales (Frederick, elder son of George II, a most popular prince, had died of pneumonia on March 31st at the early age of forty-four) and in which he gave more details on platinum and referred to Ulloa's published work. He also mentioned that it takes a high polish and suggested its use for the mirrors of telescopes.

These two letters aroused considerable interest among chemists throughout Europe as many comments to this effect confirm in the literature. One example, written many years later by the great French chemist Antoine Francoise de Fourcroy, is typical:

"These first attempts, which announced very extraordinary properties, made a great commotion in Europe, at a time when the discovery of a metal as singular as this appeared to be was a phenomenon entirely unexpected. Then the great chemists of Europe set to work on platinum and its distinctive properties." (7)

34

Scheffer's Researches in Sweden

In the middle years of the eighteenth century there was a very active interest in the sciences in Sweden. A country rich in minerals, it produced a number of distinguished chemists and metallurgists, anxious to study these deposits and to identify the metallic elements they contained. Beginning with Georg Brandt the list includes Johann Gottschalk Wallerius, Henrik Theophil Scheffer, Axel Fredrik Cronstedt, Torbern Olof Bergman, Karl Wilhelm Scheele and Johann Gottlieb Gahn, all extremely well known for their contributions to early metallurgical science.

In 1733 Brandt discovered cobalt and carried out a systematic investigation of arsenic and its compounds; in 1751 Cronstedt discovered nickel in a mineral from a cobalt mine, while Scheffer, in the same year, made a major contribution to our early knowledge of platinum.

H. T. Scheffer studied at the University of Uppsala, where he learnt mathematics from Anders Celsius, but as there was then no course available in chemistry he entered the Mining College, where his father was secretary, at the age of twenty-one and became one of Brandt's most enthusiastic pupils. He also established a private laboratory for analytical work. For ten years from 1739 he was managing a mine and a metal works producing copper and a little gold but the enterprise did not succeed and in 1749 he returned to Stockholm to work in the Mining College and also as an assayer at the Mint, while in addition he gave lectures in chemistry. He was elected a member of the Royal Swedish Academy of Sciences in 1746 and published a number of papers in their transactions. By far the most important of these, however, was his contribution to the discovery of platinum in 1751.

It has been mentioned in Chapter 2 that Don Antonio de Ulloa, after his adventures and his return to Madrid, had been commissioned by King Ferdinand VI to undertake a mission throughout Europe to study scientific developments and in the autumn of 1751 he was for some weeks in Stockholm. Here he was welcomed by the Swedish scientists and, at a meeting of the Academy on October 12th, he was proposed for membership by the secretary, Pehr Wilhelm Wargentin, also an astronomer and mathematician who would have known of Ulloa's work, and he was duly elected.

Whether or not he met Scheffer and discussed platinum with him is not known, but it is most likely that, in seeking out those who could advise him on science and industry among the small number of scientists in Stockholm, such a meeting took place.

However that may be Scheffer, undoubtedly prompted by Ulloa or by the letters from Watson to Bose or possibly both, very soon produced a paper for the Academy, submitted on November 19th and read on the 28th, with the title "The White Gold or Seventh Metal, called in Spain 'Platina del pinto', Little Silver of Pinto, Its Nature Described" (9). In this he records that in June 1750 he received a sandy specimen containing "flat triangular scales, white as silver" that were not attracted by a magnet, that the sample had been obtained from

Det hvita Gullet, eller ſjunde Metallen,
kalladt i Spanien, Platina del Pinto, *Pintos*
ſmå Silfver, beſkrifvit til ſin natur
Af
HENR. THEO. SCHEFFER.

År 1750 i Juni Månad, fick jag af Herr Aſſes-
ſor RUDENSKÖLD en mörk ſand, hvilken
Herr Aſſeſſoren hade fått i Spanien, med
den underrättelſe, at den vore ifrån Väſt-In-
dien.

Denna ſand beſtod 1 af mörka ſarn-korn. 2. af
järn-färgade Järnmalms korn, hvilka drogos af
Magneten. 3. af få gedigna gullkorn. 4. af flata
ſcalena Trianglar, hvita ſom Silfver, hvilka Magneten aldeles icke drog.

Deſſa trekantiga Metall-ſmulor ſyntes vara
Jern, hvilket af något tilfälle hade blifvit hvitt
utan uppå; men det beſynnerligaſte var, at det
icke drogs af Magneten, faſt än det var få ſmi-
digt, ſom något järn kan vara; få at det, denna
Metall, oförſkylt tillägges at vara oſmidig: i den
han-

The opening of Henrik Theophil Scheffer's paper read to the Royal Swedish Academy of Sciences on November 28th, 1751. The title reads: "The White Gold or Seventh Metal called in Spain 'Platina del Pinto', Little Silver of Pinto, its Nature Described". This records that in June 1750 he had received a sample of native platinum from the President of the Academy, Ulric Rudensköld. After removing the sandy content the sample weighed only forty grains, but Scheffer was none the less able to make a thorough examination of the new metal

Spain and that he understood it came from the West Indies. He had been given the metal by Ulric Rudensköld, then President of the Swedish Academy of Sciences, who had spent the years from 1740 to 1744 in the Swedish embassy in Madrid and must have been acquainted with many people in influential positions in Spain. After removing the sandy content he was left with only forty grains of a metallic nature; these he found could be melted readily with copper, that they were not attacked by sulphuric or nitric acid but dissolved in aqua regia, while with the addition of a small amount of arsenic the material melted easily. (His choice of arsenic must surely have occurred to him from the work of his master, Georg Brandt, on this element). His conclusions were:

"1   That this is a metal hard but malleable, but of the hardness of malleable iron.

2   That it is a precious metal of durability like gold and silver.

3   That it is not any of the six old metals; since first it is wholly and entirely a precious metal, containing nothing of copper, tin, lead, or iron because it allows nothing to be taken from it. It is not silver, nor is it gold; but it is a seventh metal among those which are known up to now in all lands."

Finally he recommended that:

"This metal is the most suitable of all to make telescope mirrors because it resists as well as gold the vapours of the air, it is very heavy, very dense, colourless and much heavier than ordinary gold, which is rendered unsuitable for this particular use by lacking these two latter properties. There remains only to be found the manner of giving white gold unity and a proper state and a mixture that can aid to melt it and to make it capable of receiving a polish."

36

This was the first accurate examination of platinum, carried out on an extremely small quantity, but immediately afterwards, and also dated November 28th, 1751, Scheffer read a short paper to the Academy "An Addendum on the Same Metal" (10) in which he refers to another sample of material that he had received from Brandt, who had also had it from Rudensköld. In this supplementary communication he reported that the metal, unlike gold, was not precipitated from solution in aqua regia by ferrous sulphate, but that it was precipitated by alkalies and by ammonia in the form of a red powder. This last observation, as will be seen, was a very important one that led on to much useful research.

William Lewis of London

Before William Watson's communications were published by the Royal Society and therefore well before Scheffer had started his work, a London physician and lecturer had begun a major series of experiments with platinum. This was Dr. William Lewis (1708–1781) of Kingston-upon-Thames, but as he did not publish his findings until 1754 Scheffer must be accorded priority as the first follower of Watson in the publication field.

LXXXVI. *Experimental Examination of a white metallic Substance said to be found in the Gold Mines of the* Spanish West-Indies, *and there known by the Appellations of* Platina, Platina di Pinto, Juan Blanca. *By* William Lewis, *M. B. F. R. S.*

PAPER I.

Read May 30, 1754.

Experiment 1.

THE substance brought into England under the name of *platina* appears a mixture of dissimilar particles.

The most conspicuous, and by far the largest part of the mixt, are, white, shining grains, of seemingly smooth surfaces, irregular figures, generally planes with the edges rounded off. Upon examining these with a microscope, the surface appear'd in some parts irregular; the prominencies smooth, bright, and shining; the cavities dark-colour'd and roughish. A few of them were attracted, tho' weakly, by a magnetic bar.

The grains above describ'd are the true *platina*. The heterogeneous matters intermingled among them, in the several parcels, were,

1. A

Before Scheffer's researches were known to him William Lewis had embarked on a long series of researches on platinum. Within five weeks from late May to early July in 1754 he read four papers to the Royal Society, with a further two in 1757. This is the opening page of his first paper from the *Philosophical Transactions*

Lewis was a skilled experimenter as well as a prolific author and editor, and in his time he was the undisputed authority on any subject on which he wrote. He had, in fact, a considerable influence on the chemical technology of the Industrial Revolution. Lewis studied medicine first at Christ Church, Oxford, and later at Emmanuel College, Cambridge. He then settled in London and soon became established as a public lecturer on chemistry and on the improvement of pharmacy and the manufacturing arts. In 1745 he was elected a Fellow of the Royal Society and two years later he moved to Kingston where he equipped a large laboratory. An excellent account of the life and work of William Lewis, until then almost a forgotten man, was provided by the late F. W. Gibbs in 1952 (12).

Lewis was acquainted with Brownrigg, and he was permitted to select a small sample from the specimens that had been presented to the Royal Society. On these he made a few preliminary observations but it was not until early in 1754 that he obtained a sufficient quantity for his extensive research. This he secured from General Richard Wall (1694–1778), an Irish Roman Catholic who, unable to hold any public office because of his faith, had served the Spanish King Ferdinand VI in a number of posts, including one in the South American colonies. In 1747 he was appointed Spanish ambassador to London and was elected to the Royal Society in 1753, clearly becoming known to Lewis at that time. Through Wall's influence, Lewis was sent one hundred ounces of platinum from Spain, with a further quantity a year or two later.

The results of Lewis's long and exhaustive series of experiments were read to the Royal Society in 1754 in four long papers over the course of only five weeks (13). He went about his work and reported his findings in a manner so logical, clear and almost modern in character that his papers present the first authoritative and comprehensive account of the properties of platinum.

In the first paper Lewis summed up his conclusions:

"That the pure platina is a white metallic substance, in some small degree malleable; that it is nearly as ponderous as gold, equally fix'd and permanent in the fire, equally indestructible by nitre, unaffected by sulphur. That it is not to be brought into fusion by the greatest degree of fire procurable in the ordinary furnaces, whether expos'd to its action in close vessels, or in contact with the fuel; by itself, or with the addition of inflammable, saline, vitreous or earthy fluxes."

The second paper recorded the effects of acids upon platinum freed from its contaminating dust, mercury and gold. Its insolubility in sulphuric, hydrochloric and nitric acids was noted, and of course its solubility in aqua regia, but his most important discovery was that:

"The spirits of sal-ammoniac added to solutions of platina (in aqua regia) diluted with distilled water, precipitated a fine red sparkling powder; which, exsicated and expos'd to the fire in an iron ladle became blackish; without at all fulminating, which calces of gold, prepar'd in the same manner, do violently."

Papers III and IV described a long series of experiments in the melting of platinum alloyed with practically every other metal then known and some of their alloys, but curiously he did not then try the effect of arsenic. The idea was that the best hope of bringing platinum into a malleable form that could be fabricated was by alloying it with another metal, but little progress resulted from these experiments except his finding that if an alloy of lead and platinum was thoroughly oxidised at a high temperature it yielded a spongy mass that could be forged.

After the publication of Papers I to IV Lewis obtained a copy of Scheffer's communication to the Swedish Academy in which he found

> "a remarkable experiment seeming to show that platina and arsenic have some disposition to unite."

and he at once set about repeating it in several different ways without great success. He concluded that:

> "It appears upon the whole, that platina does melt with arsenic, but less perfectly than with other metals; and that it would be very difficult, if not impossible, to bring it, on this foundation to sufficient fusion for being poured into a mould." (14)

On melting platinum together with an equal part of gold Lewis obtained a brittle alloy although it could be worked after annealing, and this finding drew a letter from Brownrigg to Watson, read to the Royal Society in the following December, in which he wrote:

> "Dr. Lewis will find that Platina being mixt with Gold destroys the ductility of that metal. How Gold is affected with a very small proportion of Platina I know not, but my acquaintance Mr. Charles Wood (who was Assay Master in Jamaica) a very curious man tho' he might be mistaken in some things, showed me about a drachm weight of a substance which he said was Platina and Gold mixed by him in equal quantities. This mixture was extremely brittle." (15)

Lewis was lavish in his use of platinum in his experiments, for in most of them he took at least one ounce and sometimes even three or four ounces as starting materials. One other aspect of his work, pointed out by Professor Cyril Stanley Smith (16) is that he made one of the earliest observations of the microstructure of an alloy as an aid to interpreting its constitution when he noted the complex structure of incompletely melted alloys of platinum and gold:

> "Some appeared to have suffered no alteration; others exhibited an infinite number of minute globular protruberances, as if they had just begun to melt."

For this outstanding work Lewis was awarded the Copley Medal by the Royal Society in 1754, but he continued his experiments, prompted by Brownrigg's letter, on the alloying of platinum with gold, in an attempt to ascertain:

> "whether there is reason to credit the report of great frauds having been committed by mixing them together; how far such abuses are practicable; and what is of more importance, the means by which they are discoverable."

So Papers V and VI, read to the Royal Society in March 1757, (17) dealt largely with the methods of assaying and separation of the metals.

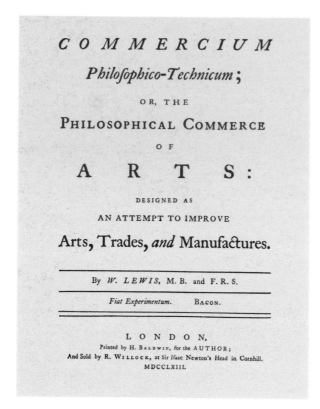

In 1763 William Lewis published his classic "Commercium Philosophico-Technicum", dedicated to George III to whom he had given private tuition when he was Prince of Wales. This included a long section running to 170 pages that formed the first authoritative and comprehensive work on the history and the properties of platinum. This work was much consulted for the next three-quarters of a century, and even as late as 1880 John Percy, Professor of Metallurgy at the Royal School of Mines in London, referred to it as "a volume rarely appreciated as it ought to be . . . it contains much information of scientific interest and of practical value "

There was yet more to come from William Lewis. Beginning in 1763, he published a massive volume in parts, the *Commercium Philosophico – Technicum*. A long section was devoted to the chemistry and applications of gold, and, following another part dealing with colours and pigments, there appeared a 170 page section on "The History of Platinum" (14). Lewis wrote in introducing this:

> "Nothing now is so much wanted, as a regular history of what has already been done, or a connected view of the experiments that have been made upon it."

and this he set out to provide. It included of course his own work as well as that of the leading chemists of Sweden, Germany and France whose work had to a large extent stemmed from his own.

William Bowles and his Researches in Spain

It was mentioned in the last chapter that in 1753 the Irish naturalist William Bowles examined a sample of native platinum from South America at the request of the Spanish Minister of the Indies and that he issued a warning on the dangers of admitting this new metal into commerce. His report on his experiments describes how, after separating the gold and the blackish sand:

"The grains of platina then resembled lead shot, still more the semi-metal which the Germans called *speiss*, which is a regulus of cobalt often included in saffre. The weight of the platina surprised me because it exceeds that of 20 carat of gold. I put several pieces of it on an anvil and hit them with the hammer. I saw they extended their diameter five or six times, remaining white as if they were silver. This determined me to send them to a gold beater to determine the extent of their malleability, but when tested between the beater's skins they broke up at once. In remarking that the sand was only malleable to a certain extent I wished to try to melt it in a furnace which a very clever Swiss used for the separation of gold by the dry way. The fire was so violent that it melted a part of the crucible and the grains of platina united themselves in a group without any loss of colour or sign of true fusion, after two hours of the most lively fire. Seeing the grains joined together, I thought that the platina might contain some portions of true sand and that this would be vitrified by the phlogiston of the metal. To convince myself, I washed a little of the platina and put it in another crucible glazed with common melted salt. After three hours nothing had melted but the grains were not so strongly united as the first time, several even remaining separate."

He then found that the platina was not attacked by sulphuric and nitric acid but there was some attack (probably on the sands) by hydrochloric acid. The platina was dissolved after "I threw on the acids a strong dose of sal-ammoniac". He noted that the grains of platina which had been united by heat broke up under a very light hammer blow. He examined the action of sulphur, lead and cupellation; he noted that the solubility in gold was limited and that the alloy with lead could be melted and could be parted in nitric acid, leaving an infusible black powder. (18)

No further research work on platinum took place in Spain for a quarter of a century and an account of this will be found in Chapter 6.

Marggraf's Work at the Berlin Academy of Sciences

Soon after his accession to the throne of Prussia in 1740 Frederick the Great, among many other activities, revived the declining Royal Prussian Academy of Sciences that had been founded in 1700 by his grandfather Frederick I, on the persuasion of Leibniz, to emulate the Royal Society of London and the French Academy of Sciences. To this end he invited to Berlin both Leonhard Euler, the Swiss mathematician, and later Pierre Louis Maupertuis from the French Academy of Sciences. Euler was sent a sample of platinum from London in about 1754, and this he passed on to his leading assistant, Andreas Sigismund Marggraf, who had studied medicine at Halle and then metallurgy at Freiberg, and who was given charge of the chemical laboratory in the Berlin Academy in 1753.

Marggraf carried out a long series of experiments reported to the Academy in 1757 (although not published until 1759) written in French, the official language insisted upon by Frederick, while they were later published in German in a collection of his papers (19).

Marggraf applied much the same procedures as the other investigators whose work has already been recorded. He again observed that at a very high temperature platinum grains sintered together somewhat but that they had not been melted and were easily broken up under the hammer. He also noted the traces of mercury among the native platinum and concluded that it was a residue from an amalgamation process. He found that hydrochloric acid dissolved some iron from the mineral and that aqua regia also dissolved the platinum itself. From this solution precipitates were formed by copper, iron, tin, lead, mercury, zinc and antimony, and with regard to that formed by zinc Marggraf remarks that "to all appearances the platinum had been precipitated" upon it. There was no precipitate by solutions of mineral fixed alkali (soda); while salt of tartar (potassium carbonate) gave a yellow precipitate. He cupelled some of this last with lead and obtained "a rough, greyish and very brittle button which was exactly like that obtained when ordinary platina was cupelled with lead". He repeated the experiment using sal-ammoniac as the precipitant and obtained a precisely similar result. The same thing happened when he evaporated his platinum solution in aqua regia to dryness, ignited the residue and cupelled it with lead. He noted that in all these cases the cupelled button always contained some lead. Next he alloyed a similar button with some pure gold and an excess of silver and parted the alloy in nitric acid. He found that in the course of this operation the platinum disappeared, the final insoluble residue being only the right weight of fine gold. He next demonstrated the presence of iron in native platinum by obtaining from its solution a precipitate of Prussian

**Andreas Sigismund Marggraf
1709–1782**

A native of Berlin, Marggraf was apprenticed to his father, the Court Apothecary, and then studied medicine at Halle and metallurgy at Freiberg. After his return to Berlin he was elected to the Academy in 1738. He was singled out as the best chemist and in 1759 he was put in charge of the chemical laboratory in the Academy, succeeding Euler as Director in 1766. He was the first to discover that the precipitate obtained from a platinum solution by means of ammonium chloride reverted to metal on heating

The beginning of the paper read to the Royal Academy of Sciences in Berlin in 1757 under the title "Experiments Concerning the New Kind of Mineral Body known under the name of Platina del Pinto". The official language of the Academy was French on the insistence of King Frederick the Great

❀ 31 ❀

◇◆◇◇◆◇◇◆◇◇◆◇◇◆◇◇◆◇◇◆◇◇◆◇◇◆◇◇◆◇◇◆◇◇

ESSAIS
CONCERNANT LA NOUVELLE ESPECE DE CORPS
MINÉRAL CONNU SOUS LE NOM DE *PLATINA DEL PINTO*.

PAR M. MARGGRAF.

Traduit de l'Allemand.

I.

Il y a déjà quelques années que l'on est parvenu en Angleterre à la connoissance de corps minéral métallique, auquel on a donné le *Platina del Pinto*. Les Auteurs Anglois qui en parlent, disent qu'on le trouve dans les Mines d'Or des Indes Occidentales Espagnoles. (Voyez les *Transactions*, Vol. 48. p. 638.) Suivant d'autres rélations, ce minéral doit se trouver en forme de sable dans les rivieres de la Province de *Quito*, & cela en très grande quantité. On ne sauroit donc dire avec aucune certitude, si c'est une matiere réellement-minérale, ou une simple raclure que l'eau entraîne de quelque veine entiere, & porte avec elle dans son cours ; ou même si ce ne pourroit point être un pur récrément métallique, d'où les Espagnols, à qui appartiennent les mines de ces contrées, auroient tiré de maniere ou d'autre ce métal parfait. Un de nos dignes Confreres (*) assure M. le Professeur *Euler*, dans une Lettre qu'il lui a écrite, qu'il tient de la bouche d'un Espagnol qui a été dans cette Province, & qui en a apporté de la *Platina*, qu'on la trouve répandue sur la campagne, près du fleuve qui traverse les montagnes du Perou auprès de *Quito*. Dans les commencemens il étoit fort difficile de se procurer quelque échantillon de cette matiere ; les Espagnols n'en vouloient point communiquer, à cause que pouvant être aisément mêlée
avec

(*) *Mr. Bertrand* de Geneve

Blue on adding "alkali that had previously been calcined with blood" (*blutlauge*, containing potassium ferrocyanide). The results of all these experiments entitle him to be called the first man to separate metallic platinum from its mineral. To say, as some of the textbooks do, that he was the first to prepare pure platinum is of course incorrect, even by the standards of purity then customary, since the metal contained all the other platinum metals, as well as other impurities.

Researches in France

France lagged behind the other European countries in applying scientific methods to the study of platinum.

In May 1751 a brief reference was made to the first letter to Bose from Watson (with his name mis-spelt) in the *Journal Oeconomique* under the heading:

METAL NOUVELLEMENT DECOUVERTE
Extrait d'une lettre ecrite de Londres par M. Guillaume Wabson
a M. ✱✱✱✱ dateé du 25 Janvier 1751 (20).

This seems to have aroused no interest whatsoever, while a further mention in the same journal in 1755, referring to the fraudulent use of platinum in the adulteration of gold and to Scheffer's work, drew very little more attention (21).

43

But in January 1758 there appeared a long letter to the editors of the *Journal des Sçavans*, the first scientific periodical that began publication in 1665, on "A New Metal called Platina", from "M. Delalande de l' Académie Royal Des Sciences". The writer was Joseph Jerome le François de Lalande (1732–1807) who was later to become the leading astronomer in his country but who was also a prolific writer on many scientific subjects. His letter opens by referring to the researches carried out on platinum in England and Germany, and goes on to justify his appearance in print:

> "Because nothing has been published in our language on the subject, except a very small extract in the *Journal Oeconomique*, I offer for your curiosity a summary of these experiments and the comparison which results between platinum and the metals already known."

Lalande, at the age of only nineteen, had been sent to Berlin in 1751 to make observations on the parallax of the moon and here, during his stay of nearly two years, he impressed both Frederick the Great and Euler and was elected to the Academy. He would have doubtless met Marggraf during this visit, but his letter refers only briefly to the latter's work on platinum while it gives a lengthy account of the "profound researches" of William Lewis.

Later in 1758 there was published in Paris a small book under the title "La Platine, l'Or Blanc, ou le Huitiéme Metal", consisting of a short preface by the author, the two letters from William Watson to Professor Bose, an account of the experiments of Charles Wood and the comments of Brownrigg upon them, Scheffer's papers to the Royal Academy of Sweden, Lewis' papers of 1754, and a somewhat alchemical letter commenting on all these from an anonymous Italian living in Venice (23). In fact, the book contained a very good summary of the work carried out on platinum in other countries up to and including 1755. It was printed anonymously but the author's name was given as M. Morin, "a man zealous in the process of science and a savant himself", by Macquer, to whom the manuscript had been submitted for approval in October 1757, in his Dictionnaire de Chimie, published in 1766. However, the true identity of this man remains somewhat in doubt. The book has often been attributed to Claude Morin, a lawyer in Dijon, but he was no scientist and it is much more likely to have been the work of Jean Morin (1705–1764), Canon and Professor of Philosophy at Chartres, who was a corresponding member of the Académie des Sciences from 1736 onwards and who wrote two or three other scientific works, mainly on the subject of electricity.

This book was given a long and quite enthusiastic review in the June issue of the *Journal des Sçavans*, and the effects of these publications on the scientists of Paris was now immediate, particularly upon Macquer himself. His colleague Antoine Baumé had obtained at the end of the summer of 1757 about a pound of platinum presented by José Arcadio Ortega, the secretary of the Academy of Medicine, Madrid, and together they set out to repeat some of the experiments carried out by their predecessors, establishing to their satisfaction that this new

The title page of the small book published anonymously in Paris in 1758 and including French translations of all the contributions so far made on platinum, with "reflections on the nature and on the essence of this singular substance". The author was given as Morin by Macquer, who had been required to approve the manuscript prior to publication, and was most probably Jean Morin, a priest and scientist of Chartres who was a corresponding member of the Académie des Sciences. This little book, together with the paper published by Lalande a few months earlier, at last prompted the scientists of Paris to take an interest in platinum

LA PLATINE, L'OR BLANC,

OU

LE HUITIEME MÉTAL;

RECUEIL d'Expériences faites dans les Académies Royales de Londres, de Suede, &c. fur une nouvelle Subftance métallique tirée des Mines du Pérou, qui a le poids & la fixité de l'Or.

Ouvrage intéreffant pour les Amateurs de l'Hiftoire naturelle, de la Phyfique & de la Chymie.

Néceffaire aux Orfévres & Affineurs, pour n'être point trompés fur des Alliages qui réfiftent aux épreuves de l'Or.

Utile dans les Arts, qui peuvent employer cette Subftance à fabriquer des Miroirs qui ne fe terniffent point à l'Air, & à ôter au Cuivre fa facilité à contracter le Verd-de-gris.

A PARIS,

Chez $\left\{\begin{array}{l}\text{LE BRETON, Imprimeur ordinaire}\\ \quad\text{du ROI, rue de la Harpe.}\\ \text{DURAND, rue du Foin.}\\ \text{PISSOT, quai de Conty.}\\ \text{LAMBERT, rue de la Comédie}\\ \quad\text{Françoife.}\end{array}\right.$

M. DCC. LVIII.
Avec Approbation & Permiffion du Roi.

body was indeed a new element. (Marggraf's paper had not yet appeared in print at this time). Their paper was read to the Académie in November 1758 and published in the Memoirs for that year although these were not printed until 1763 (24).

The paper opens with an expression of

"a kind of temerity in publishing at present a work carried out in a very short space of time and in which one has only repeated a part of the researches of M. Lewis",

but also makes the important point realised by the scientists of the time:

"As it is impossible to examine the essential properties of a metal, that is to say

45

Pierre Joseph Macquer
1718–1784

Although beginning his career as a physician Macquer was especially interested in chemistry and in 1745 he was elected to the Académie des Sciences. He began to publish papers from then on and in 1757, in collaboration with Antoine Baumé, started courses on chemistry in their laboratory. In the following year they undertook a study of platinum but although they were unsuccessful in attempts to bring about its fusion they satisfied themselves that it was a distinct metal. Macquer later became Professor of Chemistry at the Jardin du Roi and the leading French authority on the subject

those from which one can judge the usefulness that one might expect from it, such as its ductility and hardness, without melting it alone to obtain an ingot of a certain size, we have first thought it necessary to ascertain whether there is any hope of melting this metal.''

At this time Macquer, apart from his lecturing on chemistry, had just been appointed assistant to Jean Hellot, the scientific director of the porcelain factory at Sèvres, and after an abortive attempt, lasting fifteen hours, to melt platinum in a wood fire, he exposed a small quantity in a crucible in the porcelain furnace, "the greatest degree of fire known". After five days and nights no change was discernible in the platinum, but the crucible had collapsed.

After several more unsuccessful attempts Macquer and Baumé decided to determine whether platinum was essentially infusible or not by exposing it to the heat from a large concave burning mirror, a device then thought to be more powerful than any kind of fire. With this mirror, made of mercury-coated glass and 22 inches in diameter, they quickly melted iron and several mineral substances. Then, on October 16th, 1758, "the sun being perfectly clear and the air very clean", they placed at the focal length of the mirror a small piece of platinum that had already been through the porcelain furnace and that was sufficiently agglomerated to be held in a pair of pincers. They were at least partially

46

successful this time, in that glistening rounded particles of a silvery-white metal began to appear at a few points in the small mass of material. Separating the largest of these from the unmelted residue with their fingers, they found them to be readily malleable and were able to hammer them to foil without any signs of cracking. But Macquer's hopes of producing even a small ingot of platinum with which to determine its properties were not fulfilled.

They confirmed the solubility of platinum in aqua regia and its precipitation by sal-ammoniac, and attempted to melt the precipitate under a flux and found that the result, which at first sight looked like fusion, was really nothing more than an agglutination of the particles, the product being quite brittle. In the course of precipitations of the platinum with vegetable fixed alkali (potash) and volatile alkali (ammonia) they found that the colour of the precipitates varied from one occasion to another, from bright red through all shades of orange to pure yellow. Macquer thought that this depended on the strength of the solution, while Baumé attributed it to the excess or otherwise of precipitant. Lewis had commented on these conclusions but he did not pursue the matter, and the answer came only with the discovery of iridium nearly half a century later. A full

Antoine Baumé
1728–1804

A master apothecary, Baumé opened a pharmacy in Paris in 1753 and manufactured drugs on a considerable scale. In 1757 he co-operated with Macquer in conducting courses in both chemistry and pharmacy which continued for sixteen years. Baumé equipping the laboratory and preparing the experiments. He first suggested that platinum could be consolidated by heating and forging

account of all this work will also be found in Macquer's Dictionnaire de Chimie (25) but a very important deduction from it did not see the light of day until Baumé published his three volume textbook Chymie Experimentale et Raisonée in 1773 (26). Here he says of platinum:

"It combines in its admirable properties a quality still more precious which is lacking in gold, hardness. This approaches very nearly that of iron. So many excellent properties united in a single metal make it desirable that it should be introduced into commerce. There is every reason to believe that it will be possible to derive very great benefits from its use in an infinity of utensils, which made out of this metal will not be susceptible to attack by any kind of rust. For example, I have noticed that it submits to forging and welding like iron, without the introduction of any other metal. I have taken two pieces of platinum which have been cupelled in a Sèvres furnace; I have raised them to a white heat in a good forge: having placed them one on the other, and struck them with a sharp hammer blow, they have welded together just as well and solidly as two pieces of iron would have done. This property of platinum of being malleable when hot and being capable of welding in that state, leads one to hope for the greatest advantages by treating it in this way when one cannot reach finality by fusion. The case of platinum would then be the same as that of iron, with which all kinds of work can be done without our being obliged to melt it. It will be sufficient to melt this new metal a single time, either alone, or by means of lead or bismuth, and subsequently to destroy these metals in the manner we have described above."

This observation of Baumé's and the conclusion that he drew from it were responsible for the start of the fabrication of platinum for commercial uses and underlie the industry that gradually grew up during the next hundred years. The right of priority in this discovery was accorded to Baumé by his contemporaries without any hesitation and a typical example of the credit will be found in Professor Joseph Black's "Lecture on Platina, or Platinum" delivered to his students in the University of Edinburgh during the 1780s. (27)

"This method of compacting platinum and uniting the parts of it by percussion, when strongly heated, was first suggested by Mr. Baumé; and it must be employed in every case in which this metal is refined. We cannot unite the parts of it by fusion with any heat that furnaces can give."

The Iron-Gold Theory

Although Macquer and Baumé, as well as their predecessors Brownrigg, Scheffer and Lewis, were fully satisfied that platinum was an individual metal, a contrary view suddenly became apparent in France. In July 1773, at the invitation of de Morveau, the great naturalist George le Clerc, Comte de Buffon, made a visit to the Dijon Académie des Sciences, founded in 1740, to read a memoir on platinum taken from his first supplement volume to the Histoire Naturelle in which he included the findings of both de Milly and de Morveau to their great gratification (28). In the course of this he maintained:

"It is improper that chemists have regarded it as a new metal, perfect, individual and different from all the others . . . it is not believable that one may include in the

The Comte de Buffon
1707–1788

The famous naturalist George Louis le Clerc, Comte de Buffon, author of the forty-four volume "Histoire Naturelle" and Superintendant of the Jardin du Roi from 1739 as well as treasurer of the Académiè des Sciences, claimed in 1773 that platinum was not an individual metal but merely a natural alloy of iron and gold. His great reputation persuaded the Comte de Milly and Guyton de Morveau to support this view but only for a short time

George Louis le Clerc, Comte de Buffon, Intendant du Jardin du Roy, et des Académies Franç.se et des Sciences, et de celles de Londres d'Bimbourg et de Berlin.

class of metals a substance that is neither ductile nor fusible ... It is not at all a new metal but a mixture, an alloy of iron and gold formed by nature."

He went on to quote the work of the Comte de Milly, who had extracted some iron from native platinum and who also maintained that it was a natural alloy of iron and gold. Buffon further asserted that those who sold platinum were guilty of roguery.

Now Buffon, the Intendant of the Jardin du Roi, was most highly regarded for his researches in many subjects and for his massive "Histoire Naturelle" that eventually ran to forty-four volumes, and his views naturally carried a great deal of weight, although he was apt to indulge in rather hasty generalisations. During his visit to Dijon he presented a small quantity of native platinum to Guyton de Morveau and urged him to continue experiments and to seek a method of melting it, pointing out that a magnet would separate it into two parts, one containing iron and the other gold. Guyton confirmed these observations and at first supported Buffon's view. In the following July there appeared a letter from Professor Blondeau of the Naval Academy at Brest, (29) questioning both Buffon's conclusions and de Milly's reason for supporting them. Both Guyton and Milly began to have reservations, and the iron-gold theory was finally disposed of by Bergman in 1777.

**Torbern Olof Bergman
1735–1784**

Professor of Chemistry in the University of Uppsala from 1767 until his death, Bergman developed methods of quantitative analysis and wrote a treatise on chemical affinities. In 1777 he presented a paper on platinum to the Swedish Academy of Sciences, beginning by saying that "we still lack precise information about many of its longest known properties", but he disposed once and for all of Buffon's view that platinum was not an individual metal

From a portrait painted by Lorens Pasch the younger in 1778 in the possession of the University of Uppsala

The Work of Bergman

The work of Torbern Olof Bergman (1735–1784) took place considerably later than the work already recorded in this chapter, but is related to this earlier work in that it provides a summing up and clarification of what went before. As we travel on in the study of the history of platinum it will be found that on several occasions, after a certain amount of sporadic work has been carried out, there has come along a major scientific mind which has taken up the scattered results and co-ordinated and completed them: Bergman was the first of these; he was followed in the eighteen-thirties by Berzelius, curiously enough another Swede, in the eighteen-fifties by the Frenchmen Deville and Debray; so Bergman, although he did not publish until later definitely belongs to this chapter.

In 1777 he presented a paper to the Swedish Academy. This was subsequently printed in Latin in 1779 in Bergman's collected "Opuscula", in French in 1780, and in English in 1784 (30). Bergman had obtained a quantity of platinum through his friend Claude Alströmer the botanist, who had spent some time in Spain on the suggestion of Linnaeus and who was the son of the founder of the Swedish Academy of Sciences.

There had been some doubt among the ealier investigators as to whether mineral fixed alkali (soda) precipitated platinum from its solution in aqua regia or not. Bergman showed that while potash and ammonia, even in small quantities, gave a precipitate with the acid solution, soda gave it only when present in large quantities through neutralisation of all the free acid present or by direct addition. He noted that lime also precipitated platinum from its solution but did not realise that this reaction requires the influence of daylight and

50

takes place only very slowly, if at all, in the dark. He examined the precipitate produced by sal-ammoniac in a platinum solution and thought it to be a triple salt containing the constituents of sal-ammoniac as well as platinum, an observation which is quite correct. He noted that the crystalline precipitates obtained by adding to the acid solution small quantities of potassium or ammonium carbonates were red if the solution was strong and yellow if it was very dilute, an observation that remained unexplained for twenty-five years until iridium had been discovered. He also experimented on the melting of platinum and found that the sal-ammoniac precipitate appeared to be more easily melted than the native metal.

Platinum a Distinct Metal

The concluding part of Bergman's paper dealt with the question of whether platinum was in fact a distinct metal. Here he declared against Buffon's ideas and demolished his argument once and for all by both experiment and deduction:

"Since platina surpasses all metals except gold in weight, and is always found to be contaminated by iron, some scientists have believed that it could not be freed from this, that platina is nothing but a mixture of gold and iron. However, Dr. Lewis has, for several reasons, rejected this opinion. By melting together gold and iron, in whatever proportion, no such alloy is obtained which in specific gravity or other properties resembles platina in the least. Furthermore the amount of iron in the natural platina can be so reduced that it becomes hardly detectable. We do not know of any native metals that are found entirely pure. . . . When we add to this the fact that the last traces of a foreign contaminant are infinitely difficult to remove, because they are the smallest part making up the whole mass, so it is not strange if iron adheres to platina in this most obstinate manner, and rather that the difficulty in melting of platina has up to now put a special obstacle to their separation."

In a final paragraph Bergman expressed his view that it was a great pity that more platinum was not available in Europe so that it could be melted in some quantity, and that in the supplies of metal that do arrive there are many impurities that must be sought for with great attention to separate them all.

Bergman's work on classification and nomenclature is well known. He continued to make use of many of the old alchemical symbols and added new ones, including one for platinum, as illustrated here, combining the old symbols for

⊙ Gold
☉☽ Platina
☽ Silber

Bergman continued to use alchemical symbols for the elements and he devised one for the new metal platinum, based upon a combination of those for gold and silver. He also proposed the name platinum instead of the word platina used up to that time

gold and silver. He also proposed the use of the name "platinum" instead of the older "platina" in line with a number of other metals for which he had adopted the Latin ending "um".

Conclusion

From among the results of all this admirable work conducted in five countries, most of it in quite a short period of years, several facts emerged that were to prove of great importance in paving the way towards the ultimate fabrication of platinum into sheet and wire. These were:

(1) That platinum, as Macquer put it, was a particular metal as fixed, indestructible and unalterable as gold and silver, different from all other known metals.

(2) Lewis' discovery of the precipitation of platinum from solution by sal-ammoniac, which effected its separation from iron and gold.

(3) Scheffer's observation that the addition of a small amount of arsenic to a much larger quantity of platinum brought about the complete fusion of the latter at a comparatively low temperature.

(4) The discovery by Marggraf that the precipitate obtained from a platinum solution by means of ammonium chloride when heated reverted to metal. Also that the metal was thrown out of solution by metallic zinc.

(5) The observation by several people, but more especially studied by Macquer and Baumé, that the grains of native platinum and the sponge produced by the ignition of the ammonium chloride precipitate, agglutinated or sintered together when exposed to the highest possible temperature.

(6) The discovery, first by Macquer and Baumé and afterwards by Bergman, that platinum could be submitted to at least partial fusion and was then non-magnetic and malleable.

(7) The suggestion by Baumé that platinum might be consolidated by heat and percussion.

As will be seen from later chapters, all the essentials necessary for the advances towards fabrication were present in these seven points.

References for Chapter 3

1   D. McDonald, *Platinum Metals Rev.*, 1965, **9**, 20–25

2   Mr. Robert Barker, private communication, August 1981

3   W. Watson, *Phil. Trans.*, 1749–50, **46**, 584–596 (The pagination of these contributions is greatly in error)

4   L. B. Hunt, *Platinum Metals Rev.*, 1962, **6**, 28–30

5   J. Russell-Wood, *Platinum Metals Rev.*, 1961, **5**, 66–69; *Ann. Science*, 1950, **7**, 199–202

6   W. Watson, Two letters to Bose, *Physikalische Belustigungen*, 1751, (1), 107–108; (4), 285–287

7   A. F. Fourcroy, Systême des Connaissances Chimiques, Paris, 1800, **6**, 402–403

8   L. B. Hunt, *Platinum Metals Rev.*, 1980, **24**, 31–39

9   H. T. Scheffer, *Kungl. Vetensk.Akad. Handl.*, 1752, **13**, 269–275

10   H. T. Scheffer, *ibid.*, 1752, **13**, 276–278

11   A. F. Cronstedt, *"Aminnelse-tal öfver Framledne Directeuren Och Kongl, Vetensk. Acad. Ledamot, Valborne Herr Henric Theoph. Scheffer"*, Stockholm, 1760

12   F. W. Gibbs, *Ann. Science*, 1952, **8**, 122–151; *Platinum Metals Rev.*, 1963, **7**, 66–69

13   W. Lewis, *Phil. Trans.*, 1755, **48**, 638–689

14   W. Lewis, Commercium Philosophico-Technicum, London, 1763

15   W. Brownrigg, Letter to W. Watson, Royal Society Archives 1754, II, 547

16   C. S. Smith, in Powder Metallurgy, ed. J. Wolff, Cleveland, 1942, 66

17   W. Lewis, *Phil Trans.*, 1757, **50**, 148–166

18   W. Bowles, Disertacion sobre la platina, Introduccion a la Historia Natural y de la Geografia fisica de España, Madrid, 1775, 155–167

19   A. S. Marggraf, *Nouvelle Mem. Acad. Roy. Sci.*, Berlin, 1757, 31–60; Chemische Schrifter, Berlin, 1768, **1**, 1–42

20   *J. Oeconomique*, 1751, May, 93–94

21   *J. Oeconomique*, 1755, July, 148–149

22   Delalande, *J. des Sçavans*, 1758, January, 46–59

23   Morin, La Platine, l'Or Blanc ou le Huitième Metal, Paris, 1758

24   P. J. Macquer, *Mem. Acad. Roy. Sci. Paris*, 1758, 119–133

25   P. J. Macquer, Dictionnaire de Chimie, Paris, 1766, **2**, 248–263

26   A. Baumé, Chymie Experimentale et Raisonnée, Paris, 1773, **3**, 193

27   J. Black, Platinum or Platina, Lectures on the Elements of Chemistry, ed. J. Robison, Philadelphia, 1806, **3**, 388–395

28   G. L. Buffon. *Obsns. Physique (Rozier)*, 1774, **3**, 324–328; Histoire Naturelle, Paris, 1774, Suppl. **1**, 301–339

29   L. Blondeau, *Obsns. Physique (Rozier)*, 1774, **4**, 154

30   T. O. Bergman, *Kungl. Vetensk. Akad. Handl.*, 1777, **38**, 317–328; *Opusc. Phys. Chem.*, 1779, **2**, 166–183 (in Latin); *Obs. sur la Physique*, 1780, **XV**, 38–45; Physical and Chemical Essays, trans. Edmund Cullen, London, 1784, 166–183

Louis Bernard Guyton de Morveau
1737–1816

Born in Dijon, then the capital of the Province of Burgundy, Guyton
first entered the legal profession and then became a member of the
Dijon parliament, adding to his name "de Morveau" from a family
estate. He later turned to chemistry and is chiefly remembered for his
collaboration with Lavoisier in modernising chemical nomenclature but
his work on platinum, continued over many years, was of major
importance in establishing its properties and methods for its fabrication

4

Early Attempts to
Melt and Work Platinum

*"And were the properties of platinum more
fully investigated methods for working it easily
into utensils might be discovered, and then it
might justly be considered as one of the most
useful of the metallic substances."*

JOSEPH BLACK

After the experiments of a number of scientists described in the last chapter to establish the striking properties of platinum their thoughts began to turn towards making use of them, but before this could be done it was necessary for the granular native metal to be converted into malleable sheet.

The great difficulty in the way of achieving this was the presence of the iron and copper which were intimately alloyed with the platinum in the grains of the native metal. Most of the early workers made some attempt to melt these grains and found that under ordinary conditions they did nothing more than agglutinate more or less loosely together, a full fusion being prevented by the formation of surface films of iron and copper oxide at the high temperature at which the attempts were conducted. It was necessary to remove the iron before advantage could be taken of the power of self-welding at high temperatures that the metal seemed to possess, and this was the task before the workers of the seventeen seventies.

As we have seen, Baumé was the first to demonstrate that platinum could be welded when forged at a high temperature, after the removal of the iron and copper, which he brought about by cupelling the native metal with lead or bismuth at a very high temperature (1). Following Wood, Brownrigg, Lewis and others, he confirmed their observations that this process was never complete under the conditions usual at that time, the platinum remaining as a solid mass retaining a certain amount of the added base metal. The quantity of the latter left depends on the temperature available, and if the applied heat is very strong and long continued, the amount can be reduced to nothing. This no doubt explains the differing results obtained by the earlier experimenters. Lewis for instance failed to drive off all his lead because his furnace was not hot enough,

55

but in spite of this he was the first to notice that the product could be forged. Macquer and Baumé carried out the cupellation in a furnace at the Sèvres porcelain works where a very high temperature was available, with a result that the lead or bismuth that they added was entirely removed and the platinum left behind was found to have lost a sixteenth of its weight and to have become reasonably malleable (2). It was shown chemically to be free from base metals and it was this metal that Baumé was able to weld by forging. There is no evidence that he went on to apply the discovery to practical uses and this is not surprising. The product was a half-fused lump, full of cavities.

Guyton de Morveau (1737–1816), who had had the honour of collaborating with Buffon – a man already famous and thirty years his senior – in the preparation of the latter's "Histoire Naturelle des Mineraux" and had been urged by him to find a method of melting some platinum that he provided, examined Baumé's process and found that the mass could be broken up into powder and that it still contained iron; it was weakly magnetic and actually yielded iron when fused with nitre and heated with sulphuric acid (3).

So Baumé's process, valuable though it was in its indication of welding power, led no further and the thoughts of the scientists turned to the iron-free metal produced by calcining the precipitate obtained by adding ammonium or potassium chloride to the solution of native platinum in aqua regia, to see whether this could be welded more satisfactorily.

Before we turn to further attempts at the melting of platinum we must therefore consider these early examples of powder metallurgy.

The Introduction of Powder Metallurgy by de l'Isle

Apart from the rudimentary examples of the use of powder metallurgy by the natives of Colombia, described in Chapter 1, the first to achieve success in this direction on any scale was Nicholas Anne de l'Isle (1723–1780). (The name is used in several spellings such as Delisle and de Lisle). Little is known of him or of his interest in scientific matters except that he was interested in mineralogy and that he formed an important collection of minerals. In the first edition of this book he was referred to, following several earlier authorities, as the distinguished crystallographer and mineralogist Jean Baptiste Louis de Romé de l'Isle, but later this identification was queried by Dr. W. A. Smeaton in a review of Guyton de Morveau's work on platinum (4) and further investigation by the original author confirmed this and showed that the man of the same name to whom the credit is due was the one already mentioned. (5) (This confusion arose of course from the regrettable French habit of not giving initials to the authors of papers.) This de l'Isle served with the Royal Musketeers from 1739 to 1743 and was then transferred to the supply service of the French armies, serving in Germany, Flanders, Italy, Minorca and Corsica until 1769. He was then brought back to the Departement de la Guerre in Paris and, with the rank of "premier commis," took charge of supplies for the troops in Corsica, retiring in 1776.

56

It was in the August of 1775 that Lavoisier announced to the Académie des Sciences that

"M. Delisle, premier commis du bureau de la guerre, avait trouvé un moyen tres simple de fondre le platine, il a répété ses experiences avec beaucoup de succes et qu'il a obtenu un metal blanc, tres-dur, et un peu malleable" (6)

De l'Isle was fully conscious of the need to remove the sands and the alloyed iron from the platinum before working it and he set about achieving this by dissolving in aqua regia and precipitating with sal-ammoniac. More details of his procedure were given by the mineralogist and assayer Balthasar Georges Sage (1740–1824), the founder of the École des Mines, in his "Eléments de Minéralogie" published in Paris 1777:

"Sal-ammoniac dissolved in the cold in distilled water is poured into a solution of platinum made by aqua regia; there is a reddish precipitate composed of platinum and sal-ammoniac; ... this precipitate of platinum, exposed to a violent fire, melts and produces a button of malleable platinum of a whitish-grey colour resembling that of silver, and which does not alter in the air." (7)

De l'Isle discussed his procedure freely among his contemporaries, to whom he distributed a number of his small discs of undoubtedly malleable platinum, but there is no evidence that he fabricated any articles from his product or worked on any scale but a laboratory one.

There is however, one other reference to this work that should be quoted, curiously from the other de l'Isle, who was the author of a four-volume treatise on crystallography published in Paris in 1783, after the death of Nicholas Anne de l'Isle. In a footnote (8) he wrote:

"M. de Buffon has suggested that platina may be only a ferruginous substance more dense and of higher specific gravity than ordinary iron, intimately combined with a large quantity of gold. But what demolishes this theory is the fact that I possess a button of platina melted by the late M. de l'Isle and several laminae of this same platina flattened under the hammer, which have not the slightest action on the magnetic needle; which proves that all the iron which occurs interposed in platina when it is in grains can be separated from it by means of sal-ammoniac as employed by M. de l'Isle ...."

Other scientists, including Guyton de Morveau, the Count von Sickingen and the Comte de Milly, as recorded below, carried his work further and consolidated his findings, but clearly it is to this rather mysterious M. de l'Isle that the honour belongs of being the first to devise a process that still forms the basis of modern platinum refining.

The Researches of Guyton de Morveau

Before communicating de l'Isle's work to the Académie des Sciences, Lavoisier wrote to Guyton de Morveau in Dijon not only describing the new method but reporting that he had himself repeated the experiment. He had treated the

L E T T R E

DE M. DE MORVEAU

A M. LE COMTE DE BUFFON,

Sur la fusibilité, la malléabilité, le magnétisme, la densité, la crystallisation de la platine, & son alliage avec l'acier.

MONSIEUR, tout ce que vous maniez prend une nouvelle face, & produit un nouvel intérêt. Votre Mémoire sur la Platine a éveillé les Physiciens & les Chymistes : ils ont porté leurs recherches sur cette matière si singulière, si digne d'être observée, & vous avez sans doute oui parler du procédé qui a été découvert depuis peu par M. Delisle pour la fondre : ce que j'en ai appris, & qu'il a bien voulu me confirmer lui-même, m'a engagé à profiter de quelques momens de loisir pour répéter ses expériences, & reprendre celles que j'avois négligé de poursuivre depuis plus d'un an ; j'ai recommencé à traiter ce minéral dont, grace à votre générosité, il me restoit encore une assez grande quantité ; & je me persuade que vous verrez avec plaisir le récit exact de tous les phénomènes curieux & intéressans que m'a présenté ce nouveau travail, quoiqu'il ne soit pas encore possible de tous les concilier.

La première nouvelle de la découverte de M. Delisle m'avoit été donnée par M. Lavoisier, & ce savant m'avoit écrit qu'il en avoit fait l'épreuve avec succès ; qu'ayant dissous la platine dans l'eau régale, l'ayant ensuite précipitée par une dissolution très-concentrée de sel ammoniac, ce précipité traité avec mon flux réductif, lui avoit donné au bout d'une heure, un beau bouton susceptible de se polir, de se limer, mais non pas malléable ; (le flux dont parle ainsi M. Lavoisier, est celui que j'ai publié, comme devant remplacer éminemment le procédé secret de M. Bouchu pour l'essai des mines de fer). M. Delisle m'avoit marqué postérieurement, qu'il n'avoit employé aucun fondant, qu'il avoit simplement traité sa platine dans un double creuset de hesse au feu d'une forge animée par le vent de deux soufflets, & qu'il avoit eu un bouton très-bien lié, brillant, qui s'étoit laissé piler & limer, & de plus, suffisamment malléable ; les deux petites plaques qu'il avoit joint à sa lettre, en fournissoient la preuve la plus complette.

The opening of the long letter from Guyton de Morveau to the Comte de Buffon in 1775 reporting his experiments on the properties of platinum, its fusibility, malleability, magnetism and density. The practical applications of chemistry always interested Guyton and in the letter he concluded:

"If platinum one day become more common, as one would hope and desire, I do not doubt that the Arts will derive some fruits from these researches; above all they will owe an obligation to M. Delisle for it is he who established by his method that it is susceptible of being worked by the hammer, by the file and by cutting tools"

---

precipitate with a flux and had obtained a beautiful button of platinum that could be filed and polished but was not malleable. A little later de l'Isle himself wrote to Guyton to say that:

"He had not employed any flux, that he had simply treated his platinum in a double Hessian crucible in the fire of a forge animated by the wind of two blowers, and that he obtained a very compact and brilliant button which could be flattened and filed, and moreover was reasonably malleable; the two little plates which he enclosed with his letter furnish a most complete proof of this." (9)

These two letters prompted Guyton de Morveau to set about reproducing the method as it clearly represented a great improvement on the Baumé cupellation procedure. He first investigated the difference that Lavoisier had found by the use of a flux in the heat treatment stage and found that reducing fluxes (glass, borax and charcoal) never yielded a malleable metal, but that with them there was frequently real fusion into beads (no doubt due to the reduction of small amounts of phosphorus, silicon, etc., from the fluxes and charcoal ash which lowered the melting point of the platinum and at the same time destroyed its malleability). But when the sal-ammoniac precipitate was sufficiently heated in a crucible entirely by itself, as originally recommended by de l'Isle, a better result was obtained. The volume of the material was considerably reduced and the red precipitate changed into a piece of grey metal looking rather like badly melted silver, rough and granular in structure.

He reported his results to Buffon in a long letter published in *Observations sur la Physique* for July 1775 (9), reporting among other findings that:

"there was no trace of the salts with which the platinum was combined in the precipitate; the crucible was clean, and instead of the least adherence, the metallic matter appeared only to have taken the shape of the vessel; I judged by this first inspection that the platinum had simply been regenerated but that it had not undergone complete fusion, no doubt because the fire had not been lively enough or continued long enough: but what was my surprise when having put this material on the anvil, I saw it flatten almost as easily as silver, and the file and the knife make an impression on it in the same way; moreover, the magnet has no longer any influence on the pieces which I detached from it, any more than on those which M. de l'Isle had sent me."

A determination of the specific gravity of the material showed that before forging it was only 10.045 whereas in the fully hammered state it had become 20.170. It was evident therefore that the product of the heat treatment was not fully consolidated and required the subsequent forging to complete the process. Macquer was under no illusion as to what had really happened in de l'Isle's process and in his Dictionary of Chemistry (10) he makes this statement:

"It is M. de l'Ille (sic) who has made this discovery and I have verified it. The experiment consists in exposing to a good ordinary forge or furnace fire platina precipitated by sal-ammoniac from its solution in aqua regia. This precipitate appears to melt just as easily as an ordinary metal, into a metallic mass quite compact and dense, but it completely lacks malleability, when it has only been exposed to a moderate heat, and only assumes it, although always imperfectly, when subsequently subjected to a much greater degree of heat. Particles of platina being infinitely divided in the precipitate, it is not surprising that the heat penetrates such very small molecules much more effectively than the ordinary grains of platina which in comparison are of enormous size; and their softening occurring in proportion, they should show the extraordinary effect on their agglutination in the proportion of their points of contact; moreover, these points being infinitely more numerous than can be those of much greater molecules, solid masses result which have all the appearance of quite dense metal, melted and solidified by cooling, but they are really nothing but the result of a simple agglutination among an infinite number of infinitely small particles and not that of a perfect fusion as with other metals."

This looks like the first scientific contribution to the study of powder metallurgy.

In the course of his research de Morveau not only tried the effect on the precipitate of straight reducing fluxes but also of a flux containing arsenic on native platinum. He reported to Buffon (9):

"I put into a crucible a mixture of 1 gros (72 grains) of crude platinum, 2 gros of powdered neutral arsenical salt, 1 gros of charcoal dust and 2 gros of the dust of bone charcoal burned in closed vessels, the whole covered with 2 gros of powdered white glass."

He subjected this to the greatest heat that could be got from a special blast furnace designed by Macquer, and found, under a clear green slag, three well-shaped white metallic buttons, weighing in all 74 grains. These buttons were brittle but not appreciably magnetic, so that the iron had been removed but evidently "the metallic earth contained in the neutral arsenical salt has been revivified in contact with the carbonaceous matter and alloyed with the platinum". This observation had important results, to be described in the next chapter.

Guyton also pointed out that the more malleable the metal the less it was attracted by the magnet, but he was still unable to agree with the majority of his contemporaries that the magnetism of some samples was due to impurities.

The Work of Count Karl von Sickingen

At much the same time that de l'Isle was perfecting his powder metallurgy process in Paris, another worker was busily engaged in attempts to produce malleable platinum by the aqua regia method. This was the Graf Karl von Sickingen who held the post of Ambassador of the German Princedom of the Palatinate at the court of Louis XV. His long series of experiments, no less than ninety-seven are recorded, carried out in his private laboratory sometimes with the help of distinguished friends including the statesman Turgot and the Duc de la Rochefoucauld d'Enville, both active patrons of the sciences, suffered many interruptions because of his diplomatic duties and although begun in 1772 they were not reported to the Académie des Sciences until 1778. His two extremely lengthy memoirs were not, however, published by the Académie and although many of Sickingen's contemporaries were aware of his work it was not until 1782 that Professor Georg Suckow of Heidelberg, anxious that Sickingen's valuable work should be presented in his native German, made a translation available under the title "Versuche über die Platina" (11). At the same time Lorenz Crell, the well known editor and publisher, included both a summary of the work and a review of Suckow's translation in his *Neueste Entdeckungen in der Chemie* (12). Further publication of Sickingen's work was made by Jan Ingen-housz, of whom more will be said later, in both German and French (13). For a full study of these findings reference should be made to these original sources, as numerous misinterpretations and errors have crept into later text-book summaries. This is not altogether surprising, as the Count's description of his work is highly involved and detailed and only his essential findings can be summarised here. He was fully conscious of the need to remove the sand from the native mineral by dissolving in aqua regia, and also of the need to get rid of the alloyed iron that went into solution. He effected this by adding a solution of *blutlauge*, of which

The Count Karl Heinrich von Sickingen 1737–1791

The last descendant of a long line of noblemen, and the son of an alchemist, the Count von Sickingen carried out a long series of experiments on platinum in his laboratory in Paris while serving as Ambassador of the Palatinate to the French court. Several of his distinguished friends assisted him from time to time, including the statesman Turgot, the Duc de la Rochefoucauld d'Enville, and Etiénne Mignot de Montigny, the President of the Académie des Sciences and a collaborator also of Lavoisier's. Although his procedure was difficult and expensive he was the first to produce platinum in the form of wire and sheet

From a portrait by Johann Gerhard Huck, by courtesy of the Kurpfälzisches Museum, Heidelberg

the active constituent was potassium ferrocyanide. This first precipitated the iron as Prussian blue but afterwards began to precipitate the platinum also as potassium chloroplatinate. He filtered at this stage and then precipitated most of the rest of his platinum from solution by a considerable addition of oil of tartar (deliquesced potassium tartrate or, in the original German, zerflossenes Weinsteinsalz.) Thus von Sickingen used potassium salts as his precipitants rather than ammonium chloride. From this metal he was able to produce platinum in the form of sheet, and according to one of his contemporaries, the author and traveller Georg Forster, "he possessed a piece of platinum sheet more than one foot square that looked like silver and was quite pliable" (14). The red precipitate of potassium chloroplatinate so obtained was filtered off, placed in a crucible and subjected to a white heat until all the potassium chloride and other fumes had disappeared. There was left behind

"a kind of metallic flake of a silver-white colour which when heated to a white heat welds under the hammer and can be forged. In this state the platinum is perfectly ductile and lends itself to almost all the operations of the arts and does not seem to act noticeably on the magnetic needle".

Von Sickingen found it necessary for the best results to forge and alternately heat the metal to the highest possible temperature, and the forging took place on

61

Sickingen's two long memoires on platinum read to the Académie des Sciences in 1778 were never printed, but four years later Professor Georg Adolph Suckow of the University of Heidelberg, feeling that this important work should be made available in Sickingen's own language, published a German translation of which this is the title page. Lorenz Crell also published much of the work in his *Neueste Entdeckungen in der Chemie*, and in introducing the author he wrote:

"so far in this nation we have but rarely had an example of a man of noble birth, great fortune and high office who could find in chemical researches as delightful an entertainment as in the usual brilliant pastimes of the great world"

a well-polished piece of steel. On one occasion the temperatures were so high that the inside of his furnace melted down.

His early work was on four ounces of native platinum which he had been given by the Baron Holbach, but Lavoisier was so interested in the results that he presented him with a further eight ounces. Working with this larger amount he found it profitable to calcine his precipitate to metal before putting it into the crucible for the heat treatment. This enabled him to produce a little ingot which he was able to draw into wire "through all the holes of a drawing machine, whose least hole had the diameter of 19/140ths of a ligne" (0.0125 inch). This was undoubtedly the first occasion on which platinum had successfully been drawn into wire.

"The resistance of the platina to this treatment did not seem to be any greater than that of gold, which became like gold more brittle under the drawing and had to be annealed more often the smaller and more fragile the wire became. It is well known that all metals, with very few exceptions, behave in this way. . . . One piece of platina wire broke for the first time in die 54 whose diameter was half a ligne" (0.040 inch).

But it was noticed that the fracture had occurred at a place where the bar had not been completely welded. Discussing this, von Sickingen points out how necessary it is in the case of iron that the pieces being welded must hold the heat as long as they are being forged. With platinum the pieces handled were so small that this was impossible and hence the dangers of imperfect welding were considerable. In spite of this von Sickingen carried on with the drawing of wires down to about 1/16th of a ligne (0.0055 inch) and found that this wire could be rolled thin without any tendency to break.

Later he worked on eight pounds of native platina and with this large amount he found it necessary to boil his calcined platinum with distilled water to remove residues of chloride before welding. From one six ounce lot he obtained an ingot weighing over $3\frac{1}{2}$ ounces which was completely without cracks. It is interesting that after dissolving these considerable quantities of platinum he collected together the insoluble matter and attempted to melt it under borax. He obtained a rough unmelted mass but was prevented by a journey from examining this, otherwise he might have anticipated the discovery of iridium by thirty years. Unwittingly, however, von Sickingen was the first man to produce platinum comparatively free from iridium. By the addition of potassium ferrocyanide, a strong reducing agent, to his original platinum solution, he brought about the reduction of the iridic salts present to the iridous state and so prevented their precipitation by the potassium salts. Hence his platinum contained less iridium than that of any of his predecessors and this accounts for the fact that he was able to draw it down to such fine dimensions, a feat not repeated by his successors for many years.

He remained, however, modestly uncertain of the importance of his work. In a letter to Crell in 1782 he wrote:

"As platinum by itself is infusible and not ductile, but becomes so when treated with acids, is a strange phenomenon as you remark. I do not pursue this point in my work because I am no friend of hypotheses and find no ground for them here." (15)

Nicolas Christiern de Thy, Comte de Milly

Another Paris aristocratic figure who became deeply interested in platinum at this time was the Comte de Milly. He had served in the army as colonel commandant of the dragoons but after the battle of Minden in 1759 resigned his commission and entered the service of the Duke of Württemberg, returning to Paris at the conclusion of the Seven Years' War in 1763. Here he took up the study of chemistry, equipping a laboratory at Chaillot, and like many of his contemporaries became involved with the newly created porcelain industry as well

Nicolas Christiern de Thy, Comte de Milly 1728–1784

After a distinguished career in the French army and then in the service of the Duke of Württemberg, the Comte de Milly returned to Paris in 1763 and took up the study of chemistry, becoming especially interested in platinum. For a time he supported Buffon's view that this was an alloy of iron and gold, but then carried out his own investigations in continuation of the work of de l'Isle and Count Sickingen on the ammonium chloride precipitate. His work was not published in France, but in 1778 he communicated his results to the scientists of Spain, with important consequences in that country

as with platinum, a supply of which he had obtained from the Spanish Government. For a time, as we have seen, he supported Buffon in his view that this consisted merely of an alloy of iron and gold, but the work of de l'Isle convinced him that this theory was untenable, and he then set about some investigations of his own in continuation of that work as well as that of the Count von Sickingen with a view to simplifying the latter's procedure.

As with de l'Isle's and Sickingen's memoires to the Académie des Sciences, no account of Milly's paper was printed, but it is possible to establish the approximate date as sometime before 1778 from other sources.

In a paper on the assaying of platinum-gold alloys read to the Académie in November 1778 (16) Mathieu Tillet (1714–1791), the Royal Commissioner for Assays and Refining at the Paris Mint, mentioned that:

"I had read this memoire to the Academy when M. le Comte de Milly gave me a piece of ductile platinum which he had obtained by his experiments on this metal and which he had used successfully for various articles of jewellery."

In another reference in the same paper Tillet says that

"Platinum, whether forged or in its natural state, may be rendered ductile by the processes which M. le Comte de Sickingen has made known, and by those which M. le Comte de Milly is to publish."

This last reference to publication by Milly was the cause of a somewhat acid note by Sickingen in the preface to his "Versuche über die Platina". After

64

writing that some of his contemporaries who were aware of his work did not hesitate to claim a share in the discovery he continued:

> "The Count de Milly, following an early discovery of the two excellent refiners, Lewis and Baumé, that sal-ammoniac precipitates platinum out of its solution, made use of this information and adapted it to the method given in this present work, obtaining after calcination a workable and ductile platinum. He immediately sent a description of the process to the Spanish Academy of Sciences. There has not yet been sufficient experience to enable one to say whether this method of freeing platinum from its iron is as good as the one here described, the author doubts it but only experience can decide."

No record has been found of a communication from Milly to any Spanish academy of the time, but no doubt he felt under a certain obligation to report his results at an early stage to the country that had given him his raw material. At all events his action prompted the chemists of Spain to undertake some highly successful work, to be described in Chapter 6.

Fortunately we have virtually a first-hand account of Milly's procedure given by a rather curious individual who acted as an effective correspondent and go-between among the scientists of his time throughout Europe. This was Joao Jacinto de Magelhaens (1722–1790), more generally known as Magellan, a Portuguese who had devoted himself to science after an early career as a monk and who was friendly, among others, with Priestley, Lavoisier, Watson, Ingenhousz and Banks. Among his many activities in 1788 he edited, with most lengthy additions of his own, an English translation of A. F. Cronstedt's "Essay towards a System of Mineralogy" and in the course of this he gives the following account:

> "Platina may be reduced otherwise to a metallic state. The method by which the late Count de Milly employed in Paris, and which he was so obliging as to do in my presence and at my request, is as follows:
>
> "First he separated all the sand and other heterogeneous particles, by blowing them out whilst the grains of platinum were letting down from one paper to another paper. He put the metal in a matrass (flask) with twenty times its weight of aqua regia on a strong heat; the next morning he decanted it from a sediment composed of some whitish particles, of a metallic appearance, mixed with blackish matter, which he told me was a molybdenic substance. He then mixed with it an equal quantity of distilled water; precipitated the platinum by a solution of ammoniacal salt, and he filtered the liquor through blotting paper; this and the residue being dried in a plate over the fire was put on a Hessian crucible which he guarded within another large crucible. This was covered with a test and put on a blast furnace until it was red hot, even to a white heat, during half an hour; he then opened the crucible where I saw the metallic substance, like a filamentous mass; this he pressed down with an iron rod whose end was formed into a flat button; he covered it again, and continued to fire for ten or twelve minutes; the crucible being taken out, the solid mass was collected at the bottom; this could be forged and beaten on the anvil with a hammer into any form, like iron." (17)

The work of the three experimenters so far dealt with represents considerable progress in the application of powder metallurgy to platinum since it demonstrated that under the best conditions the sal-ammoniac or potassium chloride precipitate of platinum could be consolidated at high temperature and forged to a fully malleable metal. Both de l'Isle and de Milly preferred to use ammonium chloride as the precipitant rather than potassium salts in spite of its higher cost because of the greater ease with which the precipitate could be calcined to metal. Von Sickingen and de Milly both diluted their platinum solution before precipitation and reduced the possibility of base metal impurities being entrained. Von Sickingen in his later experiments found it helpful to calcine his precipitate separately and then to wash it thoroughly with water to remove the last trace of undecomposed chlorides before putting it into the crucible for the heat treatment.

As our story unfolds, it will be found that all these devices later became parts of the standard practice for refining platinum. But no one followed von Sickingen in adding a reducing agent to his original solution and so preventing the precipitation of iridium with the platinum. The others all worked on red precipitates, and it is the ammonium chloriridate that gives this colour to the yellow ammonium chloroplatinate. This was not known until Tennant discovered iridium in 1804, and wires of the fineness achieved by von Sickingen were not drawn again for a generation. In fact no one repeated his work; his process was too complex and difficult and it is not surprising to find Lavoisier in 1790 dismissing it as *très pénible*, but he had effectively demonstrated what the possibilities of the new metal really were.

De Milly simplified the process by going back to de l'Isle's sal-ammoniac, but lost the advantage of the removal of the iridium. He made a new addition to the procedure by assisting the sintering of the metal powder by pressing it down with an iron tamper, another device that became part of standard practice. In this way, by consolidating the work of his two precursors and adding an improvement or two of his own, de Milly was able to work on a larger scale and to produce good metal in quantity sufficient for the making of jewellery. Also, as already mentioned, he paved the way for a platinum industry in Spain, but we now have to turn to the numerous attempts to bring about the successful melting of platinum.

The Great Burning Glasses

After the partially successful attempts of Macquer and Baumé in 1758 to bring about the fusion of native platinum by means of a concave burning mirror nothing more of significance took place in this direction until 1772, when a new phase of activity commenced under the auspices of the Académie Royale des Sciences.

Some indecisive experiments, again under the direction of Macquer, who was by now the Professor of Chemistry at the Jardin du Roi and highly regarded as the senior chemist of his time, and with the help of the apothecary Louis Claude Cadet (1731–1799) and Antoine Laurent Lavoisier (1743–1794) then

A *Grande Lentille à liqueur*.
B *Petite Lentille pour rassembler les raions plus près*.
C *Centre de mouvement horisontal de toute la Machine*.
D *Manivelle servant à imprimer le mouvement horisontal*.
E *Manivelle servant à imprimer le mouvement vertical par le moien des Vis 1 et 2*.
F *Vis de rappel pour éloigner de la grande Loupe la petite Lentille ou la rapprocher*.
G *Porte objet aiant le mouvement de haut en bas et de bas en haut celui d'avancer et reculer parallellement à la plateforme et de s'incliner au degré du Soleil et de s'avancer parallellement aux raions*.
H *Chariot ou Plateforme portant toute la Machine et les Opérateurs*.
I *Roues du Chariot tendantes au Centre de mouvement par leurs Axes et roulantes sur des bandes de fer incrustées circulairement sur une plateforme de pierre*.
K *Escalier pour parvenir sur le Chariot, il est soutenu de deux rouleaux excentriques*.

After unsuccessful experiments with two older burning glasses this enormous piece of equipment was built for the Académie Royale des Sciences at the expense of Lavoisier's friend Jean Charles Philibert Trudaine de Montigny. Installed outside the Louvre in 1774, it was used by a distinguished committee of scientists led by Macquer and including Lavoisier but failed to achieve the melting of platinum

only a junior member of the establishment, had been carried out on the results of heating diamonds in both the presence and the absence of air. To conduct further trials it was then decided to ask permission from the Académie to make use of a great burning glass that had been kept there as a curiosity for over fifty years.

The pioneer of the burning glass had been a rich German nobleman, Count Ehrenfried Walther von Tschirnhaus (1651–1708), a chemist and a mathematician, who had devoted some of his great wealth to building a glass works on his estate. Tschirnhaus was a frequent visitor to Paris and had been elected a member of the Académie des Sciences in 1682, and one of his burning glasses was bought in 1702 by Philippe Duke of Orleans (the nephew of Louis XIV and on the latter's death regent for the young Louis XV). This was for the use of his protegé, the chemist Guillaume Homberg (1652–1715) for whom he had equipped a splendid laboratory in the Palais Royal. It was later consigned as a museum piece into the care of the Académie.

In July 1772 Cadet, with the support of the physicist Mathurin Jacques Brisson (1723–1806), asked for the use of the great lens. Permission was at once granted by the Académie, who asked Macquer and Lavoisier to join a committee

67

to take part in the proposed experiments. The apparatus was retrieved and, together with another Tschirnhaus lens owned by the Comte de la Tour d'Auvergne, was set up in the Jardin de l'Infante, a terrace beside the Louvre, where the Académie occupied rooms allocated to them by Louis XIV in 1699, and the investigation began in mid-August.

All manner of substances were exposed to this method of heating, and then on August 14th a small piece of native platinum was exposed to the heat of the Tour d'Auvergne lens. Lavoisier described the result:

> "The sky being but little favourable because of many light clouds, the platinum exposed for 24 minutes did not melt but softened and agglomerated more than it had done in earlier experiments and was still attracted by a magnet." (18)

On the 29th of the same month a further trial again resulted only in agglomeration, while on September 5th, as Lavoisier describes it:

> "a small mass of platinum, strongly agglutinated, that had already been exposed twice in the fire, exposed for 22 minutes in strong sunlight, hardly changed".

An account of these experiments was read to the Académie by Macquer on November 14th and was immediately published in the new journal founded by the Abbé Rozier, *Observations sur la Physique*. (19)

Some further experiments were carried out in the late summer of 1773, but the two Tschirnhaus lenses were not proving altogether satisfactory. They both contained bubbles, striations and other defects and the committee felt that a better effect would be obtained with an apparatus consisting of two large pieces of glass with a curvative forming part of a sphere, joined at their circumferences and then filled with alcohol. This suggestion was taken up by Jean Charles Philibert Trudaine de Montigny (1733–1777), an older friend of Lavoisier and the Intendant General of Finances. At his expense such a piece of apparatus was constructed by the engineer de Bernières, the very large pieces of glass being made in a newly built furnace in the Paris works of Saint Gobain, who donated the glass to the Académie.

This enomous piece of equipment, illustrated here, was installed in the Jardin de l'Infante and was ready for operation at the beginning of October 1744. The great lens was four feet in diameter, against the three feet of the Tschirnhaus lenses, and was mounted on a carriage to enable the movement of the sun to be followed. The focal length was ten feet, at which point the light was so strong as to harm the eyes of the observer, and a smaller lens was arranged to concentrate the sun's rays still further. Iron was readily melted by this means, but once again no success was obtained with native platinum.

A report of the experiments by Trudaine de Montigny, Macquer, Cadet, Lavoisier and Brisson was read to the Académie on November 12th, 1774 by Brisson and printed in the Mémoires of the Académie for that year, although these were not published until 1778 (20). It contains the following paragraph:

> "Having exposed to the fire some grains of platinum in a carbon cavity it appeared to congregate and to be reduced in volume and to be about to melt. A little later it

bubbled and fumed and all the grains united in to a single mass, but without forming a spherical button as with the other metals. After this kind of semi-fusion the platinum was no longer attracted by a magnet as it had been before being exposed to the action of the sun."

Thus platinum had still not been melted, at least partially on account of its admixture in its native form with iron and sand and the consequent formation of a refractory iron oxide.

But the Count von Sickingen, who had been invited to take part in these experiments, achieved success with his precipitated platinum. The report continued:

"The Baron Sickingen, Minister of the Elector Palatine, who cultivates the sciences with as much success as sagacity, subjected to the same heat a portion of platinum that he had stripped of its iron by a particular process and that was not attracted by a magnet; this platinum was reduced in volume, gave off smoke and then united into a single mass which could be flattened under a hammer."

Lavoisier Melts Platinum With Oxygen

Now it was just at this time that Joseph Priestley had discovered oxygen – or "dephlogisticated air" as he called it – and as is well known, on a visit to Paris in the October of 1774 he acquainted Lavoisier with his findings during the course

Antoine Laurent Lavoisier
1743–1794

Realising the significance of Priestley's discovery of oxygen, Lavoisier confirmed and extended the study of the constituent of air that supported and took part in calcination and combustion. This he first called "vital air", the purest part of the air. In 1782, with a massive piece of apparatus he had designed to yield a continuous stream of oxygen, he was the first to succeed in bringing about the true melting of platinum, although only on a very small scale

69

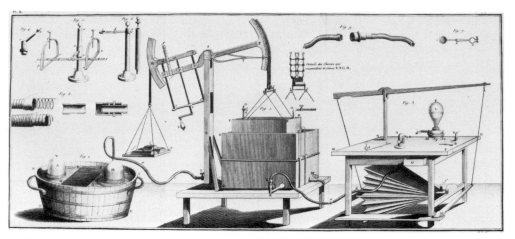

The apparatus designed by Lavoisier and built by his instrument maker Pierre Mégnié to burn
continuous streams of oxygen and hydrogen. Before this had been completed he used it in April
1782 to direct burning oxygen on to a small quantity of platinum held in a piece of charcoal.
Three months later he repeated the experiment at a meeting of the Académie Royale des
Sciences before a distinguished audience including the Grand Duke Paul of Russia and Benjamin Franklin, then in his seventy-seventh year. The drawing, as with all those illustrating his
papers, was prepared by Madame Lavoisier

of a dinner party given by him, an announcement that occasioned considerable
surprise. No thought of using this new kind of air for the melting of refractory
metals seems to have occurred to Priestley, but his friend and neighbour the
famous astronomer and geologist the Reverend John Michell (1724–1794), on
learning of the discovery "observed that possibly platina might be melted by
means of it", a comment that Priestley included in his "Experiments and
Observations on Different Kinds of Air", published in 1775 (21).

This publication, as with many others, was forwarded by Magellan to
Lavoisier, who was now directing his thoughts more and more to the role of the
air, or some part of it, in combining with metals during calcination. A few years
later, also well known to chemists everywhere, he announced that he would give
the name "oxygen" to this "most salubrious and purest part of the air".

In the meantime Franz Karl Achard in Berlin had made use of Priestley's
"dephlogisticated air" to yield higher temperatures than had hitherto been
possible, directing a stream of oxygen on to a piece of carbon. This he reported
to the Royal Academy in Berlin in 1779, although the memoires were not
published until two years later (22), but it clearly gave Lavoisier the impetus to
conduct some further experiments. He had already designed a "caisse
pneumatique" or gasometer to produce a blast of oxygen or hydrogen, or both
together, and in April 1782 he employed the first of these gasometers in a rather
spectacular experiment to direct the stream of oxygen into a hollowed-out piece
of charcoal in which he had placed a small quantity of platinum. The apparatus
is illustrated here; its construction and use were described in a long paper to the

70

Académie, "Sur un Moyen d'augmenter considérablement l'action du Feu et de la Chaleur, dans les Opérations chimiques", read in 1782 (23). This includes a reference to "this air that M. Priestley has discovered nearly at the same time as myself" and describes the melting of platinum held in a small piece of charcoal in the stream of oxygen:

> "The platinum had melted completely, and the small particles were united in a perfectly round globule; the melting was complete and easy whether I employed ordinary platinum that one finds commercially or whether the molecules attracted by a magnet had previously been removed."

This experiment, which was reported to a meeting of the Académie des Sciences on April 10, caused something of a sensation in scientific circles in Paris, and when a special meeting of the Académie was arranged three months later for the entertainment of the Grand Duke Paul of Russia and his Grand Duchess the apparatus was transported there at considerable trouble and expense. The Grand Duke, the son of Catherine the Great, and the future Tsar Paul I, was travelling incognito as the Comte du Nord although his true identity was well known, and on June 6, 1782 he was able to witness this historic melting of platinum before going on to a party given for him at Versailles by Louis XVI and Marie Antoinette.

Benjamin Franklin writes to Priestley

Another famous observer of the scene, and one of an entirely different character, was Benjamin Franklin, who had been resident in Paris since 1776 as the representative of the revolutionary United States Government. He had been elected a Foreign Member of the Académie in 1772 and had developed a close friendship with Lavoisier, whose experiments he had frequently been invited to watch at the Arsenal. He was also, of course, a close friend of Priestley, whom he had first met in London in 1765 and with whom he maintained a steady correspondence over the years.

On the day following the platinum melting demonstration Franklin concluded a letter to Priestley:

> "Yesterday the Count du Nord was at the Academy of Sciences, when sundry Experiments were exhibited for his entertainment: among them, one by M. Lavoisier to show that the strongest fire we yet know is made in a Charcoal blown upon with dephlogisticated air. In a Heat so produced, he melted Platina presently the fire being much more powerful than that of the strongest burning mirror." (24)

Two weeks later Franklin wrote a very similar letter to another of his scientific friends, Jan Ingen-housz, then in Vienna, who was already making his own contribution to the investigation of platinum and its properties.

Lavoisier carried out further experiments in the following year and in a paper to the Académie reported that:

> "Native platinum exposed to a current of 'vital air' melts in 15 or 20 minutes if the quantity does not exceed 5 to 6 grains; the fusion is quite complete, and the metal

The concluding paragraph of a letter from Benjamin Franklin, written to Priestley on June 7, 1782, the day following Lavoisier's spectacular demonstration of the melting of platinum. A similar letter was written two weeks later to another of Franklin's friends, Jan Ingen-housz

Photograph by courtesy of the Library of Congress, Washington

forms round globules, but if the quantity is a gros (72 grains) or thereabouts the fusion is very difficult" (25).

He also exposed in the same way a piece of platinum that had been forged by the Comte de Milly and concluded that the forged metal was a little more fusible than native platinum.

So platinum had at last submitted to fusion, if only in minute quantity, and the news was carried swiftly to Priestley. Lavoisier, with his ingenious piece of equipment, using the "dephlogisticated air" discovered by Priestley seven years earlier, and perhaps owing something to the suggestion made by the English parson and scientist John Michell, had achieved success where others had failed for so long. But it was to be a very long time before platinum was able to be melted on a larger or a commercial scale.

References for Chapter 4

1 A. Baumé, Chymie Expérimentale et Raisonnée, Paris, 1773, **3**, 189–194

2 P. J. Macquer, *Mem. Acad. Roy. Sci. Paris.*, 1758, 132

3 G. Bouchard, Guyton Morveau, Chimiste et Conventionel, Paris, 1938, 89–92

4 W. A. Smeaton, *Platinum Metals Rev.*, 1966, **10**, 24–28

5 D. McDonald, *Platinum Metals Rev.*, 1967, **11**, 106–108

6 A. L. Lavoisier, Oeuvres, Paris, 1868, **4**, 237

7 B. C. Sage, Elemens de Minéralogiè, Paris, 1777, **2**, 361

8 J. B. L. R. Delisle, Crystallographie, 2nd Edn., Paris, 1783, **1**, 487–490

9 L. B. Guyton de Morveau, *Obsns. Physique (Rozier)*, 1775, **6**, 193–203

10 P. J. Macquer, Dictionnaire de Chimie, 2nd edn., Paris, 1778, **4**, 197

11 K. H. von Sickingen, Versuche über die Platina, trans. G. A. Suckow, Mannheim, 1782

12 L. Crell, *Neueste Entdeckungen in der Chemie*, 1781, (3), 271–272; 1782, (5), 268–270; (6), 197–206

13 J. Ingen-housz, Nouvelle Experiences et Observations, Paris, 1785–1789, **1**, 446; **2**, 506–517

14 Letter from Georg Forster in Vienna to S. T. Sömmering, August 14, 1784; H. Hettner, Georg Forster's Briefwechsel mit S. T. Sömmering, Braunschweig, 1877, 111

15 K. H. von Sickingen, *Neueste Entdeckungen in der Chemie*, 1782, (6), 141–147

16 M. Tillet, *Mem. Acad. Roy. Sci. Paris.*, 1779, 373–377; 385–437; 545–549

17 A. F. Cronstedt, An Essay towards a System of Minerology, trans. G. von Engestrom, 2nd edn., edited J. H. Magellan, London, 1788, **2**, 574

18 A. L. Lavoisier, Manuscript, Archives of Acad. Roy. Sci., Paris; Lavoisier, Oeuvres, 1865, **2**, 612–616

19 P. J. Macquer, *Obsns. Physique (Rozier)*, 1772, December, 93–106

20 Trudaine de Montigny, Macquer, Cadet, Lavoisier, and Brisson, *Mem. Acad. Roy. Sci. Paris*, 1774, **88**, 62–72

21 J. Priestley, Experiments and Observations on Different Kinds of Air, 1775, **2**, 100–101

22 F. K. Achard, *Nouveaux Mem. Acad. Roy. Berlin.*, 1781, 20–26

23 A. L. Lavoisier, *Mem. Acad. Roy. Sci. Paris*, 1782, 457–476

24 Letter in Library of Congress, List 1048

25 A. L. Lavoisier, *Mem. Acad. Roy. Sci. Paris*, 1783, 605–606

Franz Karl Achard.

Franz Karl Achard
1753–1821

The son of a pastor, Achard studied chemistry under Marggraf in the Berlin Royal Academy of Sciences, becoming his friend and eventually his successor as Director of the Physical Class and the Chemical Laboratory in 1782. As a young man of only 26 he succeeded, where his predecessors had failed, in melting platinum with the aid of arsenic, describing his results in one of three memoires read to the Berlin Academy. In 1788 he published a little-known book, "Recherches sur les Propriétés des Alliages Metallique", the first compilation of data on alloy systems

5

The Arsenic Process and its Use by the French Goldsmiths

"But these chemical processes, which have as yet been practised upon small quantities of platina, do not prove so much with respect to the possibility of working platina in a large way and of employing it usefully in the arts, as the two pieces which I now show to the Académie made by M. Janetty."

ANTOINE LAURENT LAVOISIER

The discovery by H. T. Scheffer in 1751 that by the addition of a small quantity of arsenic to platinum and then heating to redness it "melts in the twinkling of an eye" exercised a great fascination on those seeking a method of rendering it workable. They had, however, great difficulty in confirming this effect. Lewis was unable to produce a melt fluid enough to pour into a mould; Marggraf could not achieve combination at all, and neither could Macquer and Baumé. As we have seen, Guyton de Morveau was successful in 1775, but there was still no understanding of what was happening.

The process was successfully used and explained in 1779, when Franz Karl Achard, Marggraf's successor, read a paper to the Royal Academy of Sciences in Berlin although this was not published until two years later (1). Achard heated in a luted crucible to as high a temperature as he could get a mixture of 120 grains of native platinum with a similar weight of powdered white arsenic and 180 grains of potash. On cooling and breaking open, he found that the platinum had melted into a well-rounded button weighing 120 grains, which was very hard and brittle. To see if the arsenic had combined with the platinum or had merely acted as a flux he heated a portion of the button in a muffle, and observed that as soon as it had reached a dull red heat it began to give off white arsenical fumes. Taking it out for inspection he was very surprised to find that it had softened to the consistency of tin amalgam. On raising the temperature of the muffle the metal melted, but after an hour's treatment it had become solid again and no longer gave off white fumes. Its weight had decreased from 77 grains to

75

64, and under the hammer it appeared to be fully malleable and as ductile as gold; moreover it could be filed easily. It could not be melted at the highest temperature that he could attain, but this further treatment did not produce any more loss in weight, so that he presumed that all the arsenic had been removed. From these results Achard concluded that platinum does combine with arsenic, becoming readily fusible and very brittle, and that the two can be separated by heating.

He noted that during the initial fusion with arsenic and potash, the weight of the button was the same as that of the original platinum in spite of the fact that a quite considerable quantity of arsenic had been taken up. He explains this as due not to any loss of platinum but "rather to the separation and destruction of the heterogeneous and chiefly ferruginous parts which it contained", as was confirmed by the brown colour of the slag which covered the melted material. Achard then went on to comment on the experience of his predecessors, concluding that the failure of most of them to achieve fusion was due to the fact that under the conditions of their work the arsenic volatilised and was lost before it could act upon the platinum. He contended that the alkali (potash) he added prevented this and enabled fusion to take place, and further he found by numerous experiments that, to ensure no arsenic escaped, three parts of potash had to be added for every two parts of the former. He made no reference to the work of de Morveau, the results of which can be explained in a similar way.

The First Platinum Crucibles

Achard's paper – actually three consecutive memoirs – dealt with arsenic and its combination with a large number of metals and compounds, but apparently the effects he obtained with platinum interested him more than most. He continued his work on this and in 1784, in the very first number of Lorenz Crell's *Chemische Annalen*, he published another paper under the title "Easy Methods for Making Vessels from Platinum" (2). By then he had satisfied himself that the procedure outlined above could be repeated as often as desired, and he went on to suggest that this property of arsenic of making platinum fusible and then being afterwards completely removable by heat, made it possible "for us to make from it all kinds of small vessels and especially small fusion crucibles which can be useful in certain operations". He then described a small former he had made out of clay in which such a vessel could be made by filling up the space between the inner and outer parts of the former with powdered "arsenical platina", and heating the whole to a high temperature.

"The arsenical platina melts, and when the arsenic has volatilised again solidifies and takes the shape of the space. After cooling the former is broken open and the crucible, which has been made from the platina, is hammered a little on a mandrel, and is then ready."

Achard says: "I have succeeded very well . . . in making a fusion crucible from

76

Achard's 1779 papers dealt with the melting of a wide range of metals with arsenic, but he was particularly interested in the effect obtained with platinum and in 1784 published his "Easy Methods for making Vessels of Platinum" in the first number of Crell's *Chemische Annalen*

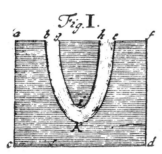

The small diagram in his paper shows his method of making a crucible by pushing powdered arsenical platinum into a clay mould and heating to a high temperature

I.

Leichte Methode, Gefäße aus Platina zu bereiten; vom Hrn. Professor Achard, Director der physikalischen Classe der Königl. Preuß. Academie der Wissenschaften.

Ich untersuchte, was der Arsenic für Würkung auf verschiedene Substanzen hätte; die Platina kam auch mit vor. Ich that zwey Quentgen davon in einen Heßischen Schmelztiegel mit gleichen Theilen Arsenic und Weinsteinsalz; welches letztere ich in der Absicht, den Arsenic zu figiren, hinzuthat. Nachdem ich den Tiegel wohl lutirt hatte; setzte ich ihn eine Stunde in einen Windofen. Die Platina war vollkommen geschmolzen; sie war sehr spröde im Bruch, noch etwas weißer, wie die reine Platina. Von dieser arsenicalischen Platina that ich ein Stück in einen Treibscherben unter der Muffel in einen Probierofen. Kaum war sie durchgeglühet; so wurde das Stück so weich, wie ein Amalgama aus gleichen Theilen Bley und Quecksilber; und bey verstärkter Hitze

A 2   kam

platina" and therefore it is to be presumed that he did manufacture at least one such vessel. If that is so, then this was the first platinum crucible ever made. The call for these crucibles came from those who were attempting to make analyses of minerals on the basis of the methods recommended by Bergman and involving fusion of the material with alkalis. For this purpose platinum offered itself as ideal and so found its first niche in laboratories. It seems unlikely, however, that Achard made many crucibles. After the volatilisation of the arsenic the shape must have been in a very fragile form and, if surviving the breaking up of the mould around it, required the most delicate handling in the earlier stages of consolidation and during the hammering that had to be applied both inside and out.

Little more was heard in fact, of Achard's crucibles but in 1785 Guyton de Morveau took up the process, finding that the flux of white arsenic and potash caused excessive swelling. This he remedied by adding common salt and powdered charcoal and he was able to show three crucibles to the Dijon Academy (3). One of these was rather spongy but the larger of the other two was sound, of equal wall thickness throughout, shining and free from blisters. This he found much more useful for the analysis of minerals than the iron crucibles recommended by Bergman although it was still attacked by metals and by nitre.

MÉMOIRE

Sur la fabrication des uftenſiles de platine.

Par M. de Morveau.

J'ai annoncé, il y a près de dix ans, l'effi-
cacité de l'arſeniate de potaſſe ou ſel neutre
arſenical pour mettre la platine en fuſion par-
faite, au point de donner un culot qui ſe
laiſſoit limer & même un peu étendre ſous le
marteau (1). Depuis ce temps j'ai fait divers
eſſais pour la faire couler, au fourneau même,
dans des moules de terre de coupelle (2);
mais ils n'avoient pour objet que de très-
petites pieces, comme des becs de chalumeau
ou des cuillers à l'uſage de cet inſtruḿent,
& l'augmentation de poids que j'avois remar-
quée dans la platine employée à ces expériences, me faiſoit craindre qu'elle ne retint
aſſez d'arſenic revivifié pour participer des
imperfections de ce demi-métal.

---

(1) *Lettre à M. le Comte de Ruſſon ſur la fuſibilité
de la platine , &c.* imprimée dans le Journal Phyſique
du mois de Septembre 1775 , expér. v.

(2) Voy. mes notes ſur les *Opuſcules chymiques, &c.
de Bergman,* tom. 2 , pag. 91 , 185 , 460 & 463.

In a paper read to the Dijon Academy in 1785 Guyton de Morveau described Achard's procedure for making a platinum crucible but considered he had improved on this method by eliminating the swelling caused by the arsenical flux. By adding common salt and charcoal to the mixture he was able to make three crucibles larger than Achard's

The King's Goldsmith Janety

The seed that Achard had sown fell upon fertile ground in an unexpected quarter and for the first time interest in platinum became evident among crafts-men as well as among scientists. In the time of great French prosperity under Louis XIV and Louis XV there had grown up in Paris and elsewhere a very able and well-organised craft of goldsmiths and jewellers and it was among these that the interest appeared. As will be seen, one or two others entered upon this development, but by far the most outstanding was Marc Etienne Janety (1739–1820), Royal Goldsmith to Louis XVI.

The name of this man is spelled variously in the literature, some examples being Janetty, Janetti, Geanty, Jeanety, Jeanetty and Gianetti, the last probably giving a clue to his origin. In this story the spelling Janety will be used throughout because that was the one used in the register of his admission to the craft as a Master Goldsmith of Paris on July 26th, 1777 (4), as well as in the signed letter to Sir Joseph Banks to be mentioned later.

Before that he had been an apprentice to Vincent Bréant, a member of an old-established goldsmithing family, and when the latter retired in 1778 Janety took over his business in the Rue de l'Arbre Sec. He continued his work on plate and jewellery, but according to Bertrand Pelletier (1761–1797) he had already turned his attention to platinum in 1786, a date fully confirmed by the one surviving piece of his, the sugar bowl illustrated here (5). By 1788 Pelletier was able to say of him, in the course of a paper read to the Académie des Sciences, that he

"has busied himself for more than two years on work in platinum and has succeeded in making it in large amounts very pure and very malleable. He has made crucibles of it, snuff-boxes, etc., which several individuals have possessed for a long time. He makes use of arsenic to melt it; but he has special methods of removing it afterwards. It is the fruit of a work not only assiduous and troublesome but also very dangerous, since he has several times been seen in an atmosphere full of arsenic fumes."

This arsenic process as applied by Janety and one or two others to be mentioned later was not merely a subterfuge for getting the platinum melted but was a real refining process for the removal of the iron and copper, producing a more or less pure platinum in a form suitable for hot forging. The difficulties were many and will be more easily understood if we examine the scientific limits within which it had to be carried out. The eutectic, that is the alloy of the arsenic-platinum series that melts at the lowest temperature, contains 13 per cent of arsenic and melts at 597°C. This fact governs the melting, or at any rate the softening, of any mixture of the two metals.

No portrait seems to exist of Janety, the Royal Goldsmith to Louis XVI. He is known to have been apprenticed to Vincent Bréant who lived and worked in the Rue de l'Arbre Sec, near the Louvre, and to have succeeded him in 1778. A year earlier he was admitted as a Master in the Paris Guild of Goldsmiths and inscribed his mark, a crowned fleur-de-lys with two grains and the initials M. E. J. on a marc weight. By 1786 he had mastered the arsenic process of making platinum malleable and of producing both articles of jewellery and pieces of chemical apparatus. The platinum coffee pot he made in the same year was shown to a meeting of the Académie des Sciences in 1790 by Lavoisier, who explained that some parts had been shaped by hammering the metal cold while other parts had been soldered. This was last seen in an exhibition held in Paris in 1933, on loan from a private collection, but unfortunately disappeared during the second world war

The only surviving piece made in platinum by Janety, now in the Metropolitan Museum of Art in New York, this beautifully designed and skilfully made sugar bowl is seven inches in length and is fitted with a dark blue glass liner against which the white brilliance of the platinum is particularly effective. It is signed and dated along the rim of the base:

PLATINA JANETY FECIT 1786

Photograph by courtesy of Mrs. Clare Le Corbeiller and the Metropolitan Museum of Art

It was therefore necessary to avoid melting since the removal of the arsenic was much easier if the surface of the material remained large and it was desirable that the temperature at which this process was conducted should not approach 600°C. Now arsenic oxidises fairly freely at quite a low temperature and the arsenious oxide produced volatilises appreciably at 300°C and freely above 450°C. The object of the refiner therefore was to obtain a rapid oxidation and volatilisation of the arsenic without melting the alloy, although later, as we shall see, he tried to keep it in a reduced state in the last stages of its removal.

Janety's interest in the working of platinum must have received a great stimulus by the arrival at his house in 1786 of Pierre François Chabaneau who, as we shall learn in the next chapter, also began in 1786 to produce malleable platinum in some quantity in Spain. He was escorted by his patron, the Spanish Ambassador to France, the Count Pablo Abarca Aranda, and brought with him

80

ingots of malleable platinum weighing 44 marcs, and together they carried out experiments in producing "coffee pots, plates, watch chains, mustard pots, tea pots and dress-coat buttons" (6). This work cannot have failed to teach Janety much about the properties of the metal in quantity, to encourage him to push forward his own difficult task and to work on a much larger scale. His work soon began to attract a great deal of attention on the part of the chemists of the capital and his products came into considerable demand.

In 1787, in a letter to Lorenz Crell from Paris, de Morveau (7) reports that

"platinum is now being fabricated very well here....M. Janett already makes excellent work from this metal and he will soon be able to make from it anything he wishes. I had advised him to quench the red hot platinum quickly in molten saltpetre in order to secure even greater purification. He assured me afterwards that by this procedure he had made malleable a piece of platinum that was brittle under the hammer; but at the same time it must not be left too long in the saltpetre because the latter begins to attack it. There is no lack of platinum so we shall soon be able to make vessels and other apparatus".

The last sentence is of interest, confirming that a supply of the metal was now being smuggled out of South America.

In 1790 another of Crell's correspondents, Professor Anton von Ruprecht of Chemnitz, writes on a visit to Paris:

"Without any doubt you will have known for a long time that some excellent chemical artists have found out how to work platinum like other metals, but it will scarcely have been believed how this work has multiplied. I can now assure you that from Janety one can get immediately at a very reasonable price snuff-boxes, watch-chains, spoons, toothpicks, little blowpipes and anything of that kind that one wants and orders, made from pure platinum in the neatest and most tasteful manner." (8)

The Effects of the French Revolution

All was not going smoothly, however, with Janety's work. The beginning of the French Revolution in July 1789 found him with a considerable stock of malleable platinum for which he could not foresee a market, and in December of that year he wrote to Sir Joseph Banks, President of the Royal Society, appealing for his help. This letter, which was located in the British Museum by Dr W. A. Smeaton (9) reads in translation:

Sir,
   Without having the honour of being known to you, I nevertheless take the liberty of writing. By the most persistent labour I have succeeded in making platinum malleable. This discovery, from which at any other time I would have greatly benefited, is of no advantage to me here, because of circumstances that are known to all Europe.

   Knowing the interests, Sir, that you take in the arts and sciences, I think I may permit myself to offer you a quantity of very malleable platinum in bars. If you are able to accept my proposition, I shall also divulge my process to you. I am not sufficiently vain to believe that if you wish to turn your attention to the subject you would need my small amount of knowledge, but much patience is required, and this would delay the full utilisation of the substance.

lumieres, mais je sais qu'il faut beaucoup de
patience, ce qui retardera encore de faire avec
cette matière ce qui est possible

Mr ingenoux qui doit etre actuelement a
Londres me fait l'honneur d'avoir de l'amitié
pour moy, comme je ne doute pas d'un instant
qu'il n'ay l'honneur de vous voir il pourra
vous dire si vous pouvez avoir confiance a
moy.

je vends icy l'once de platine 30 —
tournois mais pour me deburasser d'une
quantitée un peut considerable je ferais des
sacrifices. en me pardonnant la liberté que je
prends je vous prie de me croire avec respect

Monsieur votre tres humble
et obeissant serviteur
Janety
paris ce 13 xbre 1789 —
Md orfevre Rue de l'arbre sec
au coin de celle Bailleul

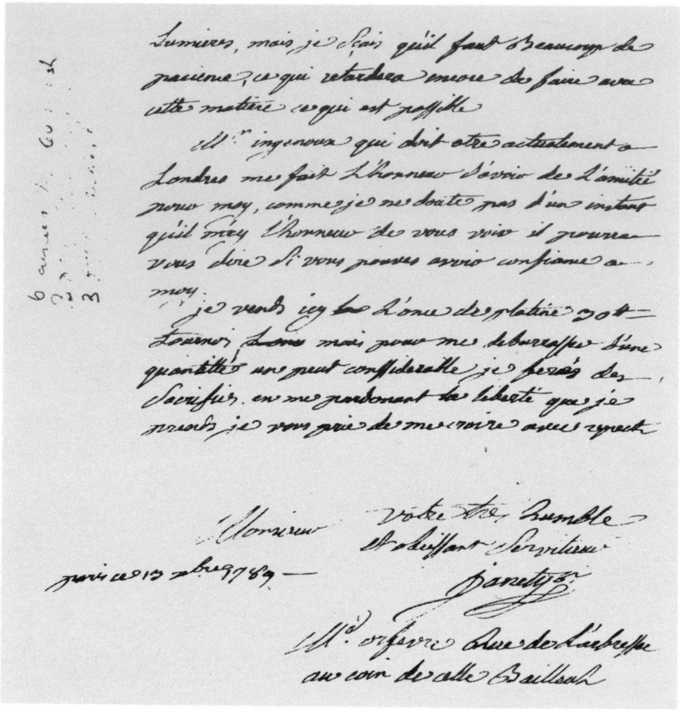

The outbreak of the French Revolution in July 1789 seriously affected Janety's business, leaving him with a quantity of malleable platinum on his hands, and on December 13th he wrote to Sir Joseph Banks, President of the Royal Society, hoping to enlist his help in disposing of it. This is the second page of the letter in the course of which Janety offers to divulge the secret of his process if Banks can find a market for the platinum

Photograph by courtesy of the Trustees of the British Museum

Mr Ingenouse, who must now be in London, does me the honour of being friendly towards me. As he will doubtless have the honour of seeing you, he will be able to tell you whether you can have confidence in me.

Here I sell an ounce of platinum for 30 *livres tournois* [about 26 shillings], but to dispose of a fairly large quantity I would make a reduction.

I ask you to excuse the liberty that I take. . . .

Janety

Master Goldsmith, at the corner of the Rue de l'Arbre Sec and the Rue Bailleul. Paris, 13 Xbre [December] 1789.

Banks was unable to dispose of any substantial quantity of this platinum. His pencil notes on the letter show that six ounces were bought by Henry Cavendish and three ounces each by Alexander Dalrymple, the hydrographer to the Admiralty, and by Alexander Aubert, an amateur astronomer who had a well-equipped observatory.

Janety's work on platinum suffered a further interruption. Early in the Revolution there was a great shortage of copper for both coinage and cannon and the large quantities of high-tin bronze contained in the bells of the many disused churches attracted the attention of the authorities. The chemist Fourcroy was called up to devise a process to separate the two metals, and in this he had the assistance of Janety who developed a satisfactory method of refining, yielding ductile copper from which several thousand cannon were made (10).

Janety need not have worried unduly about his stock of platinum because as we shall see the Académie took more than 400 ounces in 1791. Meanwhile, in 1790 he was honoured by Lavoisier himself introducing to the Académie des Sciences two pieces of his work, one of which was a coffee-pot (11). After outlining the various means available at the time for purifying platinum, Lavoisier pointed out that not only had they hitherto been employed on but very small quantities of material, but had often led to no more than imperfect results.

In his paper "Observations on Platinum", presented to the Académie des Sciences in 1790, Lavoisier summarised the several methods so far proposed for bringing platinum into a workable form and went on to refer to Janety's work in the passage quoted at the head of this chapter. He concluded:

"It depends therefore entirely on the Spanish government, which has in its possession the only mines of platina that are known, whether we shall be permitted to enjoy the advantages which might be derived to society from the use of a metal which is so unalterable, and which, for many purposes, in various arts, is preferable even to gold itself"

Enfin, j'ai fait voir que le platine, lorfqu'il eft allié à un métal volatil ou calcinable, eft fufceptible d'une efpèce d'affinage analogue à celui que reçoit le fer dans les forges.

Mais ces différens procédés chimiques, qui n'ont encore été employés que fur de petites quantités de platine, & dont plufieurs même n'ont conduit qu'à des réfultats imparfaits, ne prouvent pas autant fur la poffibilité de traiter le platine en grand & de l'employer utilement dans les arts, que les deux pièces que je mets dans ce moment fous les yeux de l'académie; elles ont été fabriquées par M. Janetty, avec du platine qu'il a traité lui-même par un procédé qui lui eft particulier, en forte que le mérite de ce travail lui appartient en entier. C'eft également lui qui avoit exécuté en platine, fous la direction de M. Chabano, un fuperbe néceffaire deftiné pour le roi d'Efpagne.

Ces pièces, & principalement le vafe qui eft fous les yeux de l'académie, prouvent qu'on peut fabriquer avec le platine des uftenfiles de toute efpèce; ce vafe contient en effet des parties planées à froid, telles que le fond, & des parties foudées. Il n'eft rien qu'on ne puiffe exécuter avec la réunion de ces deux moyens.

83

Greater possibilities of large-scale work were indicated by the two specimens which he now presented to the meeting:

"they have been fabricated by M. Janety from platinum which he has himself treated by a process of his own devising so that the whole of the credit for this work belongs to him. . . . These specimens and especially the vessel which is before the eyes of the Académie prove that utensils of every kind can be made in platinum; in fact this vessel contains cold-worked parts, such as the bottom and soldered parts. By using together these two means there is no limit to what can be made".

Further evidence of Janety making apparatus in platinum for use by the chemists of the time is given in the diary of the Scottish geologist and chemist Sir James Hall, revealed by Dr J. A. Chaldecott (12). During a second visit to Lavoisier in 1791 – he had first made his acquaintance five years earlier – Hall met Armand Seguin (1765–1835), Lavoisier's young assistant, and on July 5 records:

"Went with M. Seguin to Jannetti rue de l'arbre sec. He works in Platina. I bespoke a little spoon for the blow pipe & cup with some wire of that metal. M. S told me that as it is thus prepared the metal does not stand the action of pure caustic alkali when urged by a great heat, for by this means it becomes porous and lets the alkali through by acting as it is supposed upon the arsenic which has not been thoroughly driven off. Yet in this state it will answer many purposes."

Appreciation by the Académie des Sciences

By now Janety's work had become so well known that the Académie des Sciences appointed Claude Louis Berthollet and Bertrand Pelletier as Commissioners to investigate his process and to determine if it was deserving of

In 1791 the well known Scottish geologist and chemist Sir James Hall, during a visit to Paris, accompanied Lavoisier's assistant Armand Seguin to Janety's workshop to order several items of platinum apparatus. This entry from his diary for July 5th records the occasion

Photograph by courtesy of Dr. J. A. Chaldecott and the National Library of Scotland

84

public support. Their report was published in 1792 (13) and gives a most generous account of Janety's struggles and eventual success:

"M. Jeanety regarded it as an important aim for the goldsmiths' trade of the capital to seek means for working this metal; animated by a desire to be useful to his country he neglected his former occupations, the work in objects of gold and silver; he no longer busied himself with anything but researches on the means of working platinum. None of the scientists failed to recognise the obstinacy and courage with which this artist carried on his work; the sacrifices that he has made, the dangers he has undertaken and the success he has finally obtained are known to the whole of Europe."

They went on to record that the Académie had turned to Janety to obtain a bar of platinum no less than 14 feet long, a ball weighing 18 marks (144 ounces), two bars 19 feet long each weighing 22 marks (176 ounces), a bar 19 inches long and a pendulum bob weighing 12 marks (96 ounces) and that he had made considerable numbers of crucibles, snuff-boxes and watch-chains as well as a "set of buttons and a watch-chain of the most rare beauty for the King".

Finally the Commissioners concluded:

"You will readily agree that to have introduced into commerce a metal so precious and with such important uses is to have earned the right to a national reward."

Earlier in their report Berthollet and Pelletier gave an account of Janety's procedure – "just as M. Jeanety has submitted it to the Commissioners". In essence he first cleansed the native platinum by washing to remove as much as possible of the sand from it. He then mixed 24 ounces of it with 48 ounces of white arsenic powder (arsenious acid) and 16 ounces of refined potassium carbonate. This mixture he melted in a crucible with careful stirring with a platinum rod; after cooling and breaking up, a well-formed button was disclosed which was still magnetic. What had happened was that the mixture melted with an evolution of carbon dioxide while part of the iron and copper present in the native platinum was oxidised and passed into the slag. This came about at the expense of some of the arsenious acid with a production of free arsenic. A second reaction taking place at the same time was based on the affinity of arsenic acid for potassium and resulted in another portion of arsenious acid decomposing into arsenic acid (becoming potassium arsenate) and more free arsenic. The free arsenic produced by both these reactions alloyed with the platinum and brought about its fusion. Janety realised the importance of removing all the iron before attempting to work the metal and therefore he powdered his button and treated it in the same way a second time and, if necessary, a third. Having obtained a non-magnetic alloy he melted it with a little more arsenic and potash in such a manner as to obtain flat circular cakes of uniform thickness and about $3\frac{1}{4}$ inches in diameter, weighing about 27 ounces. He then placed these cakes in a shallow muffle leaning more or less vertically against the wall, and carefully heated them in such a way that they were equally heated all round until the arsenic began to volatilise. The temperature was very carefully maintained for six hours, the positions of the cakes being changed from time to time to ensure even heating; the greatest care had to be taken that the temperature did not rise sufficiently to cause sagging or melting. The cakes were next quenched in "common oil", replaced in the muffle and heated for another six hours at a temperature sufficient to drive off all the oil and smoke. In this part of the process the residual arsenic remained in the metallic state and was driven off in vapour. When this had ceased the temperature was raised as much as possible.

Janety's account concludes:

"If the preliminaries that I have indicated have been well followed the operation lasts only eight days. Then I clean my buttons in nitric acid and boil them in distilled water until they do not contain any more acid. I then place several of them one upon the other; I apply to them the greatest possible degree of heat and I strike them with a pestle, taking care at the first heating to bring them to redness in a crucible in order that no foreign matter shall be introduced into my buttons which are only spongy masses before this first compression. Afterwards I heat them harder and I form from them a square which I strike on every face for a more or less considerable time according to their volume."

It is highly doubtful that Janety actually received any reward for his prodigious efforts in the troubled times of the French Revolution. He continued his work in the Rue de l'Arbre Sec, but after the execution of Louis XVI in January 1793 he decided that the office of Royal Goldsmith had become rather incompatible with the revolutionary atmosphere and he retired to Marseilles where he set up again as a manufacturer of clock components. His stay here was not long, but his return on a summons from the government and his subsequent important work on platinum will have to await a later chapter.

Jacques Daumy and the Abbé Rochon

One other goldsmith who took up the arsenic process in Paris was Jacques Daumy, practising at 58 Rue de la Verrierie. Admitted a Master in February 1783, he first became well known as a maker of "doublé", gold clad on silver or copper, but soon interested himself in platinum although not on the same scale as Janety. An account of his procedure is given by Dr. Jan Ingen-housz in the second volume of his book published in Paris in 1789 (14). Describing him as Sieur Domi, Ingen-housz writes that he melts the native metal two or three times in succession with three or four times its weight of arsenic (presumably as oxide or salt), adding a few ounces of salt of tartar:

"The button which he obtains is very brittle; he reduces it to powder and exposes it, spread out to a moderate heat like that of a baker's oven (a cupellation furnace is very suitable) for the eight hours or until the powder no longer loses weight; he then subjects the powder, placed in a Hessian crucible, to a strong heat for several hours, while pressing the powder, now red hot and amenable, into a coherent mass which can then be forged with a hammer, after having been struck two or three times by a pestle. The success of this operation demands especially the avoidance of heating the powder or button high enough to melt it and being careful not to forge it without making sure that all the arsenic has gone. When the arsenic has all evaporated from

86

Alexis Marie Rochon
1741–1817

Astronomer to the French Naval Academy at Brest, the Abbé Rochon made extensive voyages and observations. His interest in mirrors for reflecting telescopes led him to the study of platinum and its methods of fabrication. He was also the first to melt optical glass in large platinum crucibles

the platinum no heat can produce a true fusion, one can then only agglutinate it forging it at red or white heat; it does, however, answer to the hammer in the cold and it is very ductile."

Daumy was called upon to employ this process by Alexis Marie Rochon who, as a young man became a priest but who managed to obtain a sinecure post of Abbé that yielded him an income sufficient for him to pursue his enthusiasm for travel and astronomy. On one of his voyages he stayed for a time in Corunna during 1770 and here he was given an ingot, weighing some eight ounces, of platinum alloyed with copper and zinc that had come from South America. This gave him the idea of making large mirrors for reflecting telescopes from "a metal whose polish will be bright and unchangeable". At first Daumy prepared the platinum for him and then it was alloyed with tin, copper and arsenic and cast into a mould. The final polishing was carried out by an engineer named Carrochez.

In a memoir read to the Académie des Sciences (15) on the occasion of his admission to membership in 1780 – but not published until 1798 on account of his long periods of travel – Rochon stated that he had constructed

"a telescope with platinum that magnifies the diameter of objects five hundred times with a degree of clearness and distinctness requisite for the nicest observations. The

87

large speculum of platinum weighs 14 pounds and is 8 inches in diameter and its focus is 6 feet."

Experience with this production of mirrors deepened Rochon's interest in platinum and he began to experiment with methods of purifying it in quantity. One was as follows: 1 pound of native platinum is mixed with 10 pounds of arsenic and 4 pounds of flux (sandifer), then melted until perfectly fluid and cast into a mould.

"Platinum in this state is exceedingly fragile and brittle; were it exposed to a red heat, the operation would absolutely miscarry. The arsenic, by disengaging itself too rapidly, would reduce it to scales which would no longer have any adhesion. It may readily be conceived that this accident must have occurred to me more than once. I have however been able to avoid it by enclosing pieces of platinum, which I have melted, in a box of plate iron, filled with sand and powdered charcoal. I then exposed them for more than a month to a fire graduated from the heat of boiling water to that which fuses silver. Platinum in this state no longer resembles a metal: it might rather be taken for a metallic calx. The particles which compose it are very close but they have only a very feeble adhesion, like that of an earthen vessel dried in the shade. It is then that the platinum must be exposed to the most violent fire; and when the metal has undergone that operation, it resumes its natural state, is sonorous, malleable, and the strongest heat gives it always new degrees of improvement."

In the course of experiments made in the famous glass works of Saint Gobain Rochon claims to have made in this way a crucible of platinum, capable of containing 30 pounds of flint glass, to which he gave more strength by covering it on the outside with a case of cast iron an inch thick. If this is true, it supports his claim to have been "the first who treated platinum in a large mass in a manner truly useful to the arts". But he generously acknowledges the assistance rendered by "the metallic talents of Daumy junior".

His experiments he says

"do not permit me to doubt of the utility of crucibles made of platina in bringing flint glass to perfection."

This last comment, completely correct, was made well before the time when this procedure became adopted in industry and foreshadowed some early work of Faraday's to be described later.

The Abbé Rochon had been allowed by the Count von Sickingen to observe his method of producing malleable platinum, and from this he benefited by giving up the use of arsenic. He goes on in his paper to relate that:

"A little time after I had read this memoir the learned Dr. Ingen-housz begged me to unite into a mass for him about two ounces of platina which he had carefully purified by means of the nitro-muriatic acid. I was obliged to enclose in a very thin foil of platinum all the fragments of this metal, which were too scattered and too minute to be subjected separately to the action of the fire and of the stamper; but, when this united, I gave them the highest degree of heat possible to be produced by charcoal excited by a pair of bellows, and I soon obtained by striking them with the stamper a ductile and malleable mass. After the success of this experiment, I employed myself in purifying in the fire, and in a crucible, platinum in grains, by means of nitre and

88

sandifer (glass-makers' flux), which must be afterwards washed in the nitric acid; and by striking at a white heat these grains contained in laminae of platinum, I procured at a small expense considerable masses of malleable platinum. This process will render unnecessary hereafter the use of the oxide of arsenic, unless the worker wishes to obtain by casting, large crucibles or muffles of platinum."

An additional detail of this process was given by Guyton de Morveau in his letter to Crell in 1789:

"The Abbé Rochon uses a hammer of 250 pounds falling from a height of $4\frac{1}{2}$. By alternate heating and forging (always by a single hammer blow each time) he produces cylinders which weigh nearly a pound and are suitable to pass the wire drawing plant." (7)

(No measure is given for the hammer drop; perhaps feet were understood.)

The translation of Rochon's paper into German – and only in this version – includes the statement

"I communicated this new method to Janety and he promised me to use it."

It seems likely, however, that Janety still adhered to his arsenic process for many years.

Pelletier's Phosphorus Process

A process with some similarity to the arsenic process was proposed at a meeting of the Académie des Sciences in 1788 by Bertrand Pelletier (1761–1797) who had studied the arsenates and phosphates of a number of metals. This consisted in treating platinum with "phosphoric glass" and carbon (5), which led to a reduction of phosphorus and the combination of the latter with the platinum to produce a melting similar to that obtained with arsenic. The phosphorus was subsequently oxidised to a slag at a high temperature and the product forged. Pelletier mentions that Janety had tried this method and by means of it had made for him a pair of balance pans which he had presented to the Académie. Pelletier claimed that the procedure was less dangerous to health but admitted that it was difficult and that it took a long time to get rid of the last traces of phosphorus. He adds that several others are using the arsenic method, which is inexpensive but dangerous for the workman.

Conclusion

Credit must undoubtedly be given to the arsenic process, and principally to Janety's practice of it, for leading to the first real exploitation of platinum in the production both of articles of jewellery and of crucibles and other apparatus. Moreover, it permitted for the first time the real properties of the metal to be appreciated not only by scientists but by the wider public, because it now became available in some quantity. Ingen-housz sums them up very well in his book (14).

"The value of this new metal is not measured solely by its scarcity but, since it has been possible to render it perfectly malleable, it is possible to begin to recognise the eminent qualities that it possesses. Its indestructibility by fire, its brilliance unalterable by the causes which tarnish even the purest silver, and its ductility, which does not appear to be less than that of gold and silver, have already placed it among the precious metals. Its specific gravity, surpassing considerably that of gold, has caused the latter metal to pass from the first to the second place among the metals with regard to this quality, which was the most certain indication of its purity, the test of Archimedes. But platinum possesses a quality or rather a perfection which is lacking in gold and silver, that is its hardness. Gold and silver when perfectly pure, cannot be put to several uses because of their flexibility and softness. It is necessary to debase them or alloy them with non-noble metals like copper, to make from them articles of jewellery, cooking utensils or plate. . . . A vessel made of pure platinum has sufficient hardness to need no alloy whatever.''

The arsenic process is of course essentially a means for removing the base metals from the native platinum and getting it into a form suitable for forging under heat. In suitable hands it proved more manageable than the working of the sal-ammoniac precipitate, but eventually as we shall see the latter caught up and the arsenic process disappeared. But this was not until it had supplied a growing demand for crucibles and other apparatus from the chemists of Europe for the best part of a generation.

Finally, the appearance of the translation of Rochon's paper in the second volume of the *Philosophical Magazine*, published in 1798, served to alert several English scientists to the special properties of platinum and to the desirability of developing an improved and economical means of rendering it workable, with important results that will become apparent in later chapters.

III. *Observations on Platina, and its Utility in the Arts, together with some Remarks on the Advantages which reflecting have over achromatic Telescopes. By* ALEXIS ROCHON, *Director of the Marine Observatory at Brest. From the* Journal de Physique, 1798.

PLATINA is a metal exceedingly refractory, unchangeable, very compact, and capable of receiving a fine polish. This singular metal has never yet been found but in the gold mines of Choco. The Spaniards gave it the name of *juan blanca*, that is to say white gold, and *platina del Pinto*, which signifies little silver of Pinto. It is brought to us from Choco, under the form of triangular grains the angles of which are rounded. These grains are irregular, ductile, and susceptible of being attracted by the loadstone. It is never pure, and always contains a black shining sand, over which an artificial magnet has great power. This sand is interspersed with gold grains and fragments of small coloured crystals. The specific gravity of platina is to that of gold as 22 to 19½*. Like that precious metal, it resists the action of simple acids,

* We have seen some platina where the difference was still greater. EDIT.

C 2   and

The memoir read to the Académie des Sciences by the Abbé Rochon was reproduced in English in Tilloch's *Philosophical Magazine* in 1798 and undoubtedly aroused the interest of Richard Knight, Tilloch himself and other English scientists in the properties of platinum and in the need to devise a process for rendering it malleable and therefore useful. This shows the opening page of the paper

90

References for Chapter 5

1   F. K. Achard, *Nouveaux Mem. Acad. Roy. Sci. Berlin*, 1781, **12**, 107–109

2   F. K. Achard, *Chem. Ann. (Crell)*, 1784, **I**, 2–5

3   L. B. Guyton de Morveau, *Nouv. Mem. Acad. Dijon*, 1785, (i), 106–112

4   H. Nocq, Le Poinçon de Paris, 1927, **2**, 353

5   B. Pelletier, *Obsns. Physique (Rozier)*, 1789, **34**, 193–197

6   V. Restrepo, "El Platino", Estudio sobre las Minas de Oro y Plata de Colombia, Bogota, 1884, 208–212

7   L. B. Guyton de Morveau, *Chem. Ann. (Crell)*, 1787, ii, 243–245

8   M. von Ruprecht, *Chem. Ann. (Crell)*, 1790, ii, 53–54

9   W. A. Smeaton, *Platinum Metals Rev.*, 1968, **12**, 64–66

10 A. F. Fourcroy, *Ann. Chim.*, 1791, **9**, 305–352

11 A. L. Lavoisier, *Ann. Chim.*, 1790, **5**, 137–141

12 J. A. Chaldecott, *Ann. Sci.*, 1968, **24**, 21–52; National Library of Scotland MS 6332, ff 49–50

13 C. L. Berthollet and B. Pelletier, *Ann. Chim.*, 1792, **14**, 20–33

14 J. Ingen-housz, Nouvelles Expériences et Observations sur Divers Objets de Physique, Paris, 1789, **2**, 505–517

15 A. M. Rochon, *J. Physique*, 1797–98, **4**, 3–15; *Phil. Mag.*, 1798, **2**, 19–27, 170–177; *Ann. Physik (Gilbert)*, 1800, **4**, 282–289 (There are significant differences in the three versions)

Pierre François Chabaneau
1754–1842

Born at Nontron in the Dordogne and self educated in Paris, at the age
of only twenty-three he left France to become Professor of Physics in
the Seminario Patriotico at Vergara in the north of Spain. In 1786, after
meeting with many difficulties and with the co-operation of Fausto de
Elhuyar, he succeeded in producing malleable platinum on a larger scale
than had ever before been achieved

6

The Platinum Age in Spain

"I hope that Europe will soon become aware of the valuable properties of this new noble metal whose worth is beyond all imagination and then that Spain, the sole possessor of this treasure, will reap useful benefits that only time will reveal."

PIERRE FRANÇOIS CHABANEAU

As we have recorded in Chapter 2, for many years platinum was regarded as worthless and as a troublesome impurity in the gold from the Spanish colony of New Granada. None the less the first researches on this newly discovered metal, collected together in Morin's book published in Paris in 1758, brought about some change of attitude among the authorities in Madrid. In the following year, for example, Juan Wendlingon (1715–1790), Professor of both Mathematics and Geography there and also the royal cosmographer for the Indies, instructed the Viceroy of New Granada to collect a substantial quantity of platinum from the heaps of discarded metal lying around the mints in Bogotá and Popayan and to despatch it to Madrid (1).

Again in 1765 the Royal Council of Commerce requested the Secretary of State for the Indies, Julian de Arriaga, to acquire further quantities and by the following January the Viceroy, Pedro Messia de la Cerdia, had replied that metal had already been shipped to Spain from the Chocó region and that it was abundant in that area (1).

A new era had opened in Spain in 1759 when Carlos III succeeded his half-brother Ferdinand VI on the throne. The new ruler was interested in promoting agriculture and industry as well as the sciences, and throughout his thirty-year reign there developed a much more enlightened and energetic atmosphere, in part under the influence of the French philosophers of the time. Among the King's initiatives were the establishment in 1771 of a Cabinet of Natural History in Madrid, based upon a large collection of mineralogical specimens formed and presented to the Government by Don Pedro Franeo Davila, and the appointment of the Irishman William Bowles to be its director. Among this collection were several specimens of platinum, and Bowles now gave more attention to its

93

properties and possible uses than he had in earlier years when he warned against the dangers of its fraudulent potential. In a long footnote to his dissertation on platinum he now concluded:

> "Finally I emphasise that platina can be available for an infinite number of uses and for making a multitude of utensils that would not be subject to rust or corrosion since this metal, with various alloys, can be worked and further submits to forging and welding like iron. See especially what M. Baumé has to say about that." (2)

The Royal Monopoly in Platinum

Then in 1774 Don José Celestino Mutis wrote home from Bogotá describing the two portrait medallions of Carlos III mentioned in Chapter 2, one made in platinum and the other in a copper-platinum alloy, that had been made by Don Francisco Benito in the mint and forwarded to the King. These were passed on to Don Miguel Musquiz, the Finance Minister, and thence to the Council of Commerce who proposed that Benito should be granted an award and that details of his procedure should be obtained. The Viceroy in New Granada, Don Manuel de Guirior, seems to have advised Benito, however, to keep his process to himself, and there is in fact no reference to it whatever in the voluminous archives of the Indies.

The consequence of this was the issue of an edict "that platinum should be worked exclusively for His Majesty as was the case with gold", and this was followed in 1778 by instructions that all platinum must be handed over to the King's representatives but without payment. Not unnaturally very little metal was brought in on these terms (3).

The event that was to have a major influence in changing this state of affairs and of bringing about the so-called "Platinum Age" in Spain had taken place some years earlier. This was the foundation in 1764, with the approval and encouragement of King Carlos III and his chief minister Count Grimaldi, of a society for the promotion of science, industry and commerce. Formed in the three Basque provinces of Viscaya, Guipuzcoa and Alava by the nobility of the region, and based in the small town of Vergara near San Sebastian, this was known as the Real Sociedad Economica Vascongada de los Amigos del Pais (the Royal Basque Economic Society of Friends of the Country) and was in fact the forerunner of a number of similar organisations in the other provinces of Spain.

The principal founder of this body, and its director, was Francisco Javier de Munibe, the Count of Peñaflorida (1723–1785), who had been educated partly in France. His elder son, Don Ramon Maria de Munibe (1751–1774), completed his studies with the society in 1768, and, in line with their policy of securing the most up-to-date knowledge from other countries, he was sent at the society's expense on a three-year tour of France, Germany, Sweden, Holland and Italy in the care of the scientist Eugenio Izquierdo, the eventual successor to William Bowles. They visited mines and iron works, attended lectures, learnt assaying from Cronstedt, and sent back regular reports to the society, some of

94

these being printed as anonymous contributions from "A Travelling Member" in their journal, *Extractos de las Juntas Generales de la Real Sociedad*. One of these, published in 1775, after the young Munibe's unfortunate death at the age of only twenty-three, contained a long account of the work of William Lewis on platinum and a review of the state of knowledge about it at that time. This awakened the interest of the society and supplies of the native metal were obtained from the government through the Marques de los Castillejos (4).

The Seminario at Vergara

The society had the intention to establish a school of university standard to make available to its members the best teaching of the time, but this naturally required several years to design and build. Their Real Seminario Patriotico finally opened its doors in the autumn of 1777, and Count Peñaflorida had some time earlier sought for suitable professors. On the recommendation of Izquierdo, who with the young Munibe had met them in Paris, the choice fell upon two very young Frenchmen, Pierre François Chabaneau (1754–1842) to teach physics

Fausto de Elhuyar
1755–1833

Appointed Professor of Mineralogy in the Seminario at Vergara after a long tour of European centres of mining and metallurgy, Elhuyar took part in the research on platinum with Chabaneau but left to become director of mines in Mexico before the work was brought to a successful conclusion

95

and the better known Joseph Louis Proust (1754–1826) for the chair of chemistry, with very handsome salaries provided by King Carlos III. Proust was to play little part during his stay in Vergara and in 1781 he returned to France, although a later period he spent in Spain did have an influence. Chabaneau (usually spelt Chavaneau in Spanish), on the other hand, became the leading figure in developing the platinum industry in his adopted country but not until several years after his appointment, being occupied first in such tasks as the analysis of the mineral waters of the nearby spa at Cestona.

In 1781 the Seminario decided to establish a chair of mineralogy and to appoint as professor a young Spaniard, Don Fausto de Elhuyar de Zubice (1755–1833), who was apparently assisted on the metallurgical side by his elder brother Don Juan José (1754–1804). They came of a good Basque family, had been educated in Paris and had been travelling in Europe, just as had Don Ramon de Munibe, spending some time at the School of Mines in Freiberg in preparation for their work at the Seminario. The King had also arranged for the elder brother to go to New Granada to supervise mining and metallurgical operations there, and to this end he had spent some time under Bergman in Uppsala.

Their first research in Vergara was on tungstic acid, resulting in their being the first to isolate metallic tungsten, a discovery they announced in the *Extractos* of the society in 1783 and which brought them considerable attention from chemists throughout Europe.

The year before this Fausto de Elhuyar had added to his duties the chair of chemistry vacated by Proust and at some time after this he began work on platinum, together with his colleague Chabaneau, drawing upon the supplies of native metal that had been obtained from the Marques de Castillejos. They met with many difficulties and a great deal of frustration, but by March 1786 they were able to announce the successful production of malleable platinum to a meeting of the council of the society. Sadly Peñaflorida, the society's main progenitor, was no longer alive to hear the news.

Not only was their procedure kept a closely guarded secret but the part which each of them played in this work is most difficult to establish, depending upon whether one relies upon Spanish or French sources, each of course according greater praise to their own compatriot.

Chabaneau's Malleable Platinum

Before the work was completed, Elhuyar had left the Seminario in September 1785 in order to visit Hungary to study the improved amalgamation process for gold devised by Baron von Born before going to Mexico as director of mining. This led to Chabaneau taking over the chair of chemistry in addition to that of physics and to his carrying on the work alone. Before Don Fausto left for Mexico, however, he paid a brief visit to Vergara in the spring of 1786, finding that considerable advances had been made, and fortunately he recorded what he

96

The Seminario, a school of university standard established by the Sociedad Vascongada at Vergara near San Sebastian in 1777. The first two professors appointed were both from France, Pierre Francois Chabaneau to the chair of physics and Joseph Louis Proust to that of chemistry. Proust remained only two years and was succeeded by Fausto de Elhuyar who, together with Chabaneau, began work on platinum. In 1785 Elhuyar was sent to Mexico and Chabaneau carried on alone, successfully producing malleable platinum in the following year.

Photograph by courtesy of Professor Francisco Aragon de la Cruz

found in two letters to his brother who was by now settled in Bogotá. Even more fortunately, Don Juan José made copies of these letters "for the interest they aroused and in case they prove useful to New Granada in the future", and these copies went into the files of his chief, José Celestino Mutis. Twenty-five years later, when the liberation revolution broke out in Colombia, these files were rescued, brought to Spain and lodged in the Botanical Gardens in Madrid as the Mutis Collection. Working on them in 1911 in connection with the compilation of his biography of Mutis, Professor A. F. Gredilla, the Director of the Botanical Gardens, came upon the two copies of Don Fausto's letters and he published their contents in his book (3). Their value is unique since, as soon as it was known that an important discovery had been made about making platinum malleable, the Minister issued the strictest orders to Chabaneau that on no

97

account was he to publish his methods. Chabaneau told the Minister that he had already informed Don Fausto about the process and the latter then received similar orders. These were duly transmitted to Don Juan José and presumably the original letters were destroyed. The letters revealed that Chabaneau had successfully used a powder metallurgy process based upon those of Sickingen and Milly. The first begins:

"Vergara, March 17th, 1786. In my former letter I told you that I was going to Hungary on behalf of the Minister for the Indies, in order to learn about the new method of amalgamation. ... As you must already know, when I went away to Madrid, Chabaneau took charge of the work on platinum to complete the research that I had promised to the Minister and he has now made some very important discoveries. The method is similar to that of von Sickingen in so far as the reduction of the precipitates and salts is concerned but in other respects it is much better. Sickingen used Prussian alkali to precipitate the aqua regia solution which is the very worst means for freeing the platinum from iron and one might almost say that it is the surest means of obtaining a mixture of the two. Chabaneau has employed the method of the Count de Mylli [sic] which consists in precipitating the said solution by sal-ammoniac, by which no iron is deposited. By this means he has obtained large precipitates and from them some very fine pieces of platinum."

Don Fausto goes on to say that Chabaneau also discovered a more economical method of dissolving the native platinum, namely, by attacking it by means of nitric acid and common salt. There follows some rather confused matter from which one gathers that they sometimes precipitated by means of other alkalis, and it is known that later on Chabaneau used potash instead of sal-ammoniac as another measure of economy. But however the platinum was separated, there is no doubt about what happened afterwards.

"The whole precipitate is placed in a crucible on an enclosed fire, when the contents diminish in size and lose their aqua regia. When crushed with an iron pestle it soon loses it brownish-grey colour, which changes to a beautiful silver-white, and with pressure gains consistency and becomes concentrated. When the precipitate has sintered, it is removed from the crucible, hammered very lightly to unite the particles and then annealed and hammered alternately until it is really firm. This mass is then exposed to more intense heating in a crucible in order to evaporate the salt occluded in its interior and is then annealed and hammered again. Finally it is exposed to the fierce heat of a forge for half an hour, when it is removed from the crucible, heated on an iron forge, forged and drawn into bars, commencing by striking very lightly in order to unite all the particles not previously affected. This is the method used by us in an operation completed today, when in a single operation we have obtained a piece weighing 13 ounces which we are going to send to the Minister in the form of a bar. Just as we have been able to correct in the course of this operation defects previously noted, so we have observed defects in this procedure which will be obviated on future occasions. It has been noticed during the last heating up and drawing of the bar that a glassy crust forms on the surface of the metal and which seems to sweat through from the interior, and we believe that this is caused by residual unreduced salts. If at the outset care is not taken to destroy those unreduced salts, the mass will crack and quickly break up into small pieces. This defect has been remedied in part by an additional treatment which consists in putting the hot casting in water. If these last

98

traces of salts are completely removed, this metal is more easily drawn into bars than silver, and one might even say that anything can be done with it! During the first operations, that is before beginning to compress the mass in the crucible, it should be well stirred with an iron bar in order to facilitate the evaporation of the salt and thus avoid inconvenience later when forging. I shall keep you advised of any further progress we may make, but it would be as well if you started to carry out experiments yourself in order to be fully prepared to install appropriate plant there, for although there is every possibility that a more economical method will be discovered, this one of course would serve very well. We estimate that the expenses would be less than 4 pesetas (8 reales or 1 Spanish dollar) per pound and we believe that the metal will find a ready market at a much higher price than that of silver. . . . I am enclosing, wrapped in paper, a small piece of platinum and although it is not the best possible sample, it will allow you to form an idea of the colour of the metal and how well it can be polished.''

The second letter, addressed from Paris on May 19th 1786, again refers to the need for secrecy:

"Chavaneau sent to the Minister for the Indies some bars of platinum produced by the method I wrote to you about in my recent letter, and he has asked him not on any account to make known his discovery before receiving instructions. He has also sent Chavaneau more native platinum with which to continue his investigations. In his reply Chavaneau had to inform the Minister that he had confided details of his new process to me, and as a consequence I later received orders not to make known to any one the secret in which I shared.

My reply, as that of Chavaneau, has been to the effect that I had already written to you about the discovery, since as you are already in the country where large quantities of the mineral can be obtained, you are therefore well placed to carry out useful work on it."

In this important correspondence it is obvious that, although Don Fausto gives most of the credit for the discovery to Chabaneau, there is very definite indication of his own participation in the work. It is also interesting that in Don Fausto's first letter there is confirmation of the Count von Sickingen's statement that de Milly sent particulars of his process to Spain.

The discovery of the way to make platinum malleable marks the end of Chabaneau's stay at Vergara. The King, wishing to have him in Madrid, created for him a special Chair there of Mineralogy, Physics and Chemistry in the School of the Natural History Museum, where Bowles had been until his death in 1780. He also installed him in one of the royal palaces with an annual salary of 2,200 Spanish dollars, in addition to a life pension of 2,800 dollars a year provided he remained in Spain, and a medal specially struck for him in platinum. He was also made Director of a Chemical Laboratory maintained by the Treasury and situated first in the Calle de Horteleza and later moved to a part of a glass warehouse in the Calle del Turco in Madrid. This was devoted to the refining and fabricating of platinum and was managed by one of Chabaneau's old assistants, Don Joaquim Cabezas (4).

Before all this took place, however, Chabaneau had been taken off to Paris in 1786 by his patron the Count of Aranda who, after a period as prime minister,

The first object made in Spain from Chabaneau's malleable platinum was this chalice, made by his silversmith Francisco Alonso, and presented by Carlos III to Pope Pius VI in 1789. It is thirty centimetres in height and weighs almost two kilogrammes. The inscription on the plinth reads:

CARLOS III HISPAN ET
IND REX PRIMITIAS HAS
PLATINAE A FR^{CO}
CHAVENEAU DUCTILIS
REDDITAE PIO VI
P.O.M.D.D. ✠

[Charles III, King of Spain and the Indies, gives as a gift the first fruit of platinum made malleable by Francisco Chavaneau to Pius VI, Supreme Pontiff of all the World (Pontifici Omnium Maximo Dono Dedit)]

Inside the cup another inscription reads:

HISPAN ELABORAVIT
ANN. R. J. MDCCLXXXVIII
✠ FRANCISCUS ALONSO

[Francisco Alonso the Spaniard fashioned (this) in the year A.D. 1788.]

The chalice is still on public view in the Treasury of St. Peter's in Rome.

had been appointed Spanish Ambassador to France in 1773. This was in order to visit the Royal Goldsmith to Louis XVI, Marc Etienne Janety, who, as described in Chapter 5, had successfully manufactured many articles in platinum by means of the arsenic process. Vicente Restrepo (3) quotes a letter from Janety recording this visit:

"The King of Spain sent one of his chemists to Paris in 1786 with 44 marcs (about 350 ounces) of very malleable platinum bar. His Embassador the Conde Aranda honoured me by accompanying him to my home for the purpose of carrying out certain experiments together. We made these experiments: we made coffee pots, plates, watch-chains, mustard pots, tea pots and dress coat buttons in my home, so many in soldered form with half an ounce of platinum in one marc of pure silver."

During the visit Janety tried hard to elicit Chabaneau's procedure from him but failed to do so. None the less a comment by Guyton de Morveau in 1787 records:

"Many sorts of useful vessels are now being made in Paris by the method of M. Chabaneau for the King of Spain in which this metal is so pure that its specific gravity is 24." (5)

On his return to Spain Chabaneau trained his own silversmith, Don Francisco Alonzo, in the working of platinum to make jewellery and instruments and provided him with a room in the laboratory in the Calle de Turco. The first object made here was a large chalice for King Carlos III, who, as recorded on the plinth, presented it in 1789 to Pope Pius VI. The chalice remained among the private possessions of the Popes until some time in the last century when Pius IX presented it to the Vatican. It is still on public view in the Treasury of St. Peter's.

Chabaneau's secret was well kept for more than a century. In 1795 he read a paper on platinum to the Royal Medical Academy in Madrid to which he had recently been elected and this was published as a pamphlet and later, in 1797, in

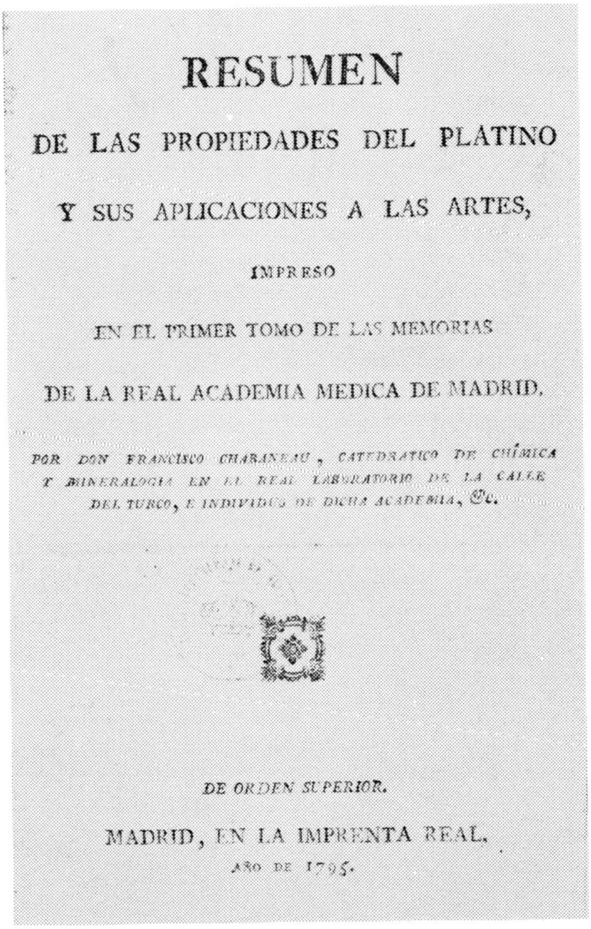

In 1795 Chabaneau at last presented a paper on the properties and potential applications of platinum but without giving the least details of his procedure in rendering it malleable. This shows the title page of his eight page pamphlet, preserved in the Bibliotheque Nationale in Paris. It did not appear in the Memorias de la Real Academia Medica de Madrid until the first issue of the journal was published in 1797.

RESUMEN

DE LAS PROPIEDADES DEL PLATINO

Y SUS APLICACIONES A LAS ARTES,

IMPRESO

EN EL PRIMER TOMO DE LAS MEMORIAS

DE LA REAL ACADEMIA MEDICA DE MADRID.

POR DON FRANCISCO CHABANEAU, CATEDRATICO DE CHIMICA Y MINERALOGIA EN EL REAL LABORATORIO DE LA CALLE DEL TURCO, E INDIVIDUO DE DICHA ACADEMIA, &c.

DE ORDEN SUPERIOR.

MADRID, EN LA IMPRENTA REAL.
AÑO DE 1795.

101

the first issue of the Academy's proceedings (6). In this communication he stated that he was still unable to describe his method of obtaining malleable platinum because a royal command issued in 1787 had forbidden him so to do. He gave only an account of the principal properties of the metal – already of course well known – and drew attention to its potential uses. Its infusibility he thought might make it a suitable substance for measuring high temperatures, while its resistance to corrosion should lead to its use in navigational and astronomical instruments. Platinum crucibles he considered might be used, as well as for analysis, for the manufacture of optical glass needed for the best telescope lenses, and he also suggested that vessels could be made of copper clad with a thin sheet of platinum. He had, in fact, already found that platinum and copper could be united so intimately that the composite material could be hammered into any shape without the two metals separating.

Chabaneau remained in charge of the platinum work until 1799 when, largely on account of poor health, he left Spain, forfeiting his pension, and retired to his native Nontron in the Perigord were he lived quietly until his death in 1842 at the age of 88. His work remained largely unknown to the scientific world, but in 1857 a local resident, Jules Delanoue, who had known him only in his declining years, published a 16-page pamphlet with the title "Notice sur Chabaneau, Chimiste Perigourdin". This was reprinted in 1862 and both editions are in the Bibliothèque National in Paris, but apart from securing a brief mention in the article on platinum by Henri Debray in Wurtz's Dictionnaire de Chimie in 1876, where he spells the name Chabanon (7), and another in 1906 in Moissan's Traité de Chimie Minérale (8) no attention was paid to it until a copy of this rare pamphlet came into the hands of Louis Quennessen, the head of the Paris firm of platinum refiners. Quennessen had earlier provided the notes for Moissan, and in 1914 he published a summary of the pamphlet in Paris and also brought it to the notice of Professor J. Lewis Howe, the American bibliographer of platinum. Professor Howe then achieved wider publication by contributing an extensive summary to *Popular Science Monthly* from which it was reproduced in *The Chemical News* of London (9, 10).

Unfortunately, this rather over-eulogistic little work contains a number of obvious errors of both fact and chronology, as well as attributing all the credit for the production of malleable platinum to Chabaneau with no mention at all of Fausto de Elhuyar. One episode that has the ring of truth, however, relates that the Count of Aranda, making one of his frequent visits to the laboratory

"Found Chabaneau in a frenzy engaged in throwing out of the doors and windows his dishes, flasks, and ores as well as all the solutions of platinum which he had prepared with so much trouble and difficulty, saying 'Away with it all! I'll smash the whole business; you shall never again get me to touch the damned metal'; and in fact he broke up all the apparatus of the laboratory."

Nevertheless, the work did go on and Delanoue records that only three months later Chabaneau showed Aranda a large cube of platinum measuring 10 cm along the sides, and weighing about 750 ounces Troy.

The Spaniards paid little attention to the work for even longer and it was not until well into the present century that comments began to appear. The first was in an address on "The Chemists of Vergara" delivered to the Spanish Academy of Sciences in June 1909 by the prominent chemist, Don Juan Fages y Virgili (4). The second arose from the discovery in 1911 by Professor A. F. Gredilla of a full account of the research in copies of two letters from Don Fausto to his brother already described. Then in 1933, the celebration of the centenary of the death of Don Fausto de Elhuyar brought forth two other papers, the first by A. de Galvez-Cañero y Alzola in the *Boletin del Instituto Geologico y Minero de España* (11), and the other a section of a Symposium in the *Anales Sociedad Española de Fisica y Quimica* mostly contributed by the same author (12). Both dealt with Don Fausto's career and, as in all of these Spanish papers, there is a tendency to decry the ability and contribution made by Chabaneau and to hold up Don Fausto as the major factor in the platinum work, although Fages (4) admits

"not only the certainty of Chavaneau's discovery but also that it was he who gave it practical application, and converted it into a practical fact useful to the progress of the sciences and arts, and certainly lucrative for himself".

The Second Royal Monopoly

The success Chabaneau had achieved in producing malleable platinum in some quantity immediately prompted the Spanish government to order the Viceroy of New Granada – now Antonio Caballero y Gongora – to collect all the platinum he could obtain while keeping its new value a secret. About 150 pounds of native metal were shipped to Spain, this time the miners being paid two or three reales a pound. Further shipments necessitated the price being raised to four reales a pound, and the authorities in Madrid recommended the importation of many more negro slaves to work the deposits and approved a scheme for the importation of tools for sale to the workers in the hope of increasing output. In 1788 it was decreed that platinum was to be sold only to the crown and penalties were established for anyone detected in hoarding the metal. By the end of that year more than three thousand pounds of platinum had been despatched from the Chocó to Cartagena for shipment to Spain, but great quantities were still smuggled out of the Chocó to be sold to other purchasers prepared to pay much more (13).

Joseph Louis Proust

The brief period of Proust's holding the chair of chemistry in Vergara has already been mentioned. He returned to France in 1781, but five years later, on the recommendation of Lavoisier to Count Aranda and on the invitation of Carlos III, he again came to Spain, first of all lecturing and carrying out research in Madrid and then in 1788 moving to the Artillery School at Segovia as professor of chemistry. His laboratory there was equipped, at the expense of the

Joseph Louis Proust
1754–1826

Born in Angers the son of an apothecary, Proust spent three years at Vergara, returned to France and then in 1786, at the invitation of the King of Spain and on Lavoisier's recommendation, came again to Spain and in 1799 succeeded Chabaneau in charge of a laboratory in Madrid equipped with an immense amount of platinum apparatus. Here Proust continued and extended the work on platinum fabrication until in 1808 his laboratory was destroyed by a mob during the siege of Madrid by Napoleon's forces

new King Carlos IV, with great luxury and an extraordinary amount of platinum apparatus, and he remained there until Chabaneau's departure from Spain in 1799 when he took over the latter's laboratory in Madrid. Here one of his first activities according to Fages y Virgili was to improve still further his equipment and he asked through his director for forty pounds of pure platinum and twenty-five pounds of the native metal in grains. This was granted, the metal being provided by Don Joaquin Cabezas, but on condition that he continued his experiments on platinum.

He had already carried out a long series of researches on native platinum while at Segovia, using relatively large quantities in his experiments, mainly concerned with the dissolution of platinum in aqua regia. The insoluble residue he described as "nothing less than graphite or plumbago", failing to grasp that it contained other metals of the platinum group. These results he published in the very first issue of the *Anales de Historia Natural* in 1799 under the title "Experimentos hechos en la Platina" and his paper was reproduced in translation in both French and English (14). Proust promised at the end of this communication to present a further contribution on platinum but this was never forth-

104

coming, apart from a short letter to Vauquelin written from Madrid in 1803 that was of no great significance (15).

Apart from the chalice presented by Carlos III to Pope Pius VI and the use of so much platinum apparatus in Proust's laboratory there is little evidence of serious applications of the metal in this period in Spain. There is a brief reference in the Archives of the Indies to the need for platinum to make a table service for the King and to use the metal in the royal chapel, while later a set of standard weights was made in a new workshop set up by Carlos IV to establish the art of making scientific instruments in Spain. As in Colombia, the counterfeiting of gold coins with platinum flourished among dishonest workers in the mints, first using solid platinum with a thin gilding and later a copper core with a thin layer of platinum, also followed by gilding. The production of these spurious pieces continued for many years. But the Napoleonic wars were now causing major disturbances; in the first French invasion of 1794 the Seminario at Vergara had been burnt and the Sociedad Vascongada had been broken up. The reign of science in Spain was coming to an end.

The Platinum Room in the Royal Palace

There, was however, one last flamboyant fling to the Platinum Age. For many years there had been a royal palace at Aranjuez, some fifty kilometres south of Madrid. This had been enlarged or embellished by successive monarchs and in 1802 Carlos IV decided to build the so-called Labourer's Cottage in the grounds, in imitation of the Petit Trianon devised by Louis XV at Versailles. Napoleon's famous architects, Charles Percier and Pierre François Fontaine, who had designed a number of elegant buildings in Paris, were called upon by the King to design the interior and to provide the richest possible decoration. The architects record with some distaste the display of ornate embellishment "contrary to the simplicity of the name of the building", consisting of mirrors and medallions not only in bronze and gold but also in platinum (16).

The second invasion of Spain by Napoleon's forces in 1808 brought about the abdication of Carlos IV. The presence of a French garrison in Madrid stirred the Spaniards to revolt and on the famous 2nd of May the crowds rioted and among other acts destroyed the laboratory in which Proust worked, leaving him destitute so that he was forced to sell his collection of minerals in order to live and then to make his escape back to France.

Conclusion

The rise and fall of the Spanish platinum industry were historically important because the metal produced seems to have been the best in quality made available up to that time. Also the quantities involved were much greater than had been the case in earlier work, as is exemplified by the 750 ounce ingot shown to Aranda. But nevertheless the whole affair was merely an episode in the history of

105

A contemporary engraving of the Platinum Room at Aranjuez. Designed by the leading French architects Percier and Fontaine for Carlos IV and his Queen Maria Luisa, the extremely ornate decoration of the otherwise simple "Labourer's Cottage" in the grounds of the royal palace included medallions and plaques made of platinum and engraved with mythological subjects and Italian landscapes. The room may still be seen by visitors to the palace

platinum. Most of the researches, and all the technique that resulted from them, were shrouded in secrecy and knowledge of them disappeared with the scattering by the Napoleonic Wars of those who employed them. They had no effect whatever on later practice, which had far out-distanced them by the time they had again been brought to light.

Chabaneau's great hopes, quoted at the head of this chapter, were not to be fulfilled. The episode, however, had one result of lasting importance, since it brought native platinum to public notice and caused it to be sought after and to acquire a value. The metal became at once an article of trade, and when the Spaniards tried to monopolise it the brisk smuggling prospered and took supplies to other countries. This continued on an increasing scale until the

106

departure of the Spaniards from New Granada, when the Republic of Colombia took over the Chocó. Despite the researches of Chabaneau, Elhuyar and Proust, and the output from Colombia running to some 500 kilograms a year as estimated by Humboldt during his visit in 1819–1820, very little use was made of platinum during this period, and when peace returned after the disturbances of the Napoleonic wars the new government of the restored King, Ferdinand VII, found a great accumulation of unrefined metal on their hands. Some part of this, as will be seen in Chapter 10, served as a basis for the further development of the platinum industry in France.

References for Chapter 6

1 Archivo General de Indias, Seville, Santa Fé, 835

2 William Bowles, Disertacion sobre la platina, Introduccion a la Historia Natural y a La Geografia fisica de España, Madrid, 1775, 155–167

3 A. F. Gredilla, Biografia de José Celestino Mutis, Madrid, 1911, 157–158; V. Restrepo, Estudio sobre las Minas de Oro y Plata de Colombia, Bogotá, 1884, 208–214

4 J. Fages y Virgili, Los Quimicos de Vergara, in Discurso del Ilmo, Madrid, 1909, 41–43; 57–61

5 L. B. Guyton de Morveau, *Chem. Ann. (Crell)*, 1787, (i), 333

6 P. F. Chabaneau, "Resumen de las Propiedades del Platino y sus Aplicaciones a las Artes", Imprenta Real, Madrid, 1795; reprinted in *Memorias de la Real Academia Médica de Madrid*, 1797, **I**, 183–188; W. A. Smeaton, *Platinum Metals Rev.*, 1978, **22**, 61–67

7 A. Wurtz, Dictionnaire de Chimie pure et appliquée, Paris, 1876, **2**, 1034

8 H. Moissan, Traité de Chimie Minérale, Paris, 1906, **5**, 661

9 L. Quennessen, *Rev. Sci.*, 1914, **52**, 553–557

10 J. L. Howe, *Pop. Sci. Mon.*, 1914, (Jan), 64–70: *Chem. News*, 1914, **109**, 229–231

11 A. de Galvez-Cañero y Alzola, *Bol. Inst Geol. y Min. Esp.*, 1933, **53**, 377–629; Apuntes Biograficos de D. Fausto de Elhuyar

12 *An. Soc. Esp. Fis. Quim.*, 1933, **31**, 115–143; El Primer Centenario de D. Fausto de Elhuyar

13 Archivo Historica Nacional de Colombia, Bogotá, 1786–1788

14 J. L. Proust, *An. Hist. Nat.*, 1799, **1**, 51–84; *Ann. Chim.*, 1801, **38**, 146–173; 225–247; *Phil. Mag.*, 1801-2, **11**, 44–55; 118–128

15 J. L. Proust, *Ann. Chim.*, 1804, **49**, 177–180

16 C. Percier and P. F. Fontaines, Residences des Souverains, Paris, 1833, 235–236

Jan Ingen-housz
1730–1799

Born in Breda in the Netherlands, Ingen-housz studied medicine and came to England to practise in 1765. Three years later, having played an important part in the early days of inoculation against smallpox, he was appointed court physician to the Empress Maria Theresa in Vienna. Here he carried out experiments on platinum and later, travelling between there and Paris and London, he encouraged other scientists to interest themselves in the subject. In his later years in England he habitually displayed a set of three platinum waistcoat buttons made for him by Matthew Boulton

7

The Widening of Interest in Platinum and its Properties

"The Value of this new metal is not measured solely by its scarcity, but since it has been possible to render it perfectly malleable it is possible to begin to recognise the eminent qualities that it possesses."

JAN INGEN-HOUSZ

The last few chapters have recounted the intensive and progressive work of a small group of men, leading gradually up to the achievements of the seventeen-eighties in France and Spain. We now come to a rather less progressive period, disturbed by revolution and war, but one in which the interest in platinum nevertheless spread more widely throughout Europe with the gaining of experience of its properties and uses.

There were several factors to account for this. One was the intense interest being taken in experimenting with magnetism and static electricity and especially in the effects of passing "the electric fire" through wires of various metals. Another was the considerable increase during the last two decades of the eighteenth century in the number of scientific journals that began publication, these not only giving a more speedy account of researches than the older transactions of the academies but also encouraging much correspondence from country to country, as well as reproducing papers from one to another, while a third was an increase in the travelling of scientists from one capital city to another and their discussions and occasionally even joint experimentation.

One other interesting example indicative of the increased interest in platinum is contained in a memoire written by A. F. Fourcroy in 1785 immediately after his election to the Académie des Sciences. The great French navigator J. F. de La Pérouse was about to embark on his voyage round the world and the Académie had been asked for advice on the work to be undertaken. Fourcroy included in his suggestions the remark that "new sources of platinum would be especially valuable" (1). Unfortunately La Pérouse and his crew were later all lost in the South Pacific and nothing was ever heard of any search he might have made for platinum.

The Influence of Jan Ingen-housz

The most fascinating figure of this phase of activity was Jan Ingen-housz, a Dutchman by birth but an inveterate traveller and a pioneer in the field of plant physiology. As a very young man he was encouraged by Sir John Pringle, then the chief medical officer of the British Army, who had met the family during his service in the Low countries in the War of the Austrian Succession. Ingen-housz took his M.D. at Louvain and then studied physics at Leyden under Musschen-broek, the inventor of the Leyden jar. After a brief period at the University of Edinburgh he settled as a physician in his native Breda until 1765 when, on the advice of Pringle, now the leading figure in London medical circles, he came over to establish his own practice there (2). Here he met Priestley and Benjamin Franklin and worked with William Watson at the Foundling Hospital in the early days of inoculation against smallpox. Watson's interest in both electricity and platinum may well have prompted Ingen-housz to pursue these matters further, but in 1768 he was chosen by Pringle, and with the approval of George III, to go to Vienna to inoculate the children of the Empress Maria Theresa. This led to an appointment as court physician with a handsome salary which he used in part to set up a laboratory and to cultivate his taste for experimentation in physics and chemistry. In Vienna he entertained a number of distinguished visitors, including Maria Theresa's son the Emperor Joseph II, as well as visiting scientists.

He was elected a Fellow of the Royal Society in 1769 and in one of his early papers, dated November 1775 (3) he described several experiments on platinum and its magnetic properties, following this with an account of his melting the metal in a narrow glass tube into which he "directed five or six electrical explosions from three very large jars." He concluded:

> "By this experiment it should seem as if platina (which hitherto could never be melted by common fire by itself, but only in the focus of a very strong burning glass such as was a little while ago made at Paris) were equally fusible if not more so than iron, by electrical fire."

Thus Ingen-housz was the first to melt platinum by means other than a burning glass. Between 1775 and 1782 he sent no fewer than nine papers to the Royal Society, although most of these concerned the respiration of plants, and he became a respected member of the scientific communities of Vienna, Paris and London. The greater part of the years 1777 to 1779 he spent in London, where he delivered the Bakerian lecture to the Royal Society, dealing with electrical phenomena and supporting Franklin's "One-fluid" theory (4). He remained in Vienna from 1780 until 1788, continuing the work on plant respiration that made him famous but also experimenting on platinum. During the visit to Vienna by the Count von Sickingen in 1784 Ingen-housz was shown the Count's method of rendering platinum malleable and they collaborated in a further experiment with the electric discharge, using a piece of wire provided by

The title page of the first volume of researches published by Ingen-housz in Paris in 1785 and dedicated to his friend Benjamin Franklin. This included a detailed account of the work on platinum carried out by the Count von Sickingen and by the Abbé Rochon, with both of whom he had collaborated, followed by a description of his own experimental work on the melting of a platinum wire by means of an electric discharge from a Leyden jar while the wire was surrounded by oxygen. His suggestions for the cladding base metals with platinum are also given

NOUVELLES
EXPÉRIENCES
ET
OBSERVATIONS
SUR DIVERS OBJETS
DE PHYSIQUE.

Par JEAN INGEN-HOUSZ, Conſeiller Aulique, & Médecin du corps de Leurs Majeſtés Impériales & Royales, Membre de la Société Royale de Londres, &c. &c.

A PARIS,

Chez P. THÉOPHILE BARROIS le jeune, Libraire, quai des Auguſtins, n° 18.

M. DCC. LXXXV.

AVEC APPROBATION, ET PRIVILÈGE DU ROI.

Sickingen. In a letter to his friend Jacob van Breda of Delft he described their experiment, and also included it in his major publication (5, 6).

"A wire of platinum of the thickness of a large knitting needle was surrounded by a steel wire of which the point was very sharp; we raised the steel wire to incandescence by means of a Leyden jar inside a flask of good quality dephlogisticated air. The steel wire burned and to all appearances communicated the flame to the platinum . . . which was found to be in perfect fusion and had lost neither its colour nor its malleability. This experiment demonstrates that platinum is a true metal endowed with all the qualities that make the other metals useful in the arts and trades; that it can be melted; that the melting destroys none of its valuable properties, and in consequence it only remains for us to take advantage of it, to see it in commerce and to find a less expensive method for melting it."

His work on platinum continued when he received a gift of six pounds of the native metal from the Marquis de Santa Cruz and at the same time obtained

111

information about the work of Chabaneau in Spain. Going to Paris in July 1778 to study Lavoisier's new chemistry he wrote that he found the chemists there were not taking very much interest in research on platinum.

"I did what I could to awaken the interest of some of the best chemists of my acquaintance and encouraged them to take up afresh their work on this metal that they had almost abandoned. I succeeded in drawing them out of their inactivities: among others M. Pelletier, whose talents and learning are well known, having resumed his researches, soon succeeded in taking a large step forward."

During this visit to Paris (which was brought to an abrupt end by the storming of the Bastille on July 14 1789) he made contact, as we have seen in Chapter 5, with Janety and with the Abbé Rochon while he also collaborated with Pelletier in preparing malleable platinum from the sal-ammoniac precipitate:

"One should ram it and press it into a red-hot-crucible and then strike it with a blow of a pestle, or one forges it with a hammer while hot and I have made from it several objects, among others a medal which was very successful."

Ingen-housz's own account of his work on platinum is bound up with reports of a number of other of his many researches in a book written for publication in French in 1781 (6). However, a German translation of it by his friend Niklas Karl Molitor appeared first in Vienna in one volume in 1782, and then in a second edition of two volumes also in Vienna two years later, both under the title of Vermischte Schriften (7). Only in 1785 did the first volume in French make its appearance in Paris, with the second delayed until 1789, all with the title of Nouvelles Expériences et Observations.

The first service that Ingen-housz rendered was to give a detailed account of von Sickingen's simplified method of refining platinum. He then describes Rochon's work on his mirrors and discloses the fact that he possesses two of them worked by Carrochez. Next he reports the use of Rochon's platinum by two prominent clockmakers, Robert Robin and Louis Berthoud, who used the metal for pendulums and movements. Then he describes Daumy's procedure, mentioning his manufacture of doublé metal, turns for a moment to Janety and from him to Pelletier. Finally comes a summary of the properties of platinum, at the end of which he points out that in its hardness it possesses a property in which it transcends both gold and silver, these two metals having to be hardened by means of base metals like copper to make them useful for jewellery or plate, while a vessel made of pure platinum has sufficient hardness to need no alloy whatever.

Ingen-housz adds to this the statement that it was he who suggested to Daumy the cladding of copper with a thin foil of platinum "in the same way as he doubles this metal in his workshop with a foil of silver or gold to make it into ornaments and plate". The attempt was successful and Ingen-housz points out the possibilities open to "copper doubled or covered with this indestructible precious metal, unalterable and untarnishable by exposure to air or by contact with substances which spoil the brilliance of the purest silver and blacken it" (6).

Thus Ingen-housz's direct addition to the knowledge of platinum was by no means negligible, but his principal contribution was probably made by his journeying from one capital to another and by his discussions in the best academic circles. He emphasised what he had to say by having a set of three waistcoat buttons made in platinum and displaying these to his friends. These were made by Matthew Boulton, whose works in Birmingham he visited during his later period in England, and when he also attended a meeting of the famous Lunar Society, and they are referred to in a letter to Boulton dated March 9, 1792 that survives in the collection of Boulton manuscripts:

> "I would have acknowledged long ago the receipt of the platina buttons you was so good as to make for me out of the piece I left under your hands, if I had not desired to know before the judgement of some of my friends about them. They all admire them very much. Several questions were made about the price of them. For my part I am much delighted with them and am not a little astonished, that they could fall out so well and the platina without mixture of other metals is susceptible of such high polish; which however I think it would not have acquired in less able hands. Receive my gratefull acknowledgement for the civility I received from you at Birmingham and nominaly for the trouble you put yourself at in manufacturing these triple beautifull and everlasting buttons, for which I am ready to pay the expenses if you let me know the amount of them, tho you was so kind as to receive it when I had the pleasure of being at your hous." (8)

Ingen-housz spent the last few years of his life in England, which he loved, as the guest of the Earl of Shelburne at Bowood and was permitted to make use of the laboratory created there for Joseph Priestley in 1773. There is no doubt that he was a prime factor in the widening of interest in platinum that was soon to bring about important advances.

Van Marum's Giant Generator

Another friend and countryman of Ingen-housz, Martinus van Marum, (1750–1837) carried out a remarkable series of experiments on the effects on a number of metals subjected to the electrical discharge. He had an enormous electrostatic generator, built by the London instrument maker John Cuthbertson and installed in the Teyler Foundation building in Haarlem. This consisted of two glass plates over five feet in diameter, which had to be turned by two workmen standing on a specially constructed table; in modern terms it was capable of discharging half a million volts. After investigating the oxidation of a number of metals and alloys van Marum recorded:

> "Finally, in April 1790 I attempted the calcination of Platinum, which I had specially made for the purpose by M. Jeanety in Paris and which I had received shortly beforehand. The wire was drawn to a thickness of 1/75 inch; and when I now examined its fusibility by the electric discharge, it seemed to me that it was almost the same as that of silver. The electric discharge also reduces platinum as readily as silver to a fine grey powder . . . this can be considered as an oxide so long as it has not been demonstrated by decisive experiments that the effect of the discharge on this metal is a totally different one" (9).

113

William Parker's Burning Lens

Before these applications of "the electric fire", however, there are several attempts to record on the melting of platinum by the much-used methods of both the burning glass and the furnace. In 1782 William Parker, the London instrument maker who had generously supplied Priestley with his burning glasses, constructed at his own expense, running to some £700, a three-foot diameter solid lens that he hoped to sell to the Royal Society. In a letter to Sir Joseph Banks, the President, Parker requested him to attend trials of this lens, but although a number of scientists attended Banks was not among them.

Parker wrote to Banks again two weeks later giving a few details of his melting of a number of metals (10).

"Yesterday and today myself with several Gentlemen have made Some Experiments with my Lenses of which I give you as follows and I shall Esteem it a Singular favour if you will favour me with your Presence on a fine Day with any Gentlemen you please to make what Experiments you think proper and am Sir

Your Most Humble Servant
Wm Parker

Fleet Street
16 July 1782

$\frac{1}{2}$ oz cast iron	15 seconds
10 Grains Fine Gold	13 seconds
10 Grains Platina	30 seconds"

Among those present was the ubiquitous Magellan, who recorded that "Mr. Parker's burning lens in London perfectly melted platina in less than two minutes" (11). A more conservative account was given later, however, by Parker's son Samuel, writing in "The Cyclopaedia" published by Abraham Rees in 1819. His account reads (12):

"Platina: The experiments evince that the specimens were in different states of approach to a complete metallic form, several of them threw off sparks, which in most cases were metallic."

(This account misprints the time for the melting of platinum as only 3 seconds)

The Activities of Lorenz Crell

Yet another attempt to make use of the high temperature of a porcelain furnace was made in 1784 when Lorenz Crell, who was to have much greater influence through his publications, paid a visit to the Fürstenberg factory that had been established by the Duke of Brunswick in 1747. Together with its director, Johann Ernst Kohl, Crell experimented with some platinum foil and wire he had been given by the Count von Sickingen as well as with some native platinum, but found that none of his samples melted (13). He concluded, however, that each new investigation taught something.

Crell had studied in Strasbourg, Paris, London and Edinburgh and had been greatly influenced by both Cullen and Black. In 1774 he became a professor in his native city of Helmstädt and three years later decided to found and edit a periodical on chemistry. At first issued quarterly, this became the monthly *Chemische Annalen* in 1784 and created a medium in which German and other chemists could exchange their findings. Even before this Crell was a tireless correspondent with scientists in other countries, among others with Joseph Black at Edinburgh, and one interesting exchange of letters, preserved in the University of Edinburgh Library, concerned the process devised by the Count von Sickingen. In April 1782 Crell informed Black that Sickingen "had melted twenty pounds of platinum and obtained a metal that was almost more ductile than gold itself". In the following September Black included in his reply

"The Count of Sickingen's success in melting Platina in such quantity as 20 lib is very remarkable, have you learnt his method? Bergman has some curious observations upon the means for giving it fusibility."

On October 24th Crell replied to Black, and an extract from this letter is reproduced over page. It must be remembered that Crell adhered to the old nomenclature and was a confirmed phlogistonist, hence the archaic nature of his comments. Black was clearly interesting himself in the properties of platinum and its fabrication, but it was not until some years later that he obtained two pounds of native platinum from one of his former students, Ignacio Ruiz de Luzuriaga, now practising medicine in Madrid and later to become physician to the King of Spain. One consquence of this interest by Black was the influence it

**Lorenz Florenz Friedrich von Crell
1745–1816**

A native of Helmstädt in the Duchy of Brunswick, Crell studied medicine and chemistry in Strasbourg, Paris, London and Edinburgh, returning to Helmstädt as a professor in the university there. Although he carried out a series of unsuccessful attempts to melt platinum in the porcelain furnace at Fürstenberg his greater contribution was the foundation of his *Chemische Annalen*, the first journal to be devoted to chemistry, and his enthusiastic editing and tireless correspondence with scientific workers throughout Europe. He was elected a Fellow of the Royal Society in 1788

115

An extract from a letter from Lorenz Crell to Professor Joseph Black of Edinburgh, dated October 24th 1782, in answer to the letter enquiring about the Sickingen process. In modern terms this reads:

> The Count of Sickingen has published his method: it seems to me very ingenious: he dissolves the platinum in aqua regia, precipitates the iron by Prussian Blue; evaporates the clear fluid, calcines it in a crucible, and puts the metal red hot on the anvil, where when very well beaten with the hammer, it unites into one mass and is to be extended in fine wires, as gold, and similarly into leaves.

had on another of his students, Smithson Tennant, who as we shall see later in Chapter 9 made a visit to Crell in 1784 and discussed Sickingen's method with him.

Thomas Willis of Wapping

Another long but unsuccessful series of experiments was carried out by a manufacturing chemist and druggist named Thomas Willis, operating in Wapping on the Thames just below the Tower of London. His attempts to melt platinum with a variety of fluxes, with charcoal and, following Pelletier, with phosphorus, were at least partly satisfactory, but in no case was his product capable of being forged. Willis also tried to melt the sal-ammoniac precipitate, but his manipulative skill was not of the highest order and he lost the precipitate in his fire. He had the assistance in this work of the younger Thomas Henry, then only twenty-one, who had been sent to London by his father, the leading figure in the Manchester Literary and Philosophical Society, and it was presumably for this reason that Willis submitted his paper to that body. His apologetic introduction is reproduced here, but he did not carry out the promise to pursue his objective. The paper received wide publicity, however, in translation into French and German periodicals (14).

Lavoisier's Larger Scale Attempts

A further cause of greater activity directed towards the melting of platinum was the report of the Académie des Sciences, prepared with the encouragement of Guyton de Morveau, to the French National Assembly in 1790 recommending the adoption of the metric system and the preparation of standards made in

116

The introductory letter from Thomas Willis, the London chemical manufacturer, accompanying the account of his many unsuccessful attempts to melt platinum and to yield a malleable product. The paper was presented to the Manchester Literary and Philosophical Society, founded in 1785, largely because the young son of Thomas Henry, one of its vice-presidents, had assisted Willis in the work. Despite the lack of success reported in the paper it was widely reproduced in French and German journals

---

EXPERIMENTS *on the* FUSION *of* PLATINA; *by Mr.* THOMAS WILLIS, *Chemist, at the Hermitage, London. Communicated by Mr.* THOMAS HENRY, F. R. S. *&c.*

READ AUGUST 13, 1789.

TO Mr. THOMAS HENRY.

LONDON, JULY 18, 1789.

SIR,

I HEREWITH inclofe you a few experiments on platina, and beg the favour of you to lay them before your truly commendable fociety. I fhould have been happy could I have fucceeded in rendering this extraordinary metal malleable, as it would have been of the greateft confequence in conftructing optical inftruments, on account of its not being affected by the air. I do not defpair of fucceeding, as my intention is to purfue this object ftill further, and I fhall with pleafure communicate my future proceffes to the public through your moft excellent inftitution.

I am, SIR,

Your moft obedient, &c.

THOMAS WILLIS.

---

platinum. The only platinum available in Paris at this time was Janety's made by the arsenic process, and this was on a relatively small scale. Lavoisier was well aware that any method for melting platinum on a larger scale would prove difficult and expensive, yet in 1789 he was proposing "to construct a very simple furnace for this purpose, of very refractory earth", and to pass oxygen through this from several of his gasometers "to produce a heat greatly more intense than any hitherto known" (15). Lavoisier was much occupied with other matters, and the work of designing and building the furnace was delegated to his assistants Armand Seguin and Jean Meusnier.

The diaries of Sir James Hall studied by Dr. J. A. Chaldecott and already referred to in Chapter 5, (16) show that in 1791 such a furnace was under construction, the objective being the melting of some 50 kilogrammes of platinum

I should not however have intruded upon you at present with a letter merely on this subject. Messrs Lavoisier & Seguin have requested some information from me, which I do not find myself very capable of supplying out of my own funds. They are constructing a furnace to be fed with vital air, in which they mean to melt 100 pounds of Platina, for the use of the commissaires appointed by the academy for the ascertaining of weights & measures. It was natural enough for them to apply to me on this occasion for clay of the most unfusible nature; but I know of none ~~of mine~~ which possesses this property in a greater degree than the Stourbridge, and I am confident that they have in France clays as refractory as this. Do you think, Sir, that these or any clays can stand the action of a fire with dephlogisticated air?

In 1791 Lavoisier and his two assistants Seguin and Meusnier began to design a furnace, to be fed with oxygen, in which to melt fifty kilogrammes of platinum and sought the advice of Josiah Wedgwood on the most suitable refractory to withstand the very high temperatures involved. Wedgwood felt unable to help and passed on the enquiry to Joseph Priestley. This is the relevant passage from Wedgwood's letter to Priestley of September 2nd 1791

for the preparation of the standards of length and weight for the Commission of Weights and Measures.

The search for a suitable refractory was however a major problem, and Lavoisier and Seguin called upon Josiah Wedgwood for advice. Wedgwood felt unable to recommend "a clay of the most infusible nature" and in turn appealed to Priestley for his views (17). The latter also felt unable to advise except to say that he considered magnesia to be the most likely substance to withstand such intense heats (18).

That these efforts of Lavoisier and his two colleagues did not meet with success is evident from the fact that the Commission turned to Janety for their needs of platinum, as will be shown in Chapter 10.

The Properties of Eighteenth Century Platinum

The early workers in platinum checked the purity of their product by determining its specific gravity and considered that the higher it was, the purer the metal must be. The method was rather rough, since at first the amounts of platinum available were very small and the accuracy of the balances of the time not high. Nevertheless, the check had its value and increased in exactitude as the quality of the apparatus available improved. From time to time, however, the figures quoted were too high, being greater than what is now generally accepted as the correct value for the pure metal (21.45). Von Sickingen, the first man to prepare malleable metallic platinum in more than minute quantities, found its gravity to be 21.061. Then came Chabaneau, claiming that his forged metal was so pure that it had a gravity of 24, but de Morveau could not make a sample of it yield a figure higher than 20.833 (19). At the same time he examined a piece of hard-worked wire bought in Paris and probably made by Janety and found its gravity to be 20.847. Ingen-housz, working before 1789, obtained a figure of 22.285 for a piece from a Chabaneau bar, whereas two medals made by himself and Janety showed 20.108 for the former and 21.0 for the latter (6). Incidentally, Ingen-housz says that Janety, like Chabaneau, claimed a gravity of 24 for his best metal, but it is interesting that in 1813, by which time the accuracy of balances had improved considerably, a sample of his metal showed a gravity of 20.01, against 21.04 given by some of the undoubtedly purer products of Wollaston. (20).

The only other physical property of platinum that came into account in the eighteenth century was its tensile strength, and the measurement of this became a practical possibility when von Sickingen produced a metal pure enough to be drawn into fine wire. For his own measurements he designed an ingenious testing machine illustrated here and used a wire of diameter 0.025 inch, obtaining a figure which, when translated into modern terms, represents a breaking strain of 24.9 tons per square inch. The test was repeated in 1795–1796 by Guyton de Morveau on platinum which probably came from Janety, and he obtained a result equivalent to 25.2 tons per square inch. De Morveau's specimen was in cold-worked condition, and so no doubt was von Sickingen's. It is interesting to compare these figures with those given by a modern product,

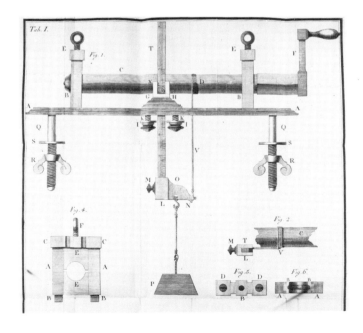

When the Count von Sickingen succeeded in producing platinum in the form of wire it became possible to determine its tensile strength. This shows the machine designed and built by Sickingen, one of the very earliest mechanical testing devices to be introduced into physical metallurgy

119

and to find that the commercial platinum of today (99 per cent platinum) can yield a figure of up to 22.3 tons per square inch in the cold-worked state. It is therefore evident that in purity the eighteenth-century metal was but little inferior to the commercial metal of today.

It should be realised, however, that with the exception of von Sickingen and to some extent of Chabaneau, all the workers in platinum up to this point had a per cent or two or possibly even more of iridium in their metal. When native platinum is dissolved in aqua regia a considerable proportion of the iridium also goes into solution as iridic chloride, which may later be reduced to iridous chloride. When ammonium chloride solution is added, if all the iridium is there as iridic chloride, most of it is precipitated as ammonium chloriridate with the ammonium chloroplatinate and the latter is thereby coloured red. It will be remembered how many of the early workers described their precipitates as red or reddish. If, however, the iridium is all present in the solution as iridous chloride, the sal-ammoniac will not precipitate any of it, since ammonium chloriridite is soluble, and the platinum precipitate will be a pure yellow in colour. Thus, in practice there will be all sorts of gradations between these two extremes according to the way in which the dissolving of the native metal has been effected. There are other factors, too, that enter into the matter. The platinum tends to be precipitated first, since the solubility of the chloroplatinate is a little less than that of the chloriridate; dilution of the solution tends to hold back precipitation of the chloriridate for the same reason; the presence of excess hydrochloric acid is said to delay precipitation of the iridium salt; and of course if the aqua regia used in the original attack on the native metal is dilute less iridium is dissolved. All these things were unsuspected by the workers before the discovery of iridium and three other metals of the group by Smithson Tennant and Wollaston in the early years of the nineteenth century (to be recounted in Chapter 9) and this was responsible for more than a few of their differences and difficulties with what they supposed to be one distinct metal.

Achard's Study of Platinum Alloys

So far only the properties of relatively pure platinum had been studied, but in 1788 there appeared a remarkable book from the hands of Franz Karl Achard in Berlin. He will be remembered as the discoverer of the arsenic process for preparing malleable platinum described in Chapter 5, but he now published the results of a laborious and comprehensive programme on the alloys of eleven metals, including platinum, with each other (21). In this he pointed out that the properties of alloys are quite different from those of the pure metals and are unpredictable. All the alloys were in the as-cast condition and on these he carried out tests for density, hardness, resistance to impact and to the file and then on the effects of exposure to air, to hydrogen sulphide and to acids on polished surfaces. He attempted to alloy platinum in the proportions of 1:2 and 2:1 with cobalt, copper, iron, lead, tin, zinc, bismuth, antimony and arsenic, and finally

The title page of the rare and almost forgotten book published in Berlin by Franz Karl Achard in 1788. Among no less than 894 samples of the binary alloys of eleven metals Achard recorded his results with alloys of platinum with most other base metals available at the time, but unfortunately few of these were sound enough to withstand his tests for hardness, ductility and resistance to impact

RECHERCHES

SUR LES

PROPRIÉTÉS

DES

ALLIAGES

MÉTALLIQUES.

PAR

MR ACHARD

Directeur de la Claſſe de Phyſique dans l'Académie Royale des Sciences & Belles-Lettres de Pruſſe, Membre de l'Académie Royale & Impériale des curieux de la nature, des Académies Royales de Turin, de Suede, d'Orléans, de Dijon, de Milan, des Académies Électorales de Baviere & Maïence, Membre de la Société Hollandoiſe de Harlem, de la Société Phyſique de Danzic, de la Société littéraire de Halle, de la Société phyſique de Berlin, de la Société Royale de Francfort ſur l'Oder, de la Société Oeconomique & Patriotique de Siléſie & de pluſieurs autres Académies, & Sociétés littéraires.

A BERLIN 1788.
IMRIMÉ CHEZ GEORGE JACQUES DECKER & FILS, IMPRIMEURS PRIVÉS DU ROI.

produced a ternary alloy of equal parts of copper, iron and platinum which he found to give considerable hardness as measured by the diameter of the flat small impression made on a small sphere falling repeatedly from successively greater heights. Not all of the binary alloys were of course sufficiently sound to withstand his series of mechanical tests, while his specific gravity figures he admitted were so low as to indicate considerable porosity. These results were published in a book "Recherches sur les Propriétés des Alliages Métalliques", written in the French language insisted upon by Frederick the Great, but unfortunately it was virtually ignored by metallurgists everywhere, while Achard himself turned to the development of the beet sugar industry in Germany. This rare work was brought to light only in recent years by Professor Cyril Stanley Smith (22) to whom the present writer is indebted for his attention having being drawn to it.

Count Mussin-Pushkin's Method

Interest in platinum was now spreading not only in those countries where scientists had already been working with it but also into other parts of Europe. The first initiative in Russia was taken by another aristocrat, the Count Apollos Mussin-Pushkin, a member of the Russian court and the founder of the Mining College in St. Petersburg. His work began in 1797 and continued until his death in 1805, the principal contribution being a method of refining platinum and rendering it malleable by the use of mercury.

He precipitated an aqua regia solution of platinum with sal-ammoniac as usual, washed the precipitate with a little cold water, dried it, calcined it to sponge and washed the latter two or three times with boiling water. After that he boiled it with dilute hydrochloric acid to remove residual iron, ignited it and then mixed it into a paste in a mortar with mercury. What he produced was not a true amalgam since the two metals did not form one, but it was a thick paste that could be moulded and handled. This so-called amalgam was then compressed

"in tubes of wood, by the pressure of an iron screw upon a cylinder of wood, adapted to the bore of the tube. This forces out the super-abundant mercury from the amalgam and renders it solid. After two or three hours I burn upon the coals, or in a

Count Apollos Mussin-Pushkin
1760–1805

An official of the Imperial Court surrounding Catherine the Great in St Petersburg and the founder of the Mining College there, Mussin-Pushkin devised a method of compressing platinum sponge with mercury, followed by heating and forging. He travelled widely in Europe on behalf of the mining authorities, meeting many other scientists, and was elected an Honorary Member of the Russian Academy of Sciences in 1796 and a Fellow of the Royal Society in 1799. He also re-organised the College of Mines into the more active Mining Cadet Corps in 1804. In 1802, after the annexation of Georgia by Russia, he was sent there to study the mineral resources and continued his work on platinum in a laboratory in Tiflis

122

crucible lined with charcoal, the sheath in which the amalgam is contained and urge the fire to a white heat; after which I take out the platina in a very solid state fit to be forged." (23)

Mussin-Pushkin published his first paper on his "platinum amalgam" in the *Annales de Chemie* in 1797 (24). Two years later, in a letter to Jeremias Benjamin Richter, the assayer and chemist to the Berlin porcelain factory who had introduced the concept of stoichiometry in 1792, he asked for an opinion on the amalgam and Richter passed the letter to Lorenz Crell for publication in his *Chemische Annalen*. In a footnote Richter, whose ideas had been sorely neglected, wrote:

"This scientist's prevailing curiosity (I will not say enthusiasm) for a new area of man's knowledge demands that I should publicly express my warmest thanks to him." (25)

Several other papers followed, written from widely different parts of Russia, concluding with a letter summarising his process written to the English chemist and Fellow of the Royal Society Charles Hatchett, which was published in 1804 in the *Journal of National Philosophy Chemistry and the Arts* founded by John Nicholson in 1797. In this he added:

"The whole of the operation seems to be governed by the pressure of the atmosphere and the laws of cohesive attraction: for the air is driven out from between the molecules of the platina, which by their solution in mercury are most probably in the primitive and consequently uniform figure. It is very visible and at the same time a very amusing phenomenon to observe how the platinum contracts every way into itself as if pressed by some external force." (26)

The only new contribution of Mussin-Pushkin was his method of getting his platinum into forgeable form, but the care he took to remove the last traces of iron from his sponge should be noted, in that he washed the original precipitate with cold water, then washed the ignited sponge and in addition boiled it in dilute hydrochloric acid. This is the only part of his process that found its way into later practice.

The Marchese Cosimo Ridolfi

No work of significance had yet been undertaken on platinum in Italy. In 1784 the Chevalier Nicolis de Robilante, the Inspector General of Mines, had presented a paper on the subject to the Académie Royale des Sciences de Turin (27) but this recorded work carried out much earlier on platinum received from Spain through the Piedmontese ambassador there and contained nothing that Lewis had not discovered by 1755 although he refers to the work of Tillet in 1779.

Much later some work was undertaken by a distinguished Italian whose early years were devoted to science but who then turned to politics. In 1815 the Marchese Cosimo Ridolfi, the son of a rich and noble family who had installed a laboratory in their palace in Florence, contributed a long paper "On the Purifi-

The Marchese Cosimo Ridolfi
1794–1865

After studying physics, chemistry and botany in the Museum of National History in Florence and writing a textbook on botany for the benefit of agriculturalists Ridolfi established a laboratory in the family palace in 1813. Here he carried out a long series of experiments on platinum, hoping to develop a simpler process that would lead to an increase in its applications. He later became Director of the Mint in Florence and for a brief period in 1848 he was prime minister to the Grand Duke of Tuscany

cation, Fusion and Economic Uses of Platinum" to the *Giornale di Scienza ed Arte*. After reviewing the procedures adopted by Janety, Pelletier and Leithner and then the mercury amalgam method of Mussin-Pushkin, Ridolfi proposed "a simpler method that would make its use more possible in the manufactures". He had observed that no-one had succeeded in combining platinum with sulphur and so conceived the idea that by converting all the other metals found in native platinum it should be easy to purify. After removing as much of the sand and iron as possible from the crude platinum and washing in hydrochloric acid Ridolfi melted his metal with lead and then granulated it by pouring into cold water. It was then pulverised, mixed with its own weight of sulphur, put into a white-hot clay crucible and heated for ten minutes. After cooling the metallic button, freed from iron and copper, was fused again with enough lead to remove the sulphur, the resulting lead-platinum alloy being heated and forged with a hot hammer until all the lead was squeezed out. Ridolfi claimed that his platinum was malleable and ductile with a density of 22.63 and that he successfully produced wire as well as foil that could be employed for the cladding of copper and brass, but his process never became anything more than a laboratory curiosity. His paper received brief summaries in Tilloch's *Philosophical Magazine* and rather later in Thomson's *Annals of Philosophy* (28).

124

Platinum in the Decoration of Porcelain

The applications of platinum in the late eighteenth century were but few. Janety produced ornamental items and one or two pieces of tableware but a growing demand for crucibles made itself felt very soon and he had to turn his attention to these. Similarly Chabaneau had confined himself to decorative objects and a few platinum medals had been struck.

There was, however, one other application – and the first in the long history of the uses of platinum in manufacturing industry – that began to develop before the close of the century.

Contemporary with the early scientific studies of platinum were those on porcelain and its decoration. In the winter of 1788–89 a paper "On the Use of Platina in the Decoration of Porcelain" was read to the Berlin Academy by Martin Heinrich Klaproth (29), the most distinguished German chemist and the leading analyst and mineralogist of his time. Beginning as a poor apothecary, he had later studied under Marggraf and in 1780 had married his wealthy niece, so being able to buy himself a laboratory, and from here there poured out analyses of hundreds of minerals as well as many papers to the Academy. His activities did not end here, however, and for some years he was retained as a consultant to the Royal Berlin Porcelain Factory.

Klaproth had not been among the numerous chemists who had worked on platinum, but he was obviously thoroughly familiar with the progress that had

**Martin Heinrich Klaproth
1743–1817**

The greatest German chemist of his time, Klaproth also took an active interest in technology and served for some years as a consultant to the Berlin Porcelain Factory. In 1788 he presented a paper to the Berlin Academy of Sciences describing his experiments on the use of platinum for the decoration of porcelain and displaying a number of specimens, so initiating the first application of platinum in manufacturing industry

so far been made. After a summary of what was then known and a reference to the work of "my worthy colleague M. Achard" Klaproth wrote

"How far platina may be employed in porcelain painting has never yet, as far as I know, been examined: I therefore thought it of considerable importance to make some experiments on this subject, which did not deceive my expectation; but on the contrary, convinced me that this object, in the hands of an ingenious artist, may be brought to perfection".

Klaproth exhibited to the meeting a number of samples of porcelain made in the Berlin factory and ornamented with platinum. He described his "simple and easy" process as follows:

"I dissolve crude platina in aqua regia, and precipitate it by a saturated solution of sal ammoniac in water. The red crystalline precipitate thence produced is dried, and being reduced to a very fine powder is slowly brought to a red heat in a glass retort. As the volatile neutral salt, combined with the platina in this precipitate, becomes sublimated, the metallic part remains behind in the form of a gray soft powder. This powder is then subjected to the same process as gold; that is to say, it is mixed with a small quantity of the same flux as that used for gold, and being ground with oil of spike is applied with a brush to the porcelain; after which it is burnt-in under the muffle of an enameller's furnace, and then polished with a burnishing tool.

The colour of platina burnt into porcelain in this manner is a silver white, inclining a little to a steel gray. If the platina be mixed in different proportions with gold, different shades of colour may be obtained; the gradations of which may be numbered from the white colour of unmixed platina to the yellow colour of gold. Platina is capable of receiving a considerable addition of gold before the transition from the white colour to yellow is perceptible."

The bound collection of papers, including Klaproth's, read before the Berlin Academy in 1788–89 appeared in 1793, and was certainly a rather exclusive publication and little attention was paid to this interesting and attractive development until his paper was reproduced in 1802 in the *Allgemeine Journal der Chemie*, founded by Alexander Nicholas von Scherer in Leipzig in 1798, and immediately afterwards in English in the *Philosophical Magazine* and again in an abridged version in *Nicholson's Journal* (30).

By 1791, however, Jean D'Arcet, Professor of Chemistry at the Collège de France and Macquer's successor as technical director of the famous Sèvres porcelain factory, had introduced platinum decoration there. He had interested himself in platinum some twenty-five years earlier and had tried unsuccessfully to melt it in the porcelain furnace at the factory established by the Comte de Lauraguais in Alençon (31).

Platinum decoration was also introduced into the Vienna porcelain factory by their colour chemist Joseph Leithner in about 1804 although an account of his work did not appear until some years later (32). He mixed platinum in the form of very fine powder with oil of turpentine and applied the paste in several layers, each coating being dried before the next was added, and then fired the porcelain at about 800°C. In a curious variation of this procedure Leithner similarly built-up successive layers of platinum on to paper so that, after

Platinum decoration on porcelain came into use in the Sèvres factory only three years after Klaproth read his paper in Berlin and before it was published. This is one of a set of Sèvres hard paste porcelain plates dated 1792 on which the decoration has been carried out in gold and platinum. All the faces and many of the architectural details are in platinum on a black ground. The gift of Lewis Einstein in 1962, this is now in the Metropolitan Museum of Art in New York

heating, a thin sheet of the metal remained that could be worked into any desired shape, It was used for a time to repair cracks and holes in crucibles but was not of course of significant importance.

English Lustre Ware

The wider publicity given to Klaproth's work by the English translations led to the British pottery manufacturers taking advantage of platinum as a decorative medium. The date of its introduction into the Staffordshire potteries in England has, however, been the subject of great debate and of many claims that cannot be upheld. Had Josiah Wedgwood still been alive and active when Klaproth's paper appeared in English the story might well have been different; he was of course a Fellow of the Royal Society and a keen reader of the scientific journals, but he had retired by 1790 and he died in 1795. Many writers on ceramics have maintained that the "silver lustre ware" as it was called was produced by the Wedgwood factory in the 1790s or early 1800s, but in fact it was not until 1806 that any real evidence can be presented for their use of platinum, and then they were following the practice of a rival pottery.

It is to one John Hancock, employed by Henry Daniel as an enameller in the Spode factory, that credit must be given for a process that made metallic decoration with platinum a commercially successful technique. Hancock had been

127

apprenticed to the Derby Porcelain Factory under William Duesbury where he acquired his knowledge of the preparation of colours. He was taken on by Daniel in 1805 under a curious arrangement by which undecorated ware was passed on to Daniel, working inside the Spode factory as an independent decorator. Hancock must have read, or had his attention drawn to, one of the English translations of Klaproth's paper, for by the closing months of 1805 he had successfully developed metallic lustre decoration. The details of this sequence of events have been established by Leonard Whiter in his study of the Spode family and factory (33). Under the arrangement just mentioned Daniel had to pay Spode a rental for the equipment used in the preparation of colours, and he had no intention of revealing Hancock's invention to his employer. Whiter continues:

"This presented him with the problem of how to enter in the record the grinding of platinum, the very mention of which would tell too much. No such secrecy was necessary for 'pale gold' (which Daniel made in the proportion of half an ounce of prepared gold to 12 grammes of prepared silver) and in his colour book he noted

November 11th 1805 half pound platina entered by the name pail gold in Mr. S.— book."

The earliest surviving pieces of Spode ware with lustre decoration show the rather dull leaden variety and is likely that several years of experimentation and failure by Hancock and by others were necessary to bring the process to full success.

The secret could not be kept however – Hancock is said to have sold the recipe for a small sum of money to anyone who wanted it (34) – and very soon Wedgwood, among others, was employing platinum decoration. By December 1806 John Wedgwood had sent jugs decorated all over with platinum and gold respectively to their friend John Leslie, the Professor of Mathematics in Edinburgh University, who had once been employed as tutor for his children by the first Josiah. Leslie, afterwards Sir John, and the great authority on heat radiation, sent a characteristic reply to Josiah II:

"26 Decr. 1806

I have to thank your brother John for the handsome present of the Jugs. They are very much admired. I wish you could inform me how the metallic coating is applied. While the Common Jug cools down in 60 minutes the platina one required 80 and the gold one 70 to come to the same point. The platina coating is therefore much thicker. I would strongly recommend it for coffee pots". (35)

Unquestionably "silver lustre", an essentially English development, became commercially successful and was exploited throughout Staffordshire and also by the potteries in Leeds, Swansea, Newcastle, Sunderland and elsewhere. When applied all over the ware it was seen as a cheap substitute for the silver and Sheffield plate tea pots, cream jugs, sugar basins and so on, and for many years, until the introduction of electroplating in the 1840s, there was a great production of such articles in rigid imitation of the shapes of the silverware. The all-over usage then declined, but the use of platinum decoration in a more artistic form had arisen in which either a pattern or its background was painted on to white

The publication of Klaproth's paper in English translations in 1804 quickly aroused the interest of the English porcelain and pottery manufacturers of Staffordshire and by the end of the following year a metallic lustre decoration had been successfully developed in the Spode factory. This tea-set is decorated with deep bands of platinum and with gold borders. It is preserved in the Spode Museum in Stoke-on-Trent

During the early years of the nineteenth century the so-called silver lustre was used extensively on earthenware by British potters. Applied all over the ware, the platinum coating was intended to simulate the appearance of silver. A little later a more artistic form arose combining enamel colours with the platinum. This typical "silver resist" jug, now in the Victoria and Albert Museum in London, is decorated in underglaze blue combined with platinum. Specimens of this kind are now collectors' pieces and very rare

129

ware with an aqueous solution of a viscous substance such as gum or honey before the platinum was applied. After drying, the "resist" was dissolved off in water, leaving the decorative platinum effect on firing.

A fuller account of the development of platinum decoration on porcelain and later on earthenware has been given elsewhere by the present writer (36).

Conclusion

This chapter has recorded a period of consolidation rather than of great progress in the history of platinum although a few scientists spread over several countries introduced minor variations in its treatment. But just as there had been a great upsurge of scientific activity in Spain a generation earlier so as the eighteenth century drew to a close another important phase opened in London. In addition, platinum was now available there in growing quantities. Together these factors led to the important work to be described in the next two chapters.

References for Chapter 7

1 W. A. Smeaton, Fourcroy, London, 1962, 28–29

2 J. Wiesner, Jan Ingen-housz, Vienna, 1905; P. Smit, *Janus*, 1980, **67**, 125–139

3 J. Ingen-housz, *Phil. Trans. Roy. Soc.*, 1776, **66**, 257–267

4 J. Ingen-housz, *Phil. Trans. Roy. Soc.*, 1778, **68**, (ii), 1027–1055

5 J. Ingen-housz, letter to Jacob van Breda of Delft, April 17, 1784. I am indebted to Dr. J. G. de Bruijn of Haarlem for this reference.

6 J. Ingen-housz, Nouvelles Expériences et Observations sur divers Objets de Physique, Paris, 1785–1789, **1**, 446; **2**, 505–517

7 J. Ingen-housz, Vermischte Schriften, trans. N. K. Molitor, 1st edition (1 vol.), 1782; 2nd edition (2 vols.), 1784, both published in Vienna

8 Letter from Ingen-housz to Matthew Boulton, Boulton MSS., Library of Birmingham Assay Office

9 E. Lefebre and J. G. de Bruijn, Martinus van Marum, Life and Work, Leyden, 1974, **5**, 209

10 W. Parker, Letter to Sir Joseph Banks, British Museum Add. MS. 33,977, ff 159–160

11 J. H. Magellan, in A. F. Cronstedt, An Essay towards a System of Mineralogy, trans. G. von Engestron, 2nd. edn., 1788, **2**, 573

12 A. Rees, The Cyclopaedia, London, 1819, **5**, Article on Burning Glass, not paginated

13 L. Crell, *Chem. Ann.*, 1784, (i), 328–334

14 T. Willis, *Mem. Lit. Phil. Soc. Manchester*, 1789, **3**, 467–480; *Obsns. Physique*, 1789, **35**, 217–225; *Chem. Ann. (Crell)*, 1790, (iii), 242–247; *Ann. Chim.*, 1791, **9**, 219

15 A. L. Lavoisier, Traité Elémentaire de Chimie, Paris, 1789, **2**, 552–558

16 J. A. Chaldecott, *Notes and Records Roy. Soc.*, 1967, **22**, 155–172; *Ann. Science*, 1968, **24**, 21–52

17 Keele University Library Wedgwood MS., 4091–5; J. A. Chaldecott, *Platinum Metals Rev.*, 1970, **14**, 24–28

18 Royal Society Misc. MSS., V, Letter 30; J. A. Chaldecott, Reference 17

19 L. B. Guyton de Morveau, *Ann. Chim.*, 1798, **25**, 1–20

20 A. F. Gehlen, *J. Chem. Physik (Schweigger)*, 1813, **7**, 309–316

21 F. K. Achard, Recherches sur les Propriétés des Alliages Métalliques, Berlin, 1788

22 C. S. Smith, Four Outstanding Researches in Metallurgical History, Philadelphia, A.S.T.M., 1963, 11–17

23 A. Mussin-Pushkin, *Allg. J. Chem.*, 1804, **3**, 450–455.

24 A. Mussin-Pushkin, *Ann. Chim.*, 1797, **24**, 205–215

25 A. Mussin-Pushkin, *Chem. Ann. (Crell)*, 1799, ii, 3–9

26 A. Mussin-Pushkin, *J. Nat. Phil. Chem. & Arts (Nicholson)*, 1804, **9**, 65–67

27 N. di Robilante, *Mem. Acad. Roy. Sci. Turin*, 1784–85, **2**, 123–147

28 C. Ridolfi, *Giornale Sci. ed Arte*, 1816, **1**, 24–35; 125–139; *Phil. Mag.*, 1816, **48**, 72–73; *Ann. Phil.*, 1819, **13**, 70–71

29 M. H. Klaproth, Sammlung der deutschen Abhandlungen König. Akad. der Wissenschaft zu Berlin in Jahren 1788 u. 1789, Berlin, 1793, 12–15

30 M. H. Klaproth, *Allg. J. Chemie (Scherer)*, 1802, **9**, 413–422; *Phil. Mag.*, 1803–4, **17**, 135–138; *J. Nat. Phil. Chem & Arts (Nicholson)*, 1804, **7**, 286–287

31 *Obsns. Physique*, 1771/2, Intro I, 108

32 A. N. Gehlen, *J. Chem. Physik (Schweigger)*, 1813, **7**, 309–316; *Ann. Phil. (Thomson)*, 1815, **5**, 20–21

33 Leonard Whiter, Spode; A History of the Family, Factory and Wares from 1733 to 1833, London, 1970, 43

34 Simeon Shaw, History of the Staffordshire Potteries, Hanley, 1829, 227

35 Wedgwood Archives, Etruria I, 315

36 L. B. Hunt, *Platinum Metals Rev.*, 1978, **22**, 138–148; *The Connoisseur*, 1979, **200**, (March), 185–189

William Allen
1770–1843

The centre of a remarkable group of scientists who not only earned a living by their professional activities but formed an extensive network of societies for experiment and discussion, Allen was producing small articles of platinum by 1805 and encouraging one of his young assistants, Thomas Cock, to develop a process for its refining and fabrication

From a portrait by H. P. Briggs in the possession of the Pharmaceutical Society

8

The Professional Scientists of London and their Societies

"Platinum is a most valuable metal; as it is not oxidisable nor fusible under common circumstances, and only with difficulty combinable with sulphur and not acted upon by common acids, it is admirably adapted for the uses of the philosophical chemist."

HUMPHRY DAVY

The thirty year period from 1790 until 1820 was a vital and productive phase in the development of both pure and applied science. Lavoisier's chemical revolution had cleared the minds of chemists, his definition of an element had opened the way to the discovery of many more, and methods of analysis, if still rather simple, had become established. The chemical industry began to expand while the scientific study of mineralogy was aiding in the development of mining and metallurgy. At the same time there existed a body of scientists, particularly in London, who were interesting themselves in technology and who were able to secure an income for themselves from their professional work by lecturing, by editing and publishing, by the manufacture of chemicals or by the making of scientific instruments. These men tended to associate into groups or small societies for joint experimentation and discussion. They were particularly interested in the metals recently discovered including titanium, molybdenum, tungsten and niobium, and although it would be a considerable overstatement to say that they were particularly concerned with platinum they none the less took an active interest in its properties and its fabrication and from their activities stemmed several important advances.

The Leading Role of William Allen

One of the first to engage in lecturing was Bryan Higgins (1741–1818) at his School of Practical Chemistry in Greek Street, Soho, started in 1774, and from which developed the Society for Philosophical Experiments and Conversations,

133

Allen's laboratory at Plough Court in the early nineteenth century. A successful pharmaceutical business was carried on here and a school of young research assistants was built up. Meetings of both the Askesian Society and the British Mineralogical Society were also held here from 1796 until 1807 when the members took part in the founding of both the Geological Society and the London Institution

founded in 1794. Another very active organisation was the Physical Society of Guy's Hospital, recognised as an important meeting place for scientists from its foundation in 1771 until the middle of the nineteenth century.

In 1792 the young William Allen, the son of a Quaker silk manufacturer who disliked his father's business, entered the famous Plough Court Pharmacy, established in 1715 in Lombard Street in the City of London by Joseph Gurney Bevan, a leading member of the Quakers. He began at once to attend the lectures given by Bryan Higgins and to devote his spare time to studying "the new system of chemistry" and in 1794 he also became a member of the Physical Society at Guy's Hospital. In that year Bevan retired, being succeeded by Samuel Mildred who took Allen into partnership but retired in 1797, leaving the business in Allen's charge (1).

Next door to the pharmacy there worked a printer named James Phillips who had two sons William (1775–1828) and Richard (1778–1851). The elder was establishing himself as a mineralogist, while the younger Richard became apprenticed to Allen at Plough Court. In the same neighbourhood there lived William Hasledine Pepys, the son of a cutler and maker of surgical instruments, and it was these very young men who took the initiative in forming the Askesian

134

Society in March 1796, the name being taken from the Greek word for exercise or training. They appointed an older man, Samuel Woods, a woollen draper also of Lombard Street, as president and were soon joined among others by Alexander Tilloch, the founder in 1798 of *The Philosphical Magazine*, and later by two distinguished surgeons, Astley Cooper (1768–1841) and William Babington (1756–1833). The latter, an older man than most of the members of the Askesian Society and resident apothecary at Guy's Hospital, had tried to melt platinum in 1797, writing in his lecture notes preserved there:

> "The most violent fires are insufficient to melt it though its parts may be made to cohere together into a solid button by the strong heat of a wind furnace." (2)

Meetings were held fortnightly at Plough Court and a collection of apparatus was assembled. In December 1798 Allen wrote in his diary:

> "I am making great progress in chemical experiments – fused platina with oxygen on charcoal." (3)

At a meeting held in the Askesian Society's room on April 2nd 1799 attended by Allen, Pepys, Tilloch, Richard Knight, the son of a London ironmonger who became an instrument maker, and William Lowry, the engraver and geologist, it was resolved that

> "those present do form themselves into a Society under the Denomination of the British Mineralogical Society." (4)

The number of members was not to exceed twenty and was to be confined to those able and willing to undertake the chemical analysis of mineral substances. It was felt that the absence in England of any school or college of scientific minerology such as the École des Mines established in France in 1783 could be remedied by the collective activities of the members in applying science to mining and metallurgy. Before the end of the year the founders were joined by William Phillips, Charles Rochemont Aikin and his brother Arthur, the latter becoming president of the society. Other members included Dr. Babington, Richard Phillips and Robert Bingley, the Assay Master at the Royal Mint, while William Henry of Manchester, Charles Hatchett and Richard Kirwan were among those elected corresponding members.

Bingley had already conducted a series of experiments on platinum and particularly on its separation from gold in the course of normal assaying. These he described in a long letter to Charles Hatchett (5) in which he wrote:

> "That it is a source of great satisfaction to be able compleatly to separate two metals which have so many properties in common, and by such means as are not in the least destructive to gold."

Meetings were held fortnightly at Plough Court, reports being read on the analyses of minerals, and on January 9th 1800 a paper was read by Richard Knight (1768–1844) entitled "A New and Expeditious Process for Rendering Platina Malleable" (6). The author, who displayed to the meeting a specimen of his malleable metal, opened his remarks in these words:

135

At a meeting of the British Mineralogical Society on January 9th 1800, Richard Knight, the son of an ironmonger in the City of London and a competent analyst and instrument maker, read a paper describing his method of producing malleable platinum. His principal advance lay in a more forceful compression of the sponge while hot. There is no evidence that Knight made more than two or three pieces of apparatus from his platinum, but it is likely that his paper served to arouse the interest of Wollaston and Tennant whose major contribution is described in the next chapter

THE

PHILOSOPHICAL MAGAZINE.

FEBRUARY 1800.

I. *A new and expeditious Proceſs for rendering Platina malleable. By Mr.* RICHARD KNIGHT, *Member of the Britiſh Mineralogical Society. Communicated by the Author.*

THE many peculiar advantages which platina in a malleable ſtate poſſeſſes over every other metal for the fabrication of a variety of inſtruments and utenſils particularly uſeful for the purpoſes of chemiſtry, together with the extreme difficulty of procuring it, being hitherto only to be obtained from Paris, of a very indifferent quality, and at a price equal to that of gold, firſt induced me to turn my attention to the ſubject. After having repeated a variety of experiments, from the different writers on this ſubſtance, without effect, I at length completed a proceſs, the ſucceſs of which has fully anſwered my expectations. By the proceſs which I follow I am able to reduce any quantity of crude platina to a perfectly malleable ſtate, entirely free from impurity, and capable of being wrought into any form whatever. As this is a circumſtance of conſiderable importance to the chemical world, and the advantages which may reſult from it to ſociety in general are perhaps incalculable, I would conſider myſelf deſerving of cenſure, could I allow any motive whatever to induce me to withhold it from the public. By ſending it for publication in a work of ſuch extenſive circulation as the

VOL. VI.   B   Philo-

"The many peculiar advantages which platina in a malleable state possesses over every other metal for the fabrication of a variety of instruments and utensils particularly useful for the purposes of chemistry, together with the extreme difficulty of procuring it, being hitherto only obtained from Paris, of a very indifferent quality, and at a price equal to that of gold, first induced me to turn my attention to the subject. After having repeated a variety of experiments, from the different writers on this substance, without effect, I at length completed a process, the success of which has fully answered my expectations. By the process, which I follow I am able to reduce any quantity of crude platina to a perfectly malleable state, entirely free from impurity, and capable of being wrought into any form whatever."

136

He dissolves the native platina in aqua regia, allows the insoluble to settle, decants the clear liquor, precipitates with the sal-ammoniac, again allows settlement, decants and washes the precipitate with cold distilled water until it is free from acid. The last water is then poured off and "the precipitate evaporated to dryness". He goes on to say:

"So far my process is in a great measure similar to that which some others have also followed; but my method of managing the subsequent, and which are indeed the principal manipulations, will be found to possess many advantages over any that has yet been made public. The best process hitherto followed has been to give the precipitate a white heat in a crucible, which in some measure agglutinates the particles; and then to throw the mass into a red-hot mortar or any similar implement, and endeavour to unite them by using a pestle or stamper. But the mass is so spongy that it is hardly possible to get a single stroke applied to it before the welding heat is gone; and though by peculiar dexterity and address some have in this way succeeded, it has been found to require such innumerable beatings and hammerings, that most of those who have attempted it have either failed entirely or given it up as being too laborious and expensive. I have succeeded in obviating all these difficulties by adopting the following simple, easy and expeditious method."

In this the dry precipitate is tamped tight into a strong, hollow inverted cone of crucible earth with a conical stopper of the same material made to fit the opening. The whole is slowly raised to a white heat with a cover resting lightly on top. Then the conical stopper is made red hot and is gradually pressed down by means of tongs until the platina becomes more solid.

"It is then repeatedly struck with the stopper, as hard as the nature of the materials will admit, till it appears to receive no farther impression."

The cone is then removed from the furnace, the platina knocked out in a metallic button which is worked as desired. This process avoids any contact between the platinum and red-hot iron, with which it readily becomes contaminated, having a great affinity for it at high temperatures. Knight claims that platinum prepared by this method is of a higher purity than hitherto, having a specific gravity of 22.26.

What Richard Knight did was essentially to reduce the process to its simplest possible form. Perhaps, in the light of a fuller knowledge, he is unduly uncomplimentary to the achievements of his forerunners, but there is no reason why his own product could not have been a good one. The principal advance brought in by him was, as he himself suggests, the stronger hot compression of the sponge, so far applied only lightly by hand. There is no evidence that Knight himself ever made great practical use of his process or manufactured more than two or three articles from it.

The publication of this paper in *The Philosophical Magazine* probably stimulated greater interest in platinum among the group comprising these two societies, and a few years later, sometime in the winter of 1804–1805, Alexander Tilloch himself contributed a paper to the Askesians, "On a New Process of Rendering Platina Malleable" (7). In this he reviewed earlier procedures and

137

then proposed his own. In this the native metal is to be dissolved and precipitated with sal-ammoniac in the usual manner. The precipitate is calcined to metallic sponge and the latter is wrapped up in a piece of platinum foil and the whole "spread out by means of a flatting mill". The foil and its contents are then exposed repeatedly to a sufficient temperature and hammered between each exposure until the whole is brought to a compact state. This it will be seen is merely a modification of a method earlier proposed and practised by Rochon.

The Work of Thomas Cock

Whether or not Tilloch's paper gave a further stimulus to Allen, he evidently and immediately took up the production of articles of platinum, the first mention of his refining and fabricating being in January 1805. In this work he was assisted by one of his young men, Thomas Cock (1787–1842) the son of a wealthy merchant who had prospered in the textile trade with Germany. His father had died when he was four years of age and his mother when he was fourteen. His elder brother was married and his guardian was his sister's husband, the rising young surgeon Astley Cooper, afterwards a very famous figure in his profession. Both Astley Cooper and William Allen were by now fellow lecturers on their respective subjects of anatomy and chemistry at the medical school of Guy's Hospital, and Cooper was a frequent visitor at Plough

Two pages from the cost sheets of the Plough Court Pharmacy, now Allen and Hanburys, for January 1805 in William Allen's handwriting show the expenses incurred in refining some 56 ounces of crude platinum. The first sheet is headed with the name of Thomas Cock, then a young man of only eighteen who later married the sister of Percival Norton Johnson's wife and afterwards devoted much of his time to the platinum refining side of Johnson's business in Hatton Garden

138

Court and for a time a member of the Askesian Society. There seems little doubt that, his young ward having shown some tendency to be interested in chemistry and metallurgy, Cooper arranged with Allen for him to become one of the "young men" at Plough Court. This is confirmed by the mention of his name as such in Allen's diary in the autumn of 1804, when he accompanied him on an extensive visit to the mines of Cornwall (8), and in the December he was proposed as a member of the British Mineralogical Society by W. H. Pepys and seconded by Allen.

The entries in Allen's cost-book, reproduced here, headed with the name "T. Cock", show that 56 ounces of crude platinum, costing 4s. 6d. an ounce, were refined by the usual precipitation method, while an entry dated February 13th 1805 is as follows:

				£	s	d
℥ 33.9/16	Small Platina		18/-	30	4	1
Deduct Cuttings	℥ 11½ at 4/6			2	11	9
				27	12	4
Making etc. 4 Crucibles				2	6	0
				£29	18	4

Wrought crucibles 30/3 oz.

		oz.		£	s	d
L. H.	Crucible	5	1/8, 1/16, 1/32 at 30/-	7	16	6
Henry	,,	4	¾, 1/5, 1/16, 1/32	7	9	0
R. Phillips		5	1/8, 1/32	7	14	8
Davy		4	½	6	15	0
				£29	15	2

(The initials L. H. represent Allen's one-time partner Luke Howard, Henry is William Henry of Manchester, Richard Phillips has already been mentioned, and Davy was of course Allen's friend Humphry Davy)

Platinum tubes were also being made by rolling up sheet and soldering with gold, as is shown by a further item in the cost books in October 1805:

		£	s	d
	Borax & Mur: Acid		1	3
	Charcoal		1	6
Platina Tubes	Gold		11	3
	Attendance		6	0
		£1	0	0

On the facing page is the following entry, initialled "T. C."

			s	d		£	s	d
Adding to platina tube 1	$\frac{1}{4}\frac{1}{8}$1/16 at 13			2		1	6	1
Borax & Muriatic Acid								6
Gold℥__∧__$\frac{1}{2}$℥							5	8
Charcoal″ $\frac{1}{4}$ B								9
Attendance 1 Day							4	0
						1	17	0
Credit bits								6
T. C.						£1	16	6

Thomas Cock continued to work at Plough Court on platinum and other subjects until his marriage in 1809 to a Miss Anna Maria Smith whose sister Elizabeth later married Percival Norton Johnson. In 1807 there was published in A Dictionary of Chemistry and Mineralogy by A. and C. R. Aikin (9) an account of the latest method for rendering platinum malleable, and one "that has been attended by compleat success." This "was invented by Mr. T. Cock, through whose liberality we are enabled to communicate it to our readers".

"The platina being dissolved in nitro-muriatic acid, the liquor is to be filtered through clean white sand, in order to separate the black powder which floats among it. The clear solution being then decomposed by sal-ammoniac the yellow precipitate is to be collected, moderately well washed in warm water and dried. It is then to be distributed into saucers which are placed in a small oven constructed for the purpose, where they are exposed for a short time to a low red heat in order to bring the platina to the metallic state, and to drive off by sublimation the greater part of the muriated ammonia. When withdrawn it is a spungy mass of a grey colour. About half an ounce of the platina in this state is to be put into a strong iron mould about $2\frac{1}{2}$ inches long by $1\frac{1}{2}$ wide and is to be compressed as forcibly as possible by striking with a mallet upon a wooden pestle cut so as accurately to fit the mould; another $\frac{1}{2}$ ounce is then added and treated in the same manner, and so on till 6 ounces have been forced into the mould; a loose iron cover just capable of sliding down the mould is then laid upon the platina, and by means of a strong screw press, almost every particle of air is forced out from among the platina. This is a part of the process that requires especial care, for if any material quantity of air is left in the mass, the bar into which it is formed is very apt in the subsequent operations to scale and be full of flaws. The pressure being duly made, the mould is to be taken to pieces, and the platina will be found in the form of a dense compact parallelopiped. It is now to be placed in a charcoal forge fire and heated to the most intense white heat in order compleatly to drive off the remaining ammoniacal muriat; this being done it is to be quickly placed on a clear bright anvil and gently hammered in every direction by a clean hammer. This is to be repeated several times, at the end of which the mass will be perfectly compact, and fit to be laminated or wrought in any other manner that the artist chuses. It is to be observed that while the platina is heating it must lie loose in the fire, for if it were held by the tongs they would infallibly become welded to the platina, and thus greatly damage it. By the time that the platina is thus drawn down to a compact bar it will be covered by

140

Arthur Aikin
1773–1854

Trained by his father's friend Joseph Priestley, Aikin moved to London in 1796 and there established a considerable reputation as a geologist, chemist and lecturer. In 1807, together with his brother Charles Rochemont Aikin, he compiled "A Dictionary of Chemistry and Mineralogy" in which he included as an original contribution a full account of Thomas Cock's process for producing malleable platinum "that has been attended with compleat success". Aikin was a close friend of Percival Norton Johnson from their early years in Stoke Newington. In his later life he was a founder member of the Chemical Society and President from 1845 to 1847

a somewhat reddish semi-vitreous crust proceeding chiefly from particles of the ashes melted down upon it and extended over its surface by the hammer. To remove this, the bar being made red hot is to be sprinkled over with pulverised glass of borax, and then kept for a few minutes at a white heat; when moderately cool it is to be plunged into dilute muriatic acid by which the borax and other vitreous matter will be dissolved, leaving the platina with a perfectly clean white surface."

Cock's contributions to the advanced technique were:

Complete removal of all material insoluble in aqua regia

The calcination of the ammonia chloroplatinate precipitate takes place rapidly and at a low temperature

High pressure is brought to bear on the cold sponge by means of a screw press before heating and forging

Forging takes place at the "most intense white heat"

The surfaces of anvil and hammer must be clean and bright.

Comment on the importance of such factors as these will be made in the next chapter, but it seems that Cock made no attempt to purify his solution from iron and base metals before precipitating it, nor did he appear to be aware of the

141

presence in his native metal of iridium, osmium, palladium and rhodium, all of which as will again be seen in the next chapter, had been discovered and their existence reported two or three years before the publication of his procedure.

The Network of Societies and Institutions

The skills and the opportunities of the London scientists were further displayed in a number of other societies and institutions, these having many members in common and the whole constituting a complex network of personal associations.

In 1807 the members of the British Mineralogical Society agreed to merge with the Askesian Society "as advantageous to each in their design of promoting Philosophical Research", but although meetings continued to be held for a time at Dr. Babington's house, both very soon decided to disband themselves, largely because of the foundation of the London Institution to which most of their members were subscribers and also as the new body offered the advantages of a school of mineralogy. Later in the same year the Geological Society was formed, the founder members including Allen, Pepys, Babington, Humphry Davy, Arthur Aikin, Richard Phillips and Richard Knight (10).

Earlier, in 1797, the Royal Institution had been founded by Benjamin Thompson Count Rumford, and Davy had been appointed lecturer in chemistry in 1801. Allen, Pepys and Richard Phillips at once became associated with the new institution and with Davy himself and by 1803 Allen had been invited by

William Hasledine Pepys
1775–1856

The son of a London cutler and surgical instrument maker, Pepys was one of the early members of the Askesian Society and, with Allen, Knight and Tilloch, a founder of the British Mineralogical Society in 1799. He took considerable interest in platinum and the attempts to bring about its fusion by the electric discharge. He was a skilful and ingenious designer of scientific instruments and apparatus and he also supervised the construction of the great battery of 2000 plates for the Royal Institution

After the isolation of the alkali metals by Humphry Davy by holding heated potash and soda in a platinum spoon and contacting them with a platinum wire connected to a powerful battery several designs of apparatus were proposed for the repetition and demonstration of these classic experiments. The most simple and successful apparatus, designed by W. H. Pepys and made by Richard Knight, was fitted with a platinum wire having two platinum discs attached to contain the alkali

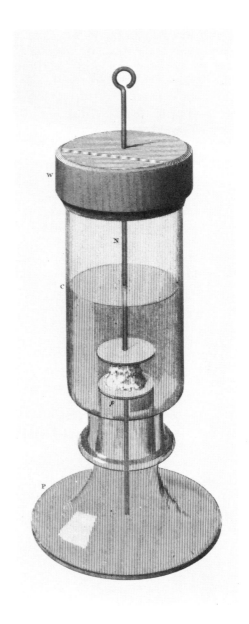

Davy to give the lectures he had been delivering jointly with Babington at Guy's Hospital to the audiences at the Royal Institution. When reporting his famous discovery of potassium and sodium in 1807 Davy described how he held pieces of caustic potash and caustic soda on an insulated disc of platinum and brought them into contact with a platinum wire connected to the negative side of his large battery, and acknowledges the assistance of both Allen and Pepys. These two went on to repeat and to demonstrate the classic experiments before the members of the Askesian Society and the British Mineralogical Society, employing the piece of platinum apparatus shown here, designed by Pepys and constructed by Richard Knight (11).

Another experiment on platinum was carried out in August 1808 by Allen, Pepys and Davy at the home of J. G. Children (1777–1852) with "a great battery" the latter had constructed. A piece of platinum wire eighteen inches in length and 1/30th of an inch in diameter was completely fused in about twenty seconds, while a three-feet length of the same wire was heated to "a bright red, visible by strong daylight"(12).

143

Earlier W. H. Pepys had also built a very large battery and in February 1803, together with a number of his scientific friends, he ignited or melted a number of metals including platinum wire 1/16th of an inch in diameter which "became red hot, white, and fused into globules at the contact" (13). Pepys extended his father's cutlery business into the manufacture of scientific instruments, but in 1805 he also began to produce cutlery in platinum. Among his products were a pair of fruit knives which he presented to King George III and Queen Charlotte through the good offices of Sir Joseph Banks (14).

Another informal group was the London Philosophical Society, initiated by Samuel Varley, an instrument and watch maker, in which Pepys and Tilloch were active together with John Francillon (1744–1818), a medical man and also a gemmologist and entomologist, who had sometimes assisted Pepys in his experiments. At a meeting of this society in the autumn of 1798 Francillon had proposed the construction of a furnace in which a stream of oxygen could be directed into a crucible containing eight ounces of platinum surrounded by charcoal so that:

Soho Square
June 16 1805

Sir

I have the pleasure to acquaint you that the two Platina Fruit Knives intended as a Present by you to the King & the Queen were received in the most Gracious manner by Both their Majesties on Thursday last from the hands of General Manners who at my desire had the honor of Presenting them. I took care that the General should be acquainted for their majesties information that the Donor is not only at the head of his art as a Cutler but that in the Science of Chemistry he is second to few

I am Sir
your most obedient servt
Jos: Banks

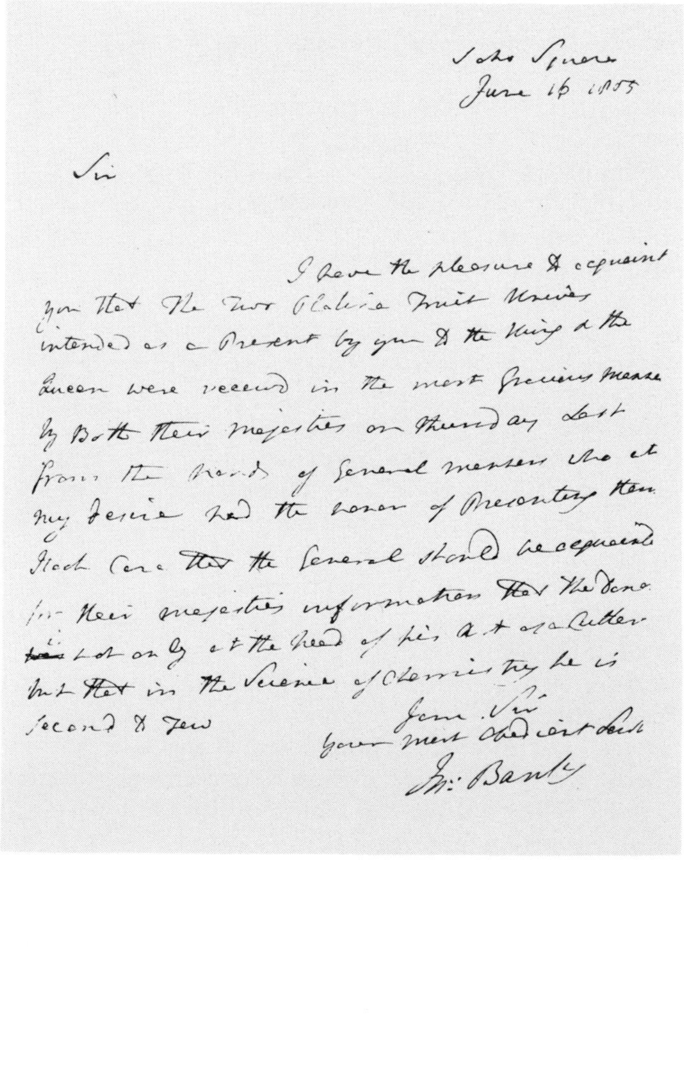

In 1805 Pepys made a pair of fruit knives in platinum which he asked Sir Joseph Banks, the President of the Royal Society, to present on his behalf to King George III and Queen Charlotte. This is the letter from Banks telling Pepys that the knives were received by their Majesties "in the most gracious manner"

Photograph by courtesy of the Royal Institution

"means might be devised to fuse large quantities of crude platina and at the same time to obtain the metal pure and malleable, an object so desirable that the Society resolved at least to make the attempt." (15)

When this experiment was carried out at a further meeting a button of platinum was found at the bottom of the crucible but it broke under the hammer and the members concluded that:

"though platina may be reduced and brought to a state of purity by oxydating the iron by means of a stream of oxygen gas, it will require a considerable degree of address to be able to apply this on a scale of any considerable extent."

There were several other interlocking societies formed during this period, including the London Chemical Society proposed by Frederick Accum in 1806 and the Chemical Club, founded in 1807 and lasting until 1826, in which all the scientists already mentioned were among the members and where guests included Dalton and Berzelius. But by now the Royal Society was absorbing the interests of the former members of the Askesian Society and other groups; Allen, Pepys and Children were elected in 1807–8 with others to follow in later years. And the rather diffuse activities of all these men on the problems of platinum were followed by much more concentrated and successful efforts at the hands of two of their distinguished friends and earlier fellows of the Royal Society whose work is now to be detailed.

References for Chapter 8

1   E. C. Cripps, Plough Court, London, 1927; D. Chapman-Houston and E. C. Cripps, Through a City Archway, London, 1954
2   W. Babington, Lectures on Chemistry, MS in Library of Guy's Hospital
3   Anon, Life of William Allen, London, 1846, **1**, 42
4   Minutes of the British Mineralogical Society, British Museum (Natural History); P. J. Weindling, forthcoming in I. Inkster and J. B. Morrell (eds.), Metropolis and Province, British Science 1780–1850
5   Letter from Robert Bingley to Charles Hatchett, Read to Royal Society, May 1, 1800, Royal Society MSS.
6   R. Knight, *Phil. Mag.*, 1800, **6**, 1–3
7   A. Tilloch, *Phil. Mag.*, 1805, **21**, 175
8   Ref. 3, **1**, 74
9   A. and C. R. Aikin, A Dictionary of Chemistry and Mineralogy, London, 1807, **2**, 233
10 I. Inkster, *Ann. Science*, 1977, **34**, 1–32
11 W. H. Pepys, *Phil. Mag.*, 1808, **31**, 241
12 J. G. Children, *Phil. Trans.*, 1809, 32–38
13 Anon, *Phil. Mag.*, 1803, **15**, 94–96
14 P. J. Weindling, *Platinum Metals Rev.*, 1982, **26**, 34–37
15 Anon, *Phil. Mag.*, 1800, **8**, 21–29; 262–266

William Hyde Wollaston
1766–1828

The outstanding figure in the history of platinum, Wollaston was the first to refine the native metal on a commercial scale and to develop its uses in industry, while in the course of his analytical work on the mineral he also discovered palladium and rhodium. During a brilliant scientific career he made other fundamental discoveries in a number of different branches of science

From a portrait by John Jackson, by courtesy of the Royal Society

9

The Partnership of Smithson Tennant and William Hyde Wollaston

"A quantity of platina was purchased by me a few years since with the design of rendering it malleable for the different purposes to which it is adapted. That object has now been attained."

WILLIAM HYDE WOLLASTON

Up to the end of the eighteenth century the attempts to produce malleable platinum had advanced mainly in the hands of practical men aiming at its preparation and fabrication rather than at the solution of scientific problems. These were now to be attacked with a marked degree of success by two remarkable but very different men who first became friends during their student days at Cambridge and who formed a working partnership in 1800 designed not only for scientific purposes but also for financial reasons. They were of the same generation and much the same background as the professional scientists of London whose work was described in Chapter 8, and to whom they were well known, but with the exception of Humphry Davy they were of greater stature and made a greater advance in the development of platinum metallurgy than their predecessors.

Their combined achievements over a relatively short span of years included the successful production for the first time of malleable platinum on a truly commercial scale as well as the discovery of no less than four new elements contained in native platinum, a factor that was of material help in the purification and treatment of platinum itself.

The junior partner, who was in fact to carry most of the burden after the first few years, William Hyde Wollaston, kept meticulous records of his experiments and of his expenditure and later of his sales of platinum but these remained undiscovered for many years. A collection of his note-books and other manuscripts relating to his life and work, including brief drafts for an unwritten biography by his friend Henry Warburton, were eventually found in 1949 by the late L. F. Gilbert in the Department of Mineralogy and Petrology in the University of Cambridge and formed the subject of a paper of his in 1952 (1). In

Wollaston's partner in the long investigations of native platinum, Smithson Tennant, became interested in the metal while still a student at Cambridge. This extract from his diary for October 4th 1784, when he was only 22, describes his visit to Lorenz Crell at Helmstädt and records the details of the Count von Sickingen's method of rendering platinum malleable

more recent years these note-books and papers, now in the care of the Library of Cambridge University, have been studied in more detail by Dr. M. C. Usselman (2), and by Dr. J. A. Chaldecott (3). Other note-books compiled by Tennant and in the same keeping have enabled Dr. A. E. Wales to publish valuable information on his part in the joint enterprise (4). This chapter therefore owes much to their respective publications, while the original author of this book also contributed a study of Tennant (5).

Taking the older man first, Smithson Tennant (1761–1815) was born at Selby in Yorkshire, the son of a clergyman. In 1781 it was proposed that he should study under Joseph Priestley, now settled in Birmingham, but this did not materialise and instead he entered the University of Edinburgh to study chemistry under Joseph Black with whom he developed a close relationship and from whom he may very well have acquired his early interest in platinum. He moved to Cambridge in October 1782, first at Christ's College and later at Emmanuel, and in 1785 while still an undergraduate he was elected a Fellow of the Royal Society. During the summer vacation of 1784 he made a journey through Denmark and Sweden, visiting mines and chemical plants and meeting Scheele and Gahn, and on his return journey visited Lorenz Crell at Helmstädt. The extract from his diary reproduced above shows that he was already interesting himself in the problem of producing malleable platinum. In the following year he paid a visit to Paris to meet Lavoisier and then to Dijon where he was the guest of Guyton de Morveau, both of these men of course having been active only a few years earlier in working with platinum. (An amusing side light on these visits is to be found in a letter from Guyton de Morveau to Crell's *Chemische Annalen* in 1786 (6). The writer refers to "one of my chemical friends the Englishman Herr Tennant", to which Crell adds in a footnote: "I also have the pleasure of counting Herr Tennant among my chemical acquaintances").

In his later period at Cambridge Tennant became friendly with his younger fellow student in medicine, Wollaston, who was at Gonville and Caius from 1782

148

until 1789 and who was particularly interested in the older man's accounts of his visits to the leading chemists in Europe.

Wollaston took up medical practice after leaving Cambridge, first in Huntingdon and then in Bury St. Edmunds, but in 1797 he moved to a practice in London where his friends considered he would have greater scope for the high ability he had already shown as a physician. Tennant had already settled there four years earlier, also after a brief spell as a physician, work for which he found himself temperamentally unsuited. In this same year he presented a short paper to the Royal Society "On the Action of Nitre on Gold and Platina" (7), quoting the earlier work of Lewis and Marggraf on the native metal before describing his own results on the corrosive attack on his more or less pure platinum.

In 1799, Wollaston, by now devoting most of his time to scientific researches, purchased small amounts of platinum and began to study the most efficient means of its dissolution in varying formulations of aqua regia. By the end of the following year, undoubtedly stimulated by the similar leanings and knowledge of his friend Smithson Tennant and probably after much discussion of the problem, the two men entered into an informal and unpublicised partnership directed towards various chemically-based commercial endeavours, one of which was the production and marketing of malleable platinum, and Wollaston abandoned his medical practice. On Christmas Eve 1800 they invested £795 in the purchase of the very large quantity of 5959 ounces of alluvial platinum from one Hutchinson of whom nothing is known but who was almost certainly a London agent connected with the Jamaican end of the smuggling trade from Cartagena. In the following February they bought a further 800 ounces from Richard Knight who had apparently not put his process to commercial use.

The underlying reasons for their undertaking this joint enterprise, one that was to lead to long years of chemical research and frustration before success was achieved, goes back to their time at Cambridge but a sense of urgency was given them by the prospects opened up by the two publications referred to earlier, the Abbé Rochon's paper that appeared in English translation in the first volume of *The Philosophical Magazine* in 1798 and Richard Knight's paper read to the British Mineralogical Society in January 1800 and also published in *The Philosophical Magazine* in the following month. An interesting and possible link is that just opposite Richard Knight's establishment in Foster Lane was the Church of St. Vedast in which his parents had been married in 1766 and where the Rector from 1779 until 1815 was in fact Wollaston's father, the Reverend Francis Wollaston, F.R.S. In any case it was becoming very clear that if a process that was technically sound could yield malleable platinum in quantity and at a reasonable price a handsome financial return might well be expected.

Tennant was a wealthy man by inheritance. It has often been said that Wollaston needed to earn a living from his researches but Dr. Usselman in the course of a detailed review of one episode of Wollaston's career has revealed that in 1799 his elder brother George had made him a present of securities valued at some £8000 (8), making it possible for him at the age of 34 to abandon the medical practice he disliked and to devote the remainder of his career to the scientific pursuits he so much preferred.

Wollaston had a great admiration for Tennant, who as we have seen had made personal contacts with several of the French and German workers on platinum and who may thus have influenced the choice of project, while his brilliant if rather erratic mind would have been of great value in the early years of the partnership.

The Discovery of Iridium and Osmium

Before we can consider Wollaston's principal work, the laborious investigation of the means of producing malleable platinum in quantity and its success after five years of frustrating experiments, we have to turn to the parallel investigations carried out by him and by Tennant that showed the presence, unsuspected by the earlier workers, of four other elements in native platinum. It will be remembered that when crude platinum was dissolved in aqua regia there was always a black insoluble residue left. Sickingen had tried to melt it under borax but was prevented by his diplomatic duties from investigating it further. Proust noted that it formed some two to three per cent of the original mineral but considered that it was "nothing else than graphite or plumbago". Early in their joint researches it was decided that while Wollaston should pursue the study of the aqua regia solution of platinum Tennant should concentrate on the insoluble residue. His curiosity was aroused and by the summer of 1803 he had begun to investigate this insoluble matter, fortunately mentioning this at the time to Sir Joseph Banks, President of the Royal Society, writing that "it did not, as was generally believed, consist of plumbago, but contained some unknown metallic ingredients". In this way he established his priority because almost simultaneously work on the same problem had begun in France.

It had been observed by many of the earlier workers that the precipitate formed by sal-ammoniac in the solution of native platinum in aqua regia differed from time to time in colour, being sometimes a pale yellow, sometimes showing various orange tints and sometimes a red that might even be a very dark one. The first to investigate this was the French mining engineer, H. V. Collet-Descotils (1773–1815), who read a paper to the Institut National on September 26, 1803, published in the *Annales de Chimie* and reproduced in Nicholson's *Journal* in the following year (9). He reported that the black insoluble powder is gradually attacked by aqua regia, especially in the presence of extra nitric acid, and he was able to trace a connection between this and the colour of the precipitate produced by sal-ammoniac in the liquor. The more completely the black powder was dissolved the redder was the precipitate. Next he found that if a red precipitate was calcined and the sponge treated with aqua regia, there always remained a certain amount of insoluble black powder, and the solution gave with sal-ammoniac a precipitate a little less red than before. Finally, he noticed that the presence of reducing agents in the platinum solution resulted in the

150

formation of yellow precipitate only, and that the red colours were obtained only after the solution had been oxidised. He concluded that

"these properties appear to me not to belong to any one of the known metals and force me to regard as a new substance a metal which colours red the salts of platinum."

At the same time as Collet-Descotils was carrying on this work a parallel research was taking place at the hands of Antoine Francoise de Fourcroy (1755–1809) with, as his partner and assistant, Nicholas Louis Vauquelin (1763–1829) of whom more will be said in the next chapter, and the first part of their paper was read to the Institut immediately after Descotils' (10). They set out to find what influence "the foreign bodies that accompany platinum in its ore, might have on the working of this metal on a large scale". They identified and removed all the minerals and base metals and then came upon Proust's insoluble black powder. As acids had so little effect on it, they tried alkalies and found that it was attacked, eventually completely, by molten caustic potash. The melt yielded a green solution with water and a green flocculent residue. Chromic acid was suspected in the former, and the solids after dissolving in acid were tested in various ways, the results leading the investigators to the theory that there was present the same substance that causes "the diversity of colours assumed by the precipitates of platinum formed by sal-ammoniac." They believed that this was a new metal but wished to isolate it in a metallic state before being convinced.

Next they found that if a solution of native platinum was precipitated in two stages with sal-ammoniac, the first was a pale yellow or orange in colour, while the second was dark red. They confirmed Collet-Descotils' discovery that the latter, if calcined and treated with aqua regia, always left some insoluble black powder. In this they recognised a perfect resemblance to the metal they had discovered in the insoluble matter left behind when native platinum is attacked by aqua regia. All this metal was not removed by aqua regia from the platinum coming from the red salt because the new solution also yielded a precipitate that was reddish though less so than before, but if these operations were repeated several times with the same platinum then eventually the strange metal was removed entirely. They also found this metal in the platinum purified by Janety in almost as great a quantity as in native platinum. At the end of their paper they proposed to go on with their experiments to obtain a larger quantity of the new metal, to examine its properties further and to seek better methods for the purification of platinum.

They were not, however, left alone in the field. Tennant continued his work during the winter of 1803–4 and carefully studied the papers of Collet-Descotils and of Fourcroy and Vauquelin. (The great advantage he had on the French chemists lay in the large amount of residue he could work with, a by-product of Wollaston's researches on the purification of platinum.) He soon realised, however, that whereas they suspected the presence of only one new metal in the black powder, there were in fact two. For breaking up the material he used a

151

XVI. *On two Metals, found in the black Powder remaining after the Solution of Platina.* By Smithson Tennant, Esq. F. R. S.

Read June 21, 1804.

Upon making some experiments, last summer, on the black powder which remains after the solution of platina, I observed that it did not, as was generally believed, consist chiefly of plumbago, but contained some unknown metallic ingredients. Intending to repeat my experiments with more attention during the winter, I mentioned the result of them to Sir Joseph Banks, together with my intention of communicating to the Royal Society, my examination of this substance, as soon as it should appear in any degree satisfactory. Two memoirs were afterwards published in France, on the same subject; one of them by M. Descotils, and the other by Messrs. Vauquelin and Fourcroy. M. Descotils chiefly directs his attention to the effects produced by this substance on the solutions of platina. He remarks, that a small portion of it is always taken up by nitro-muriatic acid, during its action on platina ; and, principally from the observations he is thence enabled to make, he infers, that it contains a new metal, which, among other properties, has that of giving a deep red colour to the precipitates of platina.

M. Vauquelin attempted a more direct analysis of the substance, and obtained from it the same metal as that discovered by M. Descotils. But neither of these chemists have observed,

Early in their joint investigations it was agreed that Tennant should examine the black insoluble powder remaining after the dissolution of platinum in aqua regia while Wollaston should concentrate on the soluble portion. Beginning work in the summer of 1803 Tennant identified two new elements which he named iridium – "from the striking variety of colours which it gives while dissolving in marine acid" – and osimium, from the pungent smell of its volatile oxide. This shows the opening of his paper read to the Royal Society in 1804

method similar to that employed by Vauquelin, namely, the alternate action of caustic alkali and of acid. The second metal he found in the alkaline solution which had been suspected by Vauquelin to contain chromium. Tennant was unable to confirm this latter conclusion, but further examination showed that the solution contained a volatile oxide which could be separated by acidification and distillation. This was a colourless body, condensing first to an oily liquid and then solidifying into a semi-transparent mass. In all stages it had a strong and very characteristic smell. It was this that caused him to confer on the metal the name of "Osmium", from a Greek word meaning smell. With regard to the other metal, Tennant's work confirmed and extended the observations of Fourcroy and Vauquelin on the subject, and this one he named "Iridium from the striking variety of colours which it gives while dissolving in marine acid". Tennant read his paper, a masterpiece of clarity and conciseness, to the Royal Society on June 21st, 1804 (11) and was awarded the Copley Medal for that year. It is a tribute to him that the French workers accepted the priority of his discovery without question in 1806 in a paper on the new metals discovered in crude platinum (12).

152

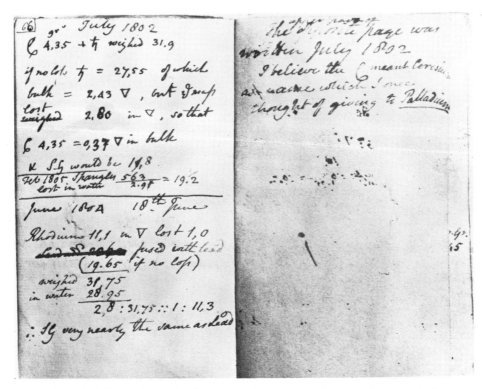

One of Wollaston's note-books open at the page containing the first record of palladium as "C". A note on the facing blotting paper reads: "The upper part of opposite page was written July 1802. I believe the C meant Ceresium a name which I once thought of giving to Palladium"

Photograph by courtesy of the Syndics of Cambridge University

Another worker had also noticed the varying colours of the sal-ammoniac precipitate. Luigi Brugnatelli, professor of chemistry in the University of Pavia and a colleague of Volta, wrote to his friend Baptiste Van Mons of Brussels, the editor of a chemical journal there, to say that he had seen the French papers and that he had obtained several years before "the substance that gave colour to solutions of platinum", enclosing a specimen of the red precipitate (13).

The Discovery of Palladium

Early in his work on the solution of platinum in aqua regia Wollaston suspected that something was present after the precipitation with sal-ammoniac that was neither platinum itself nor Tennant's iridium. The first reference to this is in one of his note-books, dated July 1802, in which the properties of an alloy of lead with "C" are described. On the facing page he later wrote:

153

"The upper part of the opposite page was written July 1802. I believe the C meant Ceresium, a name which I once thought of giving to *Palladium*".

Palladium is first mentioned by that name in August. The planet Ceres had been discovered by the German astronomer Heinrich Olbers in January 1802; he discovered another minor planet which he named Pallas at the end of March in the same year and this doubtless caused Wollaston to change his mind.

Wollaston proposed to do some further work on the solution and did not therefore publish his results, but in order to establish his priority he resorted to a most unorthodox procedure that was to involve him in an unpleasant and embarrassing controversy. Towards the end of April in 1803, some nine months after the note-books record the discovery of palladium, and by which time he had been able to isolate a reasonable quantity of metal, members of London's scientific community received in their post a small hand-bill describing the properties of a new noble metal known as "Palladium or New Silver". The leaflet, illustrated here in its actual size, announced that this new metal was to be obtained only from the shop of Mr. Forster of Gerrard Street in London's Soho in samples at five shillings, half a guinea or one guinea each. This shop has

In April 1803 a number of small leaflets were distributed anonymously among scientists in London announcing the discovery of a new noble metal called Palladium or New Silver. This described the major properties of the new metal, which was to be obtained only from the shop of Jacob Forster, a well known collector of minerals, in London's Soho. This unusual procedure aroused great controversy and caused considerable embarrassment to Wollaston. The leaflet is reproduced here in its actual size

Photograph by courtesy of the Syndics of Cambridge University

PALLADIUM;

OR,

NEW SILVER,

HAS these Properties amongst others that shew it to be

A NEW NOBLE METAL.

1. IT dissolves in pure Spirit of Nitre, and makes a dark red solution.

2. Green Vitriol throws it down in the state of a regulus from this solution, as it always does Gold from *Aqua Regia.*

3. IF you evaporate the solution you get a red calx that dissolves in Spirit of Salt or other acids.

4. IT is thrown down by quicksilver and by all the metals but Gold, Platina, and Silver.

5. ITS Specific Gravity by hammering was only 11.3, but by flatting as much as 11.9.

6. IN a common fire the face of it tarnishes a little and turns blue, but comes bright again, like other noble metals on being stronger heated.

7. THE greatest heat of a blacksmith's fire would hardly melt it;

8. BUT if you touch it while hot with a small bit of Sulphur it runs as easily as Zinc.

IT IS SOLD ONLY BY

MR. FORSTER, at No. 26, GERRARD STREET, SOHO,

LONDON.

In Samples of Five Shillings, Half a Guinea, & One Guinea each.

J. Moore, Printer, Drury Lane

154

been identified by Professor Clifford Frondel of Harvard as belonging to Jacob Forster (1739–1806), a well known collector and dealer in minerals after whom the magnesium silicate mineral Forsterite was later named (14). He travelled widely and spent several years in Russia where in 1802 the Count Mussin-Pushkin purchased from him a large collection of specimens for the Mining Cadet Corps for 50,000 roubles. Mussin-Pushkin also relates that he acquired his own supplies of native platinum from Forster who, strangely enough, was a relative of Georg Forster whose letter describing the Count von Sickingen's large piece of platinum sheet was mentioned earlier.

In his absence in St. Petersburg it was Mrs. Forster who conducted the business in London and she duly received a quantity of palladium foil amounting to 420 grains together with a copy of the leaflet, the source being quite unknown to her.

John Nicholson, the founder and editor of the *Journal for Natural Philosophy, Chemistry and the Arts*, also received a copy and in an editorial comment wrote:

> "Mrs. Foster (sic) it appears is only the vendor and totally unacquainted with the person who brought the metallic substance and the printed paper to her house . . . I received a small piece by the post" (15).

The Controversy with Chenevix

Apart from attracting general attention in scientific circles this unusual means of announcing the discovery of a new element aroused the intense interest of a quite distinguished young Irish chemist, Richard Chenevix (1774–1830), a great nephew of the famous Bishop of Waterford of the same name, who had already established a reputation as an analyst and had been elected to the Royal Society in 1801. He immediately visited the shop in Gerrard Street and purchased for fifteen guineas the whole 332 grains of palladium that had not yet been sold, suspecting that this curious procedure was nothing but a fraudulent imposition. After a fortnight's hectic experimentation Chevenix read a long paper to the Royal Society on the two evenings of May 12th and 19th describing his investigations (16). He agreed that the substance possessed all the properties claimed for it on the hand-bill, but maintained that it was merely an alloy of platinum and mercury in the proportions of two to one. He admitted that he had been unable to confirm this by analysis but maintained that he had successfully synthesised palladium, although he remained puzzled by the great discrepancy in the specific gravities expected for alloys of platinum and mercury which should have been much higher than the value of 11.3 to 11.8 claimed for palladium in the handbill. He was not deterred even by an experiment he recorded, carried out by Humphry Davy in his presence, in which a slip of palladium was ignited by "the strong galvanic batteries of the Royal Institution and burned with a very vivid light and a white smoke but no mercury was separated by this operation".

155

XII. *Enquiries concerning the Nature of a metallic Substance lately sold in London, as a new Metal, under the Title of Palladium. By* Richard Chenevix, *Esq. F. R. S. and M. R. I. A.*

Read May 12, 1803.

ON the 29th of April I learned, by a printed notice* sent to Mr. KNOX, that a substance, which was announced as a new metal, was to be sold at Mr. FORSTER's, in Gerrard-street. The mode adopted to make known a discovery of so much importance, without the name of any creditable person except the vender, appeared to me unusual in science, and was not calculated to inspire confidence. It was therefore with a view to detect what I conceived to be an imposition, that I procured a specimen, and undertook some experiments to learn its properties and nature.

The nature of the anonymous announcement of the discovery of palladium aroused the suspicions of Richard Chenevix who immediately purchased the bulk of the metal and proceeded to carry out a rapid series of experiments. After only two weeks he submitted a paper to the Royal Society in which he maintained that it was not a new metal "as shamefully announced" but merely an alloy of platinum and mercury. He persisted in this view until Wollaston disclosed that he was the discoverer in 1805

This paper, of which the introduction is illustrated here, caused something of a sensation and it was quickly reproduced in the French and German journals (17). On May 4th Chenevix wrote to Vauquelin in Paris, his letter beginning:

"The scientific world here talks of nothing but palladium: everyone received a few days ago a printed note like the one I enclose. I am sending you a little piece together with the notice so that you can see for yourself what it is" (18).

Vauquelin duly checked the properties claimed in the hand-bill and found them to be correct, but he was quite unable to find either platinum or mercury in the specimen and tentatively suggested that it might in fact be a new element. A note by the editors of the *Annales de Chimie* appended to his report concluded that "all this must give rise to doubt".

Shortly after reading his paper Chenevix left for Paris to meet Berthollet and then spent over a year in Germany, visiting chemists in Mannheim, Cassel, Leipzig, Erfurt and Freiberg among other places. Here the controversey was entered into by several scientists. The two well known Berlin analysts Valentin Rose and A. F. Gehlen carried out numerous careful experiments in an attempt to reproduce the platinum-mercury alloy of Chenevix, obtaining specific gravities of 14 to 15 but failing to produce palladium (19). J. B. Richter similarly followed the procedures described by Chenevix for the preparation of "artificial palladium" but again without success (20). Klaproth, in a letter from Berlin to Vauquelin in January 1804, also wrote that he had been quite unable to repeat this synthesis (21), but perhaps the most critical comment was that made by Professor Trommsdorff of Erfurt after Chenevix had left him for a long stay in Freiberg.

"I have repeated in vain part of the research followed by Chenevix to prepare the so-called palladium: heaven knows how it comes about" (22).

156

In December 1803 Wollaston, still anonymously, wrote to Nicholson's *Journal of Natural Philosophy, Chemistry and the Arts* offering a reward of £20 to anyone who could make twenty grains of real palladium before "any three gentlemen chymists" appointed by Nicholson. The reward was, of course, never claimed

SCIENTIFIC NEWS.

Reward of Twenty Pounds for the Artificial Production of Palladium.

THE following is a copy of a paper received by me under cover, by the two-penny post. It is written in the same hand as a note which covered a small piece of the palladium mentioned to have been received by me last Midsummer. (See Philof. Journal, June, 1803. Vol. v. p. 136.) Upon inquiry, I find, that Mrs. Forster has received the sum of 20 £. with instructions conformable to this paper. This original is cut indent-wise on the margin, and has part of a manuscript flourish or paraph on each border, but no signature.

(COPY.)

December 16, 1803.

SIR,

AS I see it said in one of your Journals, that the new metal I have called palladium, is not a new noble metal, as I have said it is, but an imposition and a compound of platina and quick-silver, I hope you will do me justice in your next, and tell your readers I promise a reward of 20 £. now in Mrs. Forster's hands, to any one that will make only 20 grains of real palladium, before any three gentlemen chymist's you please to name, yourself one if you like.

It is insisted that palladium cannot be formed by art; and a reward is offered for any procefs to that effect.

That he may have plenty of his ingredients, let him use 20 times as much quicksilver, 20 times as much platina, and in short of any thing else he pleases to use: neither he nor I can make a single grain.

Pray be careful in trying what it is he makes, for the mistake must happen by not trying it rightly.

My reason for not saying where it was found, was, that I might make some advantage of it, as I have a right to do.

If you think fit to publish this, I beg you to give the names of the umpires, as I have desired Mrs. Forster to keep the money till next Midsummer, and to deliver it only in case they can assure her that the real metal is made by a certificate signed by you, and by them, on this check.

I hope a little bit of whatever is made may be left with Mrs. Forster.

Letter

Then early in 1804 there appeared in Nicholson's *Journal* a second curious anonymous notice, reproduced here in full (23), offering a reward of £20, deposited with Mrs. Forster,

"to anyone who could make only 20 grains of real palladium before any three gentlemen chymists you please to name, yourself one if you like".

Nicholson chose Charles Hatchett and Edward Howard, both distinguished chemists and Fellows of the Royal Society, to join him as assessors but there is no evidence that any claim to the reward was ever put forward.

The writer of this notice was of course Wollaston, who had repeatedly tried to convince Chenevix of his error, but further controversy, embarrassing Sir Joseph Banks and involving a second paper from Chenevix admitting that only four out of a thousand experiments had been successful (24), (and fully related in Usselman's survey of the whole sequence of events (8)) was yet to come.

The Discovery of Rhodium

On June 24th 1804, only three days after the reading of Smithson Tennant's paper on iridium and osmium, Wollaston read a paper to the Royal Society, "On a new Metal, found in crude Platina", describing his discovery of:

"another metal, hitherto unknown, which may not improperly be distinguished by the name of *Rhodium*, from the rose-colour of a dilute solution of the salts containing it" (25).

He went on to refer to the results of various experiments:

"which have convinced me that the metallic substance which was last year offered for sale by the name of Palladium is contained (though in very small proportion) in the ore of platina . . . if we consider the difficulty of producing even an imperfect imitation of palladium the failure of all attempts to resolve it into any known metals . . . as well as the number and distinctness of its characteristic properties I think we must class it with those bodies which we have most reason to consider as simple metals."

Wollaston still did not disclose that he was the mysterious advertiser, but in the following February he wrote again to Nicholson's *Journal* announcing himself as the discoverer of palladium and offering an explanation of his unusual behaviour. After referring to his project to produce malleable platinum he wrote:

This extract from Wollaston's notebook of June 14th 1804 describes the discovery and the naming of rhodium from the rose-red colour of its salts. On this occasion the discovery was immediately announced in a paper read to the Royal Society on June 24th

158

"The object had now been attained and during the solution of it various unforeseen appearances occurred; some of which led me to the discovery of palladium; but there were other circumstances which could not be accounted for by the existence of that metal alone. On this, and other accounts, I endeavoured to reserve to myself a deliberate examination of those difficulties which the subsequent discovery of a second new metal, that I have called rhodium, has since enabled me to explain, without being anticipated even by those foreign chemists, whose attention has been particularly directed to this pursuit" (26).

Finally, in a lengthy paper read to the Royal Society on July 4th 1805, "On the Discovery of Palladium; with Observations on other Substances found with Platina" (27), Wollaston put an end to the mystery. This disclosure came of course as a complete surprise to Chenevix after his long argument against the existence of this new element, but he appears to have accepted defeat gracefully, contrary to the judgments of earlier writers who have described his "reputation in ruins", and he was a guest of Wollaston's at the Royal Society Club more than once in later years. He afterwards turned to the writing of novels and plays.

Wollaston's motives in the anonymous announcement of his discovery of palladium have been the subject of many speculations, but there seems to be no need to look further than his own statement in his letter to Nicholson in February 1805. The main object of his work was to discover the best way of producing malleable platinum, but in the course of his labours he came across evidence of the presence of first one unknown metal and then another. Anxious to pursue these to a conclusion, conscious that Descotils, Fourcroy and Vauquelin were close on his heels, it does not seem unnatural that he should wish to shelve one enquiry without losing his right of priority in another. Also, as Usselman has pointed out (8), a major factor in his delay was that not until 1805 did he finally succeed in preparing really malleable platinum after five years of intensive and often frustrating research. The entry of the rather pugnacious Chenevix into the affair clearly upset his plans.

The Production of Malleable Platinum

We can now return to the main achievement of Wollaston, the painstaking development of a process for the production of malleable platinum on a considerable scale and the promotion of its industrial uses. Tennant's part in this work was the lesser of the two although he gave much practical assistance in the early years. His financial participation ceased in 1809.

The large purchase of native metal in December 1800 has been referred to already. In February 1801 Wollaston set about his purification in earnest, working in a laboratory at the rear of his house in Buckingham Street off Fitzroy Square, dissolving his material in 16 or 20 ounce lots followed by precipitation with salammoniac, heating and forging. These early trials were beset by difficulties and many of his ingots cracked or split on hammering, a problem that was much reduced after the meticulous chemical analysis of his crude

159

platinum allowed the removal of the previously unknown metals palladium and rhodium. Further, by cleverly adjusting the proportions of hydrochloric and nitric acids in his aqua regia, and later by using more dilute mixtures, he succeeded in achieving the most economical means of dissolution.

Next he turned his attention to the technique of powder metallurgy and by a number of detailed improvements on the methods of Knight and Cock he was eventually successful, after more than three years of unremitting effort, assisted by his faithful manservant John Dowse, in obtaining a saleable product. The whole process, subjected to modifications and improvements over the years, was kept a closely guarded secret and not until long after the work had ceased and Wollaston knew in 1828 that he was suffering from a fatal illness did he consent to publish any account of his process. Then he gave a full description dictated from his bed to his friend Henry Warburton and read as the Bakerian Lecture to the Royal Society on November 20th of that year, a month before his death (28).

Before discussing the details of his procedure, therefore, we should consider both his sources of supply and then the marketing of his products.

The Role of John Johnson

The amount of native platinum purchased by Wollaston, until 1815 jointly with Tennant, totalled the remarkable figure of 47,000 ounces. In a recent closer study of the manuscripts at Cambridge and of Wollaston's account at Coutts' bank Usselman (29) has confirmed and extended the findings of Gilbert (1) and the views expressed in the first edition of this book that by far the major portion of this, over 38,000 ounces, was supplied by John Johnson during the years 1808 to 1819. This man's father, also John Johnson, had established himself in Maiden Lane in the City of London in 1777 as an assayer after a short career as a goldsmith. The son, born in 1765, was apprenticed to him in 1779 and took over the business on his father's death in 1786, developing the assaying of ores and metals still further and also engaging in the buying and selling of bullion (30). He was in fact the only commercial assayer in London, the others being employed by the Royal Mint and the Worshipful Company of Goldsmiths. For nearly twenty years Johnson both lived and worked in Maiden Lane, now a part of Gresham Street, and then in 1805, his business flourishing, he moved to a country house in Stoke Newington, the centre of a number of scientists, many of them of the Quaker persuasion. Here he became friendly with both the Aikin and the Cock families among others, and he had already made contact with Wollaston for whom he had carried out assays. A confirmation of their association is contained in a letter from Johnson to his son Charles, serving with Wellington's army in Spain and dated April 22nd, 1814. In this he refers to "my friend Dr. Wollaston" (31).

Johnson had come across platinum as an adulterant in gold in his assay work and this may well have prompted him to take an interest in the metal and to enter into its importation and sale, acting as an agent for merchants in Jamaica.

160

The first recorded sales to Wollaston occur in December 1806 and January 1807, followed by a quantity of 1000 ounces at 2s 10d per ounce in October 1808 and a further 1750 ounces at 2s 9d later in the same month. A charge of 3s 6d was made for each delivery by hired horseman. Transactions of this kind, ranging from only 72 ounces up to 5,100 ounces at a time, continued over the next eleven years. Wollaston was apparently prepared to purchase all the native platinum that Johnson could obtain and held a running account with him for this purpose, the dealings yielding handsome profits to both men. Johnson retired temporarily at the beginning of 1817, leaving the business in the hands of his son Percival, the founder later in that year of the firm that became Johnson, Matthey and Co., but he returned in 1818 and continued to be active until his death in 1831.

Another indication of the Johnsons' early interest in the platinum metals is disclosed in a paper submitted to *The Philosophical Magazine* in 1812 by Percival, then only 19, on the assaying of platinum-gold alloys to which a postscript initialled by both father and son reads:

"It may also be worthy the notice of your readers that we find palladium to be such a general alloy of Brazil gold as often to alter the colour thereof. We have particularly observed it in the Brazil coin many of which were rejected at first sight, suspecting them to be counterfeits. We found it a short time since in a Brazil bar to the amount of 20 per cent, altering the colour to nearly that of palladium" (32).

In the later years of Wollaston's work other sources of supply were sometimes drawn upon, including some 2000 ounces from a rather extraordinary character Justus Erich Bollmann (1769–1821), a German physician and chemist and something of an adventurer who had emigrated to America and then divided his time between Philadelphia and London with occasional visits to Colombia and Jamaica. An account of his colourful career and his connections with Wollaston has recently been given by Dr. J. A. Chaldecott (33). One of his last purchases was of 600 ounces in April 1819 from "Hodgson", who was most probably John Hodgson of Bucklersbury, a South American merchant, and last of all a further 400 ounces from John Johnson in August of the same year.

From his total intake of 47,000 ounces of crude platinum Wollaston produced and marketed a little over 38,000 ounces of refined and malleable metal at an average price of 16 shillings per ounce. His very large purchases clearly amounted almost to a monopoly and would have made it difficult if not impossible for any potential competitor in England to have secured supplies. But by 1820 platinum became most difficult to obtain from the now independent Colombia and his activities drew to a close, leaving orders unfulfilled.

The Marketing of Wollaston's Platinum

By early in 1805 Wollaston had the production process under control and was at last in a position to offer his platinum for sale. Evidently he did not wish to involve himself in numerous small commercial transactions and he therefore came to an arrangement with one William Cary (1759–1825), a London scientific instrument maker with premises in the Strand. Cary had become well

By 1805, after long and frustrating researches, Wollaston was able to produce marketable platinum in quantity. Among his first products were crucibles and lids, marketed through William Cary, a well known London instrument maker. This shows one of his early crucibles that was retained by the Wollaston family until it was given to the Science Museum in 1932. It is about 2 inches in height and weighs 1.59 oz troy, a great deal lighter than the crucibles made at about the same time by William Allen and Thomas Cock

known to the family some ten years earlier when he had made a transit circle to the complete satisfaction of Wollaston's father, the Rev. Francis, while Wollaston himself had purchased a number of instruments from him. These and other details of their association have recently been fully recounted by Dr. J. A. Chaldecott (34). In February and March 1805 Wollaston's notebooks record the supply of a number of crucibles and of bars and rods for wire drawing to a total weight of 258 ounces. Little or no publicity seems to have been given to the arrangement although on March 22 a letter from Sir Joseph Banks to Richard Chenevix includes the statement that:

> "Wollaston has now opened a Manufactory of Platina Crucibles, etc, which are sold in his name to such as chuse to bespeak them, by Carey, the Mathematical instrument maker" (35).

The first published reference to a Wollaston crucible came in May 1806 in the course of a paper on the analysis of a new mineral found in Cornwall by Professor John Kidd of Oxford, a former student and a friend of Astley Cooper. In this investigation he reported using "a small crucible of platina prepared by Dr. Wollaston" (36).

Altogether the note-books show that approximately 1700 ounces of platinum were made into crucibles and lids – a development that greatly assisted the analysts of the time in their discipline, by now brought to a reasonable degree of accuracy. One indicative comment came from the great Berzelius, the leading chemist and analyst, in a letter to his friend and patron Baron Hisinger in October 1815:

> "I have just received from England a delicious platinum evaporating dish holding more than $\frac{1}{2}$ stop. It is a jewel" (37).

(The old Swedish unit of capacity, a stop, was equivalent to 654 millilitres).

162

The other early application for Wollaston's platinum was for the touch-holes of flint-lock pistols and sporting guns, replacing gold and having the advantages of greater hardness and higher melting point. This double-barrelled sporting gun, made in about 1810 with platinum faced touch-hole and decoration, is now in the Armoury of the Tower of London. Large quantities of platinum were also used as counterpoise weights on gravitating stops, an invention patented by Joseph Manton in 1812 to prevent the accidental discharge of cocked guns

Numerous other items of chemical apparatus, including balance pans and blow-pipes, were produced by Wollaston and sold through Cary over the period 1806 to 1822, but a much greater quantity was in the form of bars or ingots, the total amounting to over 28,000 ounces.

But by far the largest sales were made to the London gunmakers for the touch-holes and pans of flintlock pistols and sporting guns. This type of firing mechanism, well-established by the end of the seventeenth century, utilised a piece of flint held in the jaws of a pivoted dogshead or cock. When the trigger was pressed the cock swung forward and the flint struck sharply against the steel fixed on the cover that protected the priming charge, the sparks caused by the impact igniting the charge. The explosion resulted in considerable corrosion and erosion of the metal touch-hole which thus became slowly enlarged and led to a loss of power when the gun was fired. To avoid this deterioration it had become the practice to line the touch-hole of the best quality guns with gold, but the advent of malleable platinum now offered the additional advantages of a higher

163

This label from the case holding guns made by the greatest London gun maker Joseph Manton and dating from 1805 includes the words "Inventor of Platina Touch Holes"

melting point, a greater hardness and a lower cost. At this time London was the centre of fine gun-making, and among the leaders were the brothers John and Joseph Manton, both of whom introduced platinum touch-holes in the early years of the century, to be followed by almost all other English gunmakers, of whom there were at least two hundred. A full account of this phase of the history of platinum and its uses has recently been given by I. E. Cottington (38) who supports the view of Dr. Usselman (2) that no less than 27,000 ounces of Wollaston's platinum could have been absorbed in this way – the largest single application for his malleable metal. By the end of the period in question percussion ignition had replaced the flintlock but while this type of mechanism still required small amounts of platinum the Wollaston era was over and the supply had greatly diminished.

The Sulphuric Acid Boilers

While the greater proportion of his platinum was sold through Cary, Wollaston retained to himself a major product in which he took a personal initiative and in which the quantity of platinum was by far the greatest for each individual sale. In the manufacture of sulphuric acid the lead chamber process, introduced by John Roebuck and Samuel Garbett in 1746, yielded a relatively weak acid that needed concentration, a procedure that was carried out in glass retorts bedded in sand inside iron pots with considerable risk of breakage. By Wollaston's time the manufacture of sulphuric acid had increased very greatly and the cost had of course decreased. As early as December 1805 Wollaston had constructed his first boiler in platinum for a small acid manufacturer named Philip Sandman of

164

The application of platinum that most interested Wollaston lay in its great advantage as boilers for the concentration of sulphuric acid, replacing the fragile glass vessels previously relied upon. This outlet he retained to himself, designing the vessels, specifying their manner of construction by metal workers and interesting himself in their performance in service. The first platinum boiler ever made was supplied in 1805 to a small sulphuric acid maker named Philip Sandman of Southwark on the south bank of the Thames and Wollaston's sketch is shown here together with the weight of platinum involved, 406 ounces, and the cost, £282 9s 0d. After Sandman's death in 1815 this boiler was bought by the famous Glasgow chemical firm of Charles Tennant who found it so satisfactory that they very soon purchased four more

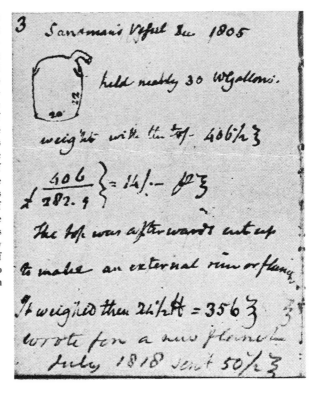

Southwark. This weighed some 406 ounces and held 30 gallons, the price being £282. Wollaston's relevant note-book entry is illustrated here. There is some evidence that Smithson Tennant played some part in the design and construction of this boiler, but it was fabricated by a Sheffield silversmith, Charles Sylvester, for a little over £50. An interesting comment on this first boiler is given by Samuel Parkes (1761–1825), himself an acid manufacturer, in his Chemical Catechism of 1807:

> "The important uses to which this precious metal may be applied can be easily conceived when it is considered that it unites the indestructibility of gold to a degree of hardness almost equal to that of iron; that it resists the action of the most violent fire and also of the most concentrated acids. Aware of these properties, a chemist in the neighbourhood of London has been induced to spend several hundred pounds in the fabrication of a single utensil for rectifying sulphuric acid" (39).

Four years passed before Wollaston succeeded in persuading a second sulphuric acid maker to purchase a platinum boiler. The firm of Richard Farmer and Son had been established at Kennington Common in 1778, and in 1809 the son Thomas ordered a larger boiler together with a siphon. For the fabrication work Wollaston now turned to a young cousin of John Johnson's named George Miles (1783–1837) who had been brought up in the Johnson family after the

165

death of his parents and who had established himself as a silversmith in Clerkenwell in 1804. Wollaston's notes for the costs incurred in making this boiler are shown here, but it seems that Miles was either inexperienced or had underestimated the work involved as in a letter to Thomas Farmer dated February 14th 1810 Wollaston wrote:

> "When I last parted from you I was in hopes that the additional charges for finishing the vessel might not require an appendix to the imperfect account which I gave you; but when I came afterwards to converse with Miles upon the subject, I found that the former sum of £14 10. 0 was not such as to satisfy him for his time and labour and actual expenditure and that he now demands an addition amounting in the whole to £19 saying at the same time that, if he could have foreseen the trouble, he would not have undertaken it for less than £30. Now, tho' I have no doubt that a little more sense would have saved a great deal of labour, we employed him such as he is and I am inclined to think he ought to be paid" (40).

Farmers large Vessel & Siphon

Platina 322½ at 16/ ... 258. 0. 0
Forging 4 Ingots ... 1. 0. 0
Flatting ... 2. 11. 6
Copper for do 21 lb / 8 pt = 2,6 8 pt 2. 6. 6
Porterage ... 9. –
Gold used 2 oz 4 dwt — 10. 10. –
Making £14. 10 – } + £4. 10 × ½ = 2. 5 } — 16. 15. 0
291. 12. –
Siphon 8¼ oz – £6. 12
gold for do 5 dwt 1. 1. –
making ... .15 8. 8 –
£300. 0. 0

Suction tube
Plat 1.6.0 1. 1. 0
Gold &c ... 9
£1. 10. 0

Pl Cast 121 111 89 102 423 oz

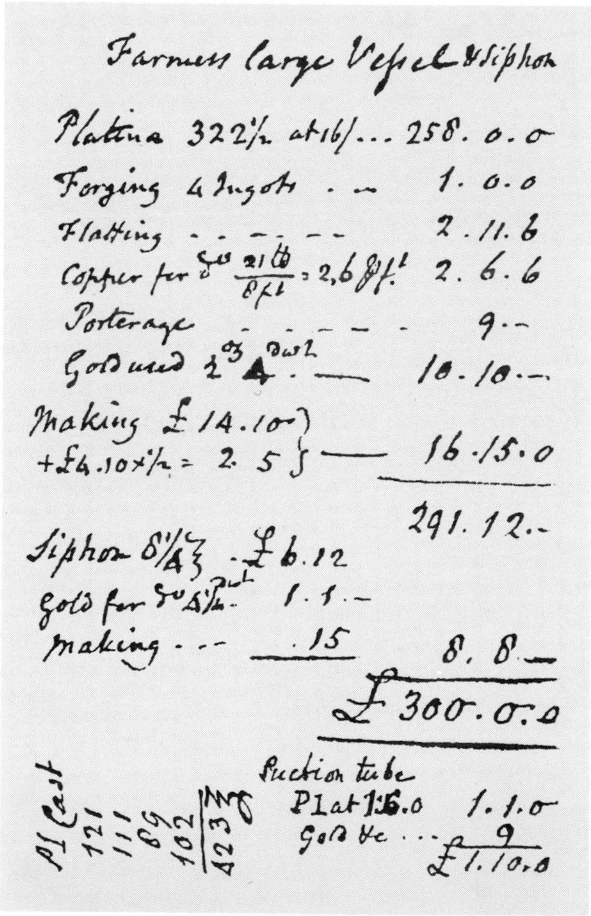

Wollaston's second boiler was ordered in 1809 by Thomas Farmer of Richard Farmer and Son of Kennington Common, London. This is believed to have been fabricated by George Miles, a silversmith in Clerkenwell who was a young cousin and protegé of John Johnson. The cost sheet shows that it was made by gold soldering from sheets that had been flattened by hammering between sheets of copper. Wollaston also designed a platinum syphon for this and later boilers to make continuous operation possible

Clearly Wollaston was not impressed with Miles, and when he secured a further order for a boiler from Pepper and Smith, sulphuric acid makers of Old Ford, east of London, he turned to his friend John Johnson for assistance. Johnson had no facilities and no space for such fabrication in his assaying business and it seems likely that, after purchasing the 413 ounces of platinum required, he supervised his young cousin Miles and ensured that this time a more satisfactory procedure was carried out according to Wollaston's specification. Miles was not particularly successful in his vocation and in 1810, perhaps not surprisingly, he abandoned it and became an assayer with the Worshipful Company of Goldsmiths.

Thirteen more boilers are shown to have been made in the note-books, and for ten of these Wollaston turned to a London coppersmith named John Kepp of Chandos Street, Covent Garden. (This business later became Richard and Edward Kepp, who described themselves as "Braziers and Copper and Platina Smiths" until 1854 when the name changed to "Benham and Froud late R and E Kepp", and the fabrication of platinum was still being carried out by them long after Wollaston's death). In 1811 Samuel Parkes, whose chemical works in Shoreditch included the manufacture of sulphuric acid, bought a Wollaston boiler holding exactly thirty gallons. An indication of the great care and interest that Wollaston took in the design and the performance of his boilers is given in a letter from him to Parkes in January 1812:

> "As it is now more than twelve months that you have been in possession of your large vessel it may reasonably be supposed that you have full trial of its merits and I hope you will not think it unreasonable in me to request a few lines upon that subject as it would gratify me to hear that it has fully answered your expectations . . . . if you would take the trouble of adding any further information on points of economy which I presume you have carefully estimated (saving of time, saving of fuel, saving of breakage, saving of labour) you would confer an additional favour on
> Sir, your most obedient humble servant
> Wm. H. Wollaston" (41).

Parkes' reply is not recorded, but in one of Wollaston's note-books under the date January 17th 1812 there is a brief entry: "S. Parkes, Platinum retort answers well" while in the second edition of Parkes' Chemical Essays published in 1823 after Wollaston had ceased to produce malleable platinum on account of the shortage of native metal he wrote:

> "I had one of these vessels and found it to be easily managed and very economical . . . It would now, however, be very difficult to procure such a vessel in consequence of the great scarcity both of crude and malleable platinum" (42).

The great Glasgow chemical firm of Charles Tennant bought Wollaston's first boiler from Sandman on his death in 1815 and its efficiency prompted the purchase of two more in the following year, and one each in 1817 and 1818. A note in Wollaston's hand in February 1816 records that Tennants:

> "boil off 3 times per day and turn out 50 bottles of 150 lb per week. They reckon to save the prime cost in 2 years, oil of vitriol being now at $3\frac{1}{4}$s. to $3\frac{1}{2}$s. per lb." (43).

The standard kilogramme weight, known as the Kilogramme des Archives, made in platinum by Marc Etienne Janety in 1798. The label on the case reads, using the French Revolutionary Calendar, "Kilogramme Conformé à la loi du 18 Germinal An 3, presenté le 4 Messidor An 7"

"The Commission, occupying itself with the choice of metal most suitable for the standards has come to the conclusion that the original standards to be kept in the premises of the National Convention should be made in a metal that is well known to be the most durable and the least alterable by time and weather and they propose to use platinum which, in this regard is greatly superior to all other known metals and which eminently possesses all the properties that could be desired for the production of invariable standards". (1)

Without Lavoisier and his furnace the Commission now turned to Marc Etienne Janety who, as recorded in Chapter 5, had felt it prudent to leave Paris in 1794, and summoned him to return from his works in Marseilles. He arrived back in September 1795, taking new premises in what had formerly been the Abbaye de Saint Germain des Prés but had been used as a saltpetre factory during the revolution, a location that was to be of importance to him later (2).

The Commission of Weights and Measures gave him 200 marcs (just under 50 kilogrammes) of native platinum, fixing the price of fabrication at 15 francs an ounce with an allowance of 25 per cent for scrap. Janety began the work in November 1795, still using his arsenic method, and over the next three years he produced four metres and four kilogramme weights, these being finally adjusted by the well known instrument maker Nicholas Fortin and the engineer Etienne

Lenoir. One of each standard was solemnly deposited in the Archives of the Republic in June 1799, these still being known as the "Metre et Kilogramme des Archives". The others are also still in existence in Paris. The kilogramme is illustrated here; the metre is a flat strip 25.3 millimetres in width and 4 millimetres thick, the length being defined by the distance from end to end, but these now show signs of wear. A final report on all this work was presented in October 1801 by the physicist Mathurin Jacques Brisson, the mathematician Andrieu Marie Legendre and Guyton, "Sur le Travail du Platine pour les Etalons des Poids et Mesures par le Citoyen Janneti" (3). In this they quote Janety's statement that he had had to make thirty kilogramme weights in order to obtain four good ones and report that they had paid him 2620 francs in cash and a further 1260 francs after he had finished repairing one of the weights that had been damaged by Fortin during its calibration.

After these labours Janety returned to his normal business as a jeweller and metal worker, achieving yet more renown for his fabrication of platinum as we shall see later in this chapter.

The Campo Formio Medal

The foundation in 1795 of the Institut National des Sciences et des Arts provided a meeting place for scientists as well as for historians and literary men, the First

The Campo Formio medal designed by Benjamin Duvivier and struck in platinum in 1798 to commemorate Bonaparte's victory in his Italian campaign. He is shown returning from the war with an olive branch held high in his right hand, his horse led by Bellona and Prudence while Victory raises a laurel wreath over his head carrying a statue of the Apollo Belvedere

181

Class being allocated to the scientists and replacing the Académie des Sciences that had been suppressed in 1793. It was to this body that the young General Bonaparte was elected in 1797 on his return from the successful Italian campaign that was concluded by the treaty signed in Campo Formio, a village near Udine. To commemorate this victory a large platinum medal was struck to the design of Benjamin Duvivier, the former medallist to Louis XVI, and presented to the Institut who in turn presented it to Bonaparte in October 1799 after his return from Egypt. The medal, illustrated here, shows him returning in triumph from the campaign in Italy. It is 57 millimetres in diameter and weighs 173 grams. The platinum has taken the design to perfection, but there is a fine crack to be seen on the reverse.

The Later Work of Guyton de Morveau

The earlier work of Guyton de Morveau on platinum and his contribution to a study of its physical properties have been described in Chapter 7. In 1791 he left Dijon for Paris on his election to the National Assembly and remained a member of the Convention which succeeded it a year later, while as a member of the Committee of Public Safety he was active in the establishment in 1794 of the École Polytechnique, the first college of technology in the world, where he became one of the professors of chemistry (4). In 1799 he was appointed Administrator of the Mints, of which there were nine altogether, and here, possibly inspired by the part played by Isaac Newton when Master of the Royal Mint from 1696 until his death in 1727 he interested himself in the activities of counterfeiters and personally investigated the possibility of platinum being used

The Platinum Pyrometer of Guyton de Morveau

The first of many applications of platinum in the measurement of high temperatures, this pyrometer was designed by Guyton de Morveau in 1803. A platinum rod 45 millimetres long supported in a groove formed in refractory clay had its free end in contact with the short arm of a bent lever, the longer arm serving as a pointer moving over a graduated scale, all made in platinum. The thermal expansion of the rod thus gave an indication of temperature. In later years Guyton modified and improved the design of his pyrometer

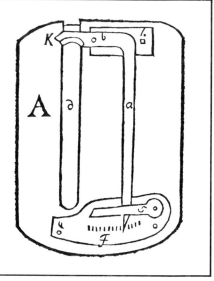

to adulterate gold. In a paper in 1803 he described a series of experiments with gold alloys to determine the effect of additions of platinum on both their colour and their specific gravity, also making the point that the use of a thin coating of gold on a platinum coin could readily be detected by treating the piece with a drop of aqua regia which would expose the grey colour of platinum (5). He concluded:

> "As fast as the counterfeiters try to perfect their pernicious art one finds the means to discredit their products."

Guyton was also interested in the measurement of high temperatures in furnaces and kilns and in the same year, 1803, he presented to the First Class of the Institut a pyrometer he had invented based upon the expansion of a platinum rod (6). This instrument has been more fully described by Dr. J. A. Chaldecott (7) and a sketch of its design is reproduced here from a German account of the pyrometer (8), no illustration being given in the original paper.

A further study of Guyton's in 1809 concerned the tensile strength of a number of metals including platinum (9). He determined the weight that a wire of each metal 2 millimetres in diameter would support before fracture, iron not breaking until a load of 250 kilogrammes was reached, copper 137 and platinum 125 kilogrammes. A comment that would have done him credit a hundred years later concluded his paper:

> "The force of cohesion is not appreciably diminished while the ductility of a metal permits its parts to slide over each other without breaking".

The Chemistry of the Platinum Metals

Reference was made in Chapter 9 in connection with Smithson Tennant's discovery of iridium and osmium to the work carried out almost simultaneously in France by Descotils and jointly by Fourcroy and Vauquelin on the constituents of native platinum. Descotils had investigated the cause of the varying colours of the precipitate formed by the addition of sal-ammoniac to its solution in aqua regia and had concluded that a new substance was present but he neither isolated it nor gave it a name (10). Fourcroy and Vauquelin had been asked by the Institut to investigate the purification of platinum and the method put forward by Mussin-Pushkin, as well as the arguments of Chenevix about the composition of palladium, and they also concluded that a new and unknown metal was present (11). They had in fact observed some of the properties of iridium but again they did not propose a name for it, while they also noticed a volatile substance obtained from the insoluble residue that affected the eyes and throat and which they thought to be an oxide of the metal already described. Smithson Tennant's paper of 1804 clearly identified and named both iridium and osmium, the metal with the volatile oxide, an announcement that prompted further work by the French chemists.

When this news reached Vauquelin in June 1805 he was about to leave for a

**Hippolyte Victor Collet-Descotils
1773–1815**

A student of Vauquelin in his early years, Descotils was invited to accompany Berthollet and Monge in the scientific party accompanying Napoleon Bonaparte in his expedition to Egypt. On his return he was appointed to the École des Mines where he spent the rest of his career. He was the first to investigate the cause of the varying colours of the salts of the platinum metals and he also devised an economical process of refining

visit to Marseilles to meet Fourcroy and from there he wrote to his colleague Bergman at the Muséum d'Histoire Naturelle, asking him to work on various problems and continuing:

"I also beg you to continue the experiments on platinum and to endeavour to discover whether it contains two new metals or only one. It is absolutely essential to clear up this question and to admit our error if we have made one." (12)

Fourcroy and Vauquelin duly continued their work, and in a paper read to the Institut in March 1806 they reviewed the properties of all four newly discovered elements, iridium, osmium, palladium and rhodium, acknowledging the more precise work of Tennant (13) although their own experimental work was accurate and detailed. In a later brief note they also acknowledged the part played by Descotils in the early stages of these discoveries (14).

Descotils had been a member of the group of scientists led by Berthollet who accompanied Bonaparte to Egypt in 1798 and on his return he was appointed to direct the laboratory of the École des Mines. His continuing friendship with Berthollet resulted in an invitation to become a member of the Société d'Arcueil founded by the latter and Laplace in 1807, a semi-formal organisation of eminent scientists who met and carried out research in Berthollet's country house outside Paris and whose work has been the subject of a full study by

184

Antoine François de Fourcroy
1755–1809

Professor of Chemistry for over twenty-five years at the Jardin du Roi, re-named the Muséum d'Histoire Naturelle after the French Revolution, Fourcroy held several other academic posts during his distinguished career. Much of his research was carried out in collaboration with Vauquelin who was appointed to the Muséum in 1804. Together they investigated the insoluble residue from native platinum but were less successful than the Englishman Smithson Tennant in identifying iridium and osmium. In this portrait by the famous French painter David he is pointing to a volume of his classic work, Système des Connaissance Chimiques, published in 1800. When Fourcroy died the titles of his principal works were engraved on a platinum plaque held by a platinum chain placed round his neck before his burial

Professor Maurice Crosland (15). In the first year of its existence Descotils read a paper, On the Purification of Platinum (16) in which he proposed both an economy in the amount of aqua regia to be used and a means of reducing the amount of iridium precipitated with the platinum. Instead of dissolving the native metal directly in the acids he proposed to alloy it with zinc and then to remove this with sulphuric acid, leaving a soft residue of the platinum metals that could be treated effectively by half the quantity of aqua regia needed for the direct attack. He also noticed that it was advantageous to pour the nitric acid on to the metal first, gradually adding the hydrochloric acid until this produced no further effect.

To separate the iridium more effectively after filtering all the insoluble black powder, he evaporated the solution "to perfect dryness". The addition of a little water left the gold and some of the palladium insoluble. The rest of the palladium was precipitated with mercuric cyanide as proposed by Wollaston and the iron and base metals by carbonate of soda. The filtrate from all this was made alkaline with more soda and then "on leaving it exposed for some time to the air, the iridium would be separated in the form of a green sediment", a

185

The lecture theatre and chemical laboratory of the Muséum d'Histoire Naturelle, formerly the Jardin du Roi, where Fourcroy and Vauquelin carried out the majority of their joint researches. After Fourcroy's death in 1809 Vauquelin continued to publish his work on the refining of iridium, osmium, palladium and rhodium and on their properties

process that could be expedited by warming to 50°C. After filtering, the platinum was precipitated with sal-ammoniac as usual, washed repeatedly with small quantities of water and reduced to metal, Descotils claiming the production of platinum "in its greatest known state of purity".

Some time after Fourcroy's death in 1809 Vauquelin, still working in the Muséum d'Histoire Naturelle, read two further papers to the Institut, the first in 1813 on palladium and rhodium (17) and the second early in 1814 on iridium and osmium (18). In the first of these he paid a graceful tribute to the discoverer of the two metals:

"Although M. Wollaston operated on only 1000 grains of native platinum and had at the most 6 or 7 grains (half a gram) of each of the new metals at his disposal, he yet recognised their principal properties, which does infinite honour to his sagacity, for the thing appears at first to be incredible. For my part, although I employed about 60 marcs (15 Kilograms) of crude platinum I found many difficulties in separating exactly the palladium and rhodium from the platinum and the other metals that are present in this mineral and in obtaining them perfectly pure".

In the second paper, devoted at great length to the separation of iridium and osmium and to their properties, he again acknowledged Tennant's discovery of these new elements which "we had taken, M. Fourcroy and I, for two modifications of one unique species". Unfortunately he attributed the blue colour of a solution he obtained in this work to iridium, so missing the discovery of the sixth platinum metal, ruthenium, which was not identified until 1844.

186

Another associate of Vauquelin's and a cousin of Fourcroy, André Laugier (1770–1832) also interested himself in osmium and read a paper to the Institut in 1813 on a new method of separating it from native platinum in greater quantities than before and at lower cost (19).

The Productions of Janety Father and Son

After completing his arduous work on the standard metres and kilogrammes, Marc Etienne Janety turned again to the manufacture of platinum jewellery and further extended his efforts in the making of chemical apparatus. At an exhibition in Paris in 1802 he displayed examples of these for which he was awarded a silver medal. His new house and factory were in the Rue Colombier, now the Rue Jacob, between the Seine and the Boulevard Saint Germain, and here in 1800 Fourcroy and Vauquelin had acquired an adjacent factory for the manufacture of fine chemicals, sharing with Janety a well and a pump in a common courtyard (20). This proximity must have interested the two professors while equally they must have encouraged Janety in his platinum work. There is in fact one reference to his being "assisted by the advice of the great chemist Vauquelin" (21).

In 1801 a new organisation was established in Paris, the Société d'Encouragement pour l'Industrie Nationale, modelled on the British Society for the Encouragement of Arts, Manufacturers and Commerce founded in 1754 and now known as the Royal Society of Arts. A committee for applied chemistry was appointed and included Fourcroy, Guyton de Morveau, Vauquelin and Descotils with Berthollet as chairman, and several reports of this committee over the next fifteen years or more provide a great deal of information on Janety's activities and those of his son François Joseph Marc Janety who joined him in the business and later succeeded him.

In 1810 a report to the Society by J. P. J. d'Arcet reads:

"M. Janety to whom we owe the best process up to now for working platinum and making it malleable has presented to the Society since its last public meeting various articles made from this metal which appear to merit your full attention; you have charged the Committee on Applied Chemistry to examine these articles and it is in their name that I now report to you.

M. Janety has presented to you:

(1) Ten platinum medals of the same diameter but of different types.
(2) A bucket shaped vessel 189 millimetres in diameter and 135 millimetres deep that has been planished and hammered until it weighs only 160 grammes.
(3) A retort with neck capable of holding 1 litre and weighing only 800 grammes.

You know with what perseverance M. Janety père has struggled for more than thirty-three years against the obstacles opposing the reduction of native platinum into malleable platinum. It is only by long labours and by the loss of his health, of his fortune, of his profession, that he has succeeded in conquering them, and the happy results that he has presented to you are without any doubt the fruit of the greatest and most willing sacrifice that has been made in the advancement of an art" (22).

Two years later d'Arcet reported again in the name of the committee, this time referring to "M. Janety fils, pupil and successor to his father", who retired at about this time. The objects presented to the Society by the son on this occasion were two still larger vessels holding respectively 22 and 16 litres and designed to line cast iron boilers for the concentration of sulphuric acid, together with a saucepan and a travelling knife with several blades. The younger Janety also announced that for more than a year he had ceased to use arsenic in the preparation of his platinum and had turned to powder metallurgy (23).

Yet again, at a general meeting of the Society in 1814 it is recorded that Janety junior presented several objects in platinum including cutlery, watch-chains, crucibles and capsules, one of the latter over 300 millimetres in diameter "of exquisite workmanship the more surprising as everyone knows the extreme difficulty of rendering platinum malleable" (24).

The progress and the praise continued. In 1818 the two Janetys were awarded a silver medal by the Société d'Encouragement, the citation going on:

"We ought to remember that M. Janety the first opened this career in which France has preceded and surpassed the other nations; that he has sacrificed to this work thirty-three years of his life, his fortune, his health; that his son has been able to value and improve this patrimony of persevering research; that in 1812 vessels and articles of jewellery, presented by him, excited a sort of surprise, and we have regarded it as a duty to mark in a solemn manner this new branch of industry" (25).

A Great Exhibition was held in Paris in the following year, 1819, and yet another silver medal was awarded to Janety junior and a new partner of his, Leonard Chatenay, by the jury chaired by Berthollet for platinum vessels and jewellery and for some standard rules in platinum made for the Royal Society of London and the Academy of Sciences of St. Petersburg (26).

Janety senior died in 1820, the business continued for a few years, and then in 1823 Janety junior died and Bréant took over the business from the widow as the "Ancienne Maison Janety". As late as 1828 it has been shown by Dr. W. H. Brock that Liebig was still buying platinum apparatus from Janety (27) but, by 1830 he had turned to Cuoq and Couturier. The activities of the Janetys, extending over more than forty years, none the less constitute an important section of this history and they made a very great contribution in their chosen field.

The Achievements of Jean Robert Bréant

The Janety's were not without a competitor. Shortly after the fall of Napoleon in 1815 and the restoration of Ferdinand VII to the Spanish throne the government in Madrid decided to dispose of a large quantity of platinum that had accumulated in their hands since the days of Chabaneau and Proust and for which they now had neither an application nor a method of treatment. This came to the knowledge of Pierre Augustine Cuoq (1778–1851) a lawyer in Lyons who travelled widely in business and turned to merchanting and exporting.

Taking into partnership one Couturier, a metal merchant, he purchased around 1000 kilogrammes of Spanish platinum and established the firm of Cuoq Couturier et Cie in Paris. Neither had any knowledge of chemistry and they turned for advice to one of the assayers at the Paris Mint, Jean Robert Bréant (1775–1850). A native of Normandy, Bréant had seen service in the mints in Rouen, Toulouse and La Rochelle before being appointed to Paris in September 1814, and he also had no expertise in the refining of platinum. He had, however, come to know Vauquelin, also a native of Normandy, and after securing his advice and spending some time in his laboratory Bréant agreed to undertake the refining of this very large quantity of metal.

At this time Wollaston's method had not been published, and it is likely that Bréant relied upon the processes of Thomas Cock and Richard Knight. In any case he achieved quite remarkable results very quickly, making large ingots, and by February 1817 Cuoq and Couturier were able to present to the Société d'Encouragement a boiler made from a single sheet of platinum refined by Bréant, weighing 15 kilogrammes and holding no less than 162 litres (28). Several other boilers had been made at the same time for the concentration of sulphuric acid.

Even before this Bréant had made a boiler from four sheets of platinum, made from ingots weighing 5 kilogrammes each, which were first riveted together and then soldered with gold. This was provided to the chemical works at Les Termes outside Paris where the younger Chaptal was in partnership with J. P. J. d'Arcet for experimental work in the concentration of sulphuric acid, but Bréant was not satisfied with this method of fabrication and so increased the size of his ingots to produce sheets as much as four feet square. He also succeeded in the autogenous soldering of platinum and displayed to the Society a tube six feet in length "of which the edges were perfectly united without the use of solder".

The same report refers to the great reduction in time in the concentration of sulphuric acid to as little as a quarter of that needed when using glass vessels as well as to the elimination of the risk of breakage. These boilers were adopted in the French sulphuric acid industry, and a little later Bréant supplied siphons in platinum to permit continuous operation.

A detailed description of one of his platinum siphons and of the economics effected in its use was given by the sulphuric acid manufacturer Auseline Payen, the partner of Cartier, in whose works at Pontoise this equipment was installed (29). This was fitted with water cooling and reduced the time required for decantation of a large boiler from half an hour to six minutes.

At the Exhibition of 1819 Cuoq and Couturier showed laboratory apparatus in platinum, a number of medals, and platinum "leaf" reduced to the same degree of thinness as gold leaf, as well as another large sulphuric acid boiler of 200 litres capacity. For these they were awarded a silver medal while Bréant received a similar medal "for having purified platinum on a large scale, having rendered it malleable and at a price much less than formerly" (26).

In 1819 also Bréant parted company with his colleagues and set up for himself at 64 Rue Montmartre while Cuoq and Couturier continued to operate in the Rue de Richelieu near the Palais Royal, later moving to the Rue Lulli where they remained until well into the 1830s, when the parties came together again in joining with a firm named Desmoutis whose sundry successors are still in operation. The further progress of this company will be described in a later chapter.

The refining process employed by Bréant was described in full detail, not by him but by Jean Pierre Baruel (1780–1838), of the École de Medicine in an English translation "at the request of a chemical gentleman from this country who wished to know the actual method now practised at Paris for preparing the great masses of platinum of which Couturiere forms his beautiful alembics" (30).

Just as in Janety's procedure, the native metal was first washed in a stream of water to remove all sand. Any mercury present was then driven off by heating. Next there was a short treatment with weak aqua regia to remove gold and copper, with usually a little platinum. The undissolved material was washed with water and ammonia and boiled in a retort with strong aqua regia, a process that was repeated five times. The insoluble "black powder" was filtered off and the filtrate evaporated down, a distillate containing osmium being collected. The concentrated platinum solution was diluted with water, precipitated with sal-ammoniac solution and the precipitate filtered and calcined. The resulting metal sponge was treated with dilute aqua regia which dissolved the platinum but left most of the iridium. After filtering off the latter, the solution was again treated with sal-ammoniac and the precipitate filtered and calcined. Here, therefore, for the first time in a published commercial process, an attempt was made not only to diminish the solution of the iridium but also to remove that which had gone forward into the first precipitate by redissolving in dilute aqua regia the sponge yielded by the latter. Consequently a much more uniform and more malleable metal was obtained. To consolidate it the sponge was charged into a crucible, heated and, as it contracted and sintered together under compression, more was added 'even to the amount of 20 or 30 pounds'. When this had been attained, the crucible was covered and heated up to whiteness, and the platinum was then transferred quickly to a steel mould in which it was compressed several times in a strong coining screw press.

This process was continued five or six times in the mould under a fly press and by then the metal was sufficiently consolidated to be transferred to an open charcoal fire and alternately heated and forged about thirty times. The ingot was then ready for rolling out into the sheet which was subsequently fashioned by the hammer into "the beautiful alembics of Couturiere". Baruel also gives detailed directions for the recovery from the various insolubles and mother liquors of palladium, rhodium, iridium and osmium in a more or less pure form. Of the chemical part of the process, the English translator remarks that it is "fundamentally the same as that published by M. Vauquelin in 1813 in the 88th volume of the *Annales de Chimie*" (17).

The process as described by Baruel had several new features. The dissolution

190

in aqua regia and the subsequent boiling-down were carried out in retorts or similar distillation apparatus, so that a distillate could be collected and a considerable part of the osmium which was volatilised as its tetroxide could be recovered. Next, as has already been pointed out, the French in order to obtain platinum reasonably free from iridium, were not content, like Wollaston, with minimising the amount taken into solution by using dilute aqua regia, but set about separating it physically from the first platinum sponge by treatment of the latter with weak aqua regia. Their recovery of palladium was based on Vauquelin's procedure of precipitating "an ammonia proto-sub-muriate of palladium" (dichloro-diammino-palladium) and not on Wollaston's equally effective use of mercuric cyanide. They also took steps to isolate and collect the rhodium and osmium. In all previous commercial work, most of these metals had either remained in the final product or had been wasted in mother liquor or fumes.

To some extent these remarks apply even to the work of Wollaston, but as regards the working qualities of the platinum produced there seems to have been little to choose between the English and French metal. The translator of Baruel's paper remarks that they seem "to be at present equally pure, malleable and ductile, and the price is nearly the same". The French technique of fabrication shows considerable advance on the British in that the production of bigger and bigger, but still sound, ingots enabled larger and larger sheets to be prepared. This in turn meant less and less recourse to soldering and therefore sounder vessels. Not only did this represent a true advance, but the introduction by Bréant of forge-welding made possible for the first time an autogenous platinum joint.

The Appeal of Palladium

In the course of his refining of the 1000 kilogrammes of Spanish platinum mineral Bréant was able to extract and purify some 900 grammes of palladium, an achievement of which he seems to have been especially proud. His superior at the Paris Mint, Aimé Puymaurin, who was deputy-director to his father, the Baron Puymaurin, succeeding him in 1830, encouraged Bréant in this connection and in July 1823 contributed a paper on palladium to the *Bibliothéque Universelle* (31). In this he refers to the rarity of this metal being so great that it was unlikely that so favourable an opportunity of obtaining it would occur again. He then details the properties of palladium as determined by Bréant and concludes:

> "Should this unoxidisable metal become sufficiently common to add to our industrial materials it might well be used successfully for medals and for chemical apparatus; its ductility and its lustre might well make it a substitute for silver in articles of jewellery, while it could be used for military braid as its lustre is never tarnished".

A month earlier Bréant, accompanied by the Baron Puymaurin, had been

From Bréant's small stock of palladium he made two cups, a large one presented to King Charles X on his accession in 1824 and a smaller replica shown here for himself. This he kept on his desk at the Paris Mint as a souvenir of his success in extracting and refining palladium

received in audience by Louis XVIII to whom he presented a medal struck in palladium (32). This carried the bust of the King on one side, and on the other the inscription:

1823.
Sous le régne de LOUIS-LE-DESIRÉ
Protecteur des lettres, des sciences et des arts,
Pour la premiére fois, le palladium purifié
Par M. Bréant
A servi à la fabrication des médailles.

Médaille en palladium présentée au Roi,
Par M. Bréant.

Another medal struck from Bréant's palladium commemorated the opening of the Museum of the Département des Monnaies et Médailles by King Louis Philippe and Queen Amélie in 1833.

Earlier, from his small stock of palladium, Bréant had made two cups, a larger one, 44 centimetres in diameter, that he presented to King Charles X on his succeeding his brother Louis XVIII in 1824 which is still preserved in the Trianon at Versailles, and a smaller one for himself. In both cases only the bowl is in palladium, the base being made of silver. The smaller cup, shown here, Bréant kept on his desk at the Mint where he became director in 1846, until his death six years later, as a constant reminder of his work in the refining of the platinum metals and the fabrication of large ingots of platinum sound enough to be rolled into large sheets and to yield better vessels for the French chemical industry.

References for Chapter 10

1   Procès-Verbaux du Comité D'Instruction Publique de la Convention Nationale, Ed. J. Guillaume, Paris, 1894, **2**, 638–646

2   G. Kersaint, *Rev. Hist. Pharm.*, 1959, **47**, 25–30

3   M. J. Brisson, A. M. Legendre and L. B. Guyton de Morveau, quoted in C. J. E. Wolf, *Ann. Chim.*, 1882, **25**, 66

4   W. A. Smeaton, *Platinum Metals Rev.*, 1966, **10**, 24–28

5   L. B. Guyton de Morveau, *Ann. Chim.*, 1803, **47**, 300–302

6   L. B. Guyton de Morveau, *Ann. Chim.*, 1803, **46**, 276–278

7   J. A. Chaldecott, *Ann. Science*, 1972, **28**, 347–368

8   L. B. Guyton de Morveau, *Französ. Ann. allg. Naturgesch. Phys. Chem.*, 1803, **3**, 28–31

9   L. B. Guyton de Morveau, *Ann. Chim.*, 1809, **71**, 189–199

10   H. V. Collet-Descotils, *Ann. Chim.*, 1803, **48**, 153–176

11   A. F. Fourcroy and N. L. Vauquelin, *Ann. Chim.*, 1803, **48**, 177–183; *ibid*, **49**, 188–224; 1804, **50**, 5–26

12   Letter from Vauquelin to Bergman, June 18, 1805, Bibliothèque Nationale Archives, quoted by G. Kersaint, *Bull. Soc. Chim.*, 1958, 1603

13   A. F. Fourcroy and N. L. Vauquelin, *Ann. Mus. Hist. Nat.*, 1806, **7**, 401–409

14   A. F. Fourcroy and N. L. Vauquelin, *Ann. Mus. Hist. Nat.*, 1806, **8**, 248

15   M. Crosland, The Society of Arcueil, London, 1967

16   H. V. Collet-Descotils, *Mém. Phys. Chim. Soc. d'Arcueil,* 1807, **1**, 370–378; *Ann. Chim.*, 1807, **64**, 334–335; *Phil. Mag.*, 1811, **37**, 65–69

17   N. L. Vauquelin, *Ann. Chim.*, 1813, **88**, 167–198

18   N. L. Vauquelin, *Ann. Chim.*, 1814, **89**, 150–181; 225–250

19   A. Laugier, *Ann. Chim.*, 1814, **89**, 191–198

20   G. Kersaint, *Rev. Hist. Pharm.*, 1959, **47**, 25–30

21   H. Vever, Histoire de la Bijouterie Française au XIXe Siècle, Paris, 1906, 119–120

22   J. P. J. d'Arcet, *Bull. Soc. Enc. Ind. Nat.*, 1810, **9**, 54–57

23   J. P. J. d'Arcet, *ibid,* 1812, **11**, 207–208

24   Anon, *ibid,* 1814, **13**, 75

25   Anon, *ibid,* 1818, **17**, 388–389

26   Exposition de 1819, Rapport du Jury Centrale sur les Produits de l'Industrie Française, 170

27   W. H. Brock, *Platinum Metals Rev.*, 1973, **17**, 102–104

28   J. F. L. Mérimée, *Bull. Soc. Enc. Ind. Nat.,* 1817, **16**, 33–36

29   A. Payen, *Bull. Soc. Enc. Ind. Nat.*, 1827, **26**, 20–22

30   J. P. Baruel, *Quart. J. Sci. Lit. Arts,* 1822, **12**, 246–292; *Phil. Mag.*, 1822, **59**, 171–179

31   A. Puymaurin, *Bibliotheque Universalle*, 1823, **23**, 235; *Bull. Soc. Enc. Ind. Nat.,* 1823, **22**, 163–164

32   Le Moniteur Universel, 1823, June 22, 76

Percival Norton Johnson
1792–1866

Apprenticed to his father, the assayer John Johnson, in 1807 Percival
set up for himself as an assayer and gold refiner in 1817, taking up the
refining of platinum and its allied metals after Wollaston had
abandoned this work. This portrait by G. J. Robertson in the possession
of Johnson Matthey shows him reading an assay report dated 1830 and
wearing a chain made from platinum he had refined. He was elected a
Fellow of the Royal Society in 1846

11

Progress in England after Wollaston

"This beautiful, magnificent and valuable metal is very remarkable in many points besides its known special uses."

MICHAEL FARADAY

When Wollaston found that he could no longer obtain supplies of native platinum and so ceased to offer his products to industry in 1820 only Percival Norton Johnson remained as a refiner and fabricator, although for some years on a smaller scale. This dominant position he maintained for many years although for a time the French were important competitors.

At the age of twenty-five Johnson had separated from his father's long established practice as an assayer in the City of London and on January 1st 1817, with a capital of £150, had set up for himself as "Assayer and Practical Mineralogist", initially in the City and then in 1822 in Hatton Garden (1). On the very first day of his independence he married Elizabeth Lydia Smith, one of whose older sisters had earlier married Thomas Cock (1787–1842), a wealthy young man who until then had been an assistant in the famous laboratory of Willima Allen at Plough Court and who had developed a process for producing malleable platinum as related in Chapter 8. Cock carried on his chemical work in a laboratory in his house after leaving Plough Court on his marriage and when Johnson embarked on the refining of platinum it was Cock's process that he employed and it was Cock himself who supervised operations, spending much of his time in the Hatton Garden laboratories for the rest of his life.

Johnson, as we have already seen, was keenly interested in platinum and he maintained the contacts with Wollaston that his father had enjoyed. At first, however, he became more heavily involved in palladium with which he and his father had become acquainted in Brazilian gold as early as 1812. In the course of his gold assaying work he achieved an accuracy much greater than anything known before and this brought him into conflict with the bullion dealers who had been accustomed to wider margins and therefore easier profits. His results were challenged, and his immediate response was to offer to buy any gold on the basis of his own assays. With a substantial injection of capital provided by one of his wife's brothers, Lieutenant W. R. B. Smith, R.N., a gold refinery was built behind the Hatton Garden house and from the first this was called upon to treat considerable amounts of the palladium-bearing Brazilian gold as Johnson was the only refiner in London capable of handling it. His process remained undisclosed for many years, but in 1846, when he was a candidate for election to the Royal Society, he at last gave full details of his procedure in a letter to the then President, the Marquis of Northampton (2). This begins:

"My Lord,

I am requested to lay before the Society the process I discovered and have adopted since 1817 for the extraction of Palladium which exists in combination with the Gold of the Gongo Soco Mine in the Brazils and also of the Candonga Mine in the same country, with some observations as to the nature of the rock and gangue of the formation in which it is found and the applications of the metal for useful purposes".

The letter goes on to describe the process of melting the gold with excess of silver, parting with nitric acid, precipitating the silver with sodium chloride and

water on a filter is dried and heated
by which the pure Metallic Sponge of Palladium

This Metallic Sponge is then subjected to an
excessive pressure in an iron box and carefully
forged as with Platinum and drawn into
wire or flattened into Sheet as with that Metal
Finding that Palladium has the property of
protecting Silver from the effect of Sulphuretted
hydrogen and for the more economical use
and greater facility of manufacture for Dental
purposes I make an alloy of equal parts
of these metals
An alloy is also made of 70 parts Palladium
20 of Silver & 10 of Copper which has the elasticity
of Steel and is used for the points of "everlasting"
Pencils and might be usefully introduced in
Surgical Instruments

I have the honor to remain
Your Lordship's
Most obedient Servant
Percival N Johnson

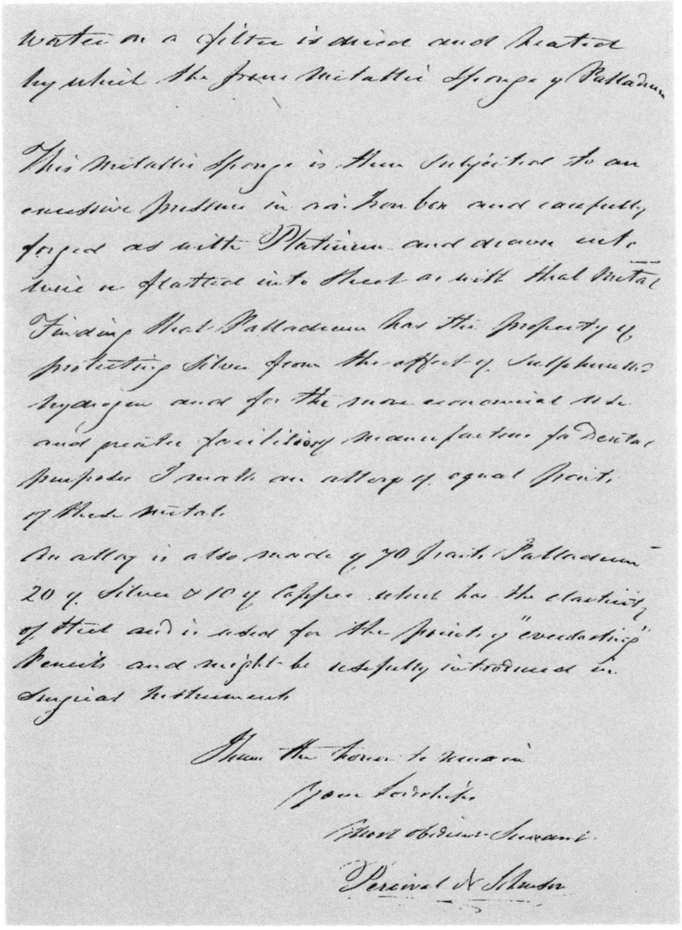

The concluding part of the letter from Percival Norton Johnson to the President of the Royal Society describing the process for the extraction of palladium from Brazilian gold that he had been employing since 1817, and the properties of this metal and some of its alloys

Photograph by courtesy of the Royal Society

The buildings in Hatton Garden, now demolished, to which Percival Norton Johnson moved in 1822. As his business developed additional houses on either side of the original location at No 79 were taken over and a refinery was set up behind the offices and laboratories

then throwing down the palladium with zinc, redissolving in nitric acid. The palladium, amounting to some 4 per cent of this native gold, was finally precipitated with ammonium chloride, followed by heating to reduce it to a metallic sponge. As with platinum, this was compressed in an iron box, carefully forged and then either drawn into wire or rolled to sheet.

The final paragraph of Johnson's letter, reproduced here, described his suggestions for the uses of palladium as a protective coating for silver, as an alloy with silver for dental purposes and in a ternary alloy with copper and silver for the construction of instruments.

A Palladium Chain for King George IV

As Wollaston had found, Johnson met with considerable difficulty in disposing of the considerable amounts of palladium he was accumulating and he sought to promote its applications. One of his early ideas was to present a massive ceremonial chain in palladium to King George IV, the offer being made through Sir Astley Cooper, surgeon to the King and of course a relative of

197

In June 1826 Johnson sought to publicise the merits of palladium by presenting this massive ceremonial chain to King George IV. The offer was made through Sir Astley Cooper, the distinguished surgeon who had married the sister of Thomas Cock in 1798 and who had exercised his influence in a number of directions, first by securing an appointment for Cock at Plough Court and later by taking an interest in Johnson's activities. The great palladium chain is still preserved in Windsor Castle

Photograph by permission of
Her Majesty the Queen

Johnson's by marriage. This chain is still preserved at Windsor Castle and is illustrated here by the kind cooperation of the Surveyor of the Queen's Works of Art, Mr. Geoffrey de Bellaigue. The letter from Sir Astley Cooper to Sir William Knighton, the King's physician who had been appointed Keeper of the Privy Purse, is also in the archives at Windsor Castle and was recently located there by Mrs. Shirley Bury of the Victoria and Albert Museum (3).

The amounts of Brazilian gold extracted began to increase substantially when a new mining company started operations in 1826 but because of the difficulties in refining their unsaleable stocks built up in their London warehouse. This situation came to Johnson's knowledge in 1832 and he at once set about treating the gold bars by the process just described and over the next twenty years he refined over a quarter of a million ounces of Brazilian gold, recovering large quantities of palladium. By the November of 1835 he had in fact extracted 2600 ounces, and he continued his promotional campaign, advocating the use of

198

The letter from Sir Astley Cooper to Sir William Knighton who was Keeper of the Privy Purse to George IV. It reads:

> "Mr. Johnson who is a great Metallurgist has a Chain of Palladium which is the only one which has been ever made, and which he wishes to present to our beloved Monarch. From his taste and talent I believe his Majesty would be pleased to accept it"

Photograph by permission of Her Majesty the Queen

My dear Sir William.

Mr Johnson who is a great Metallurgist has a Chain of Palladium which is the only one which has been ever made, and which he wishes to present to our beloved Monarch.

From his taste and talent I believe his Majesty would be pleased to accept it –

I am yours very truly

June 25th 1826 –

Astley Cooper

palladium for the pans and beams of chemical balances, for rust-free surgical instruments, for lighthouse reflectors and as a substitute for steel in many small applications.

In a letter to one of the first issues of *The Mining Journal* in 1835, addressed from "Assay Office and Metal Works, 79 Hatton Garden", he wrote:

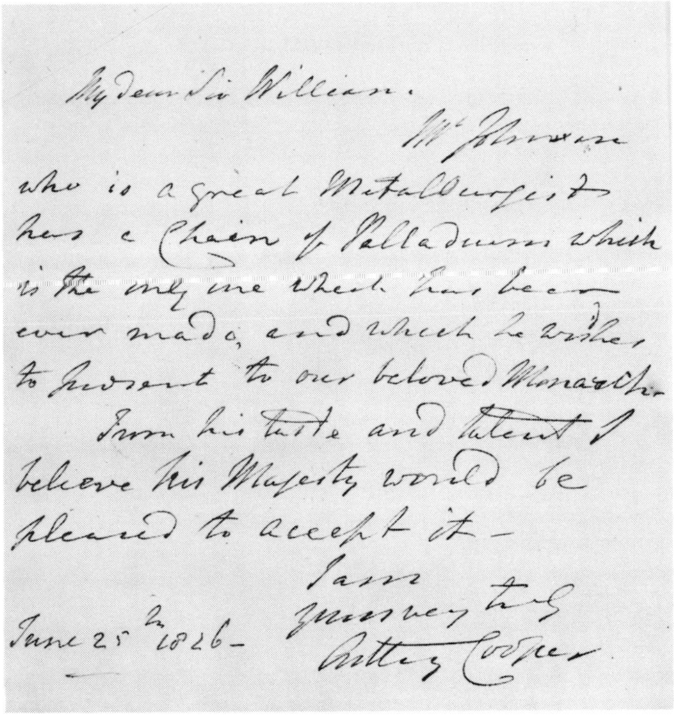

> "Palladium has not until within the last few years been an object of attention from its great scarcity, the ore of platinum being the only source from which this metal was derived and in which Dr. Wollaston first discovered it in 1803. I noticed the existence of palladium in the Brazilian gold in the year 1812 but until my engaging with the Imperial Brazilian Mining Company its extraction to any extent in a state of purity had not been effected . . . The properties of this metal render it intrinsically valuable in the arts, having most of the characters of platina, which it also resembles in colour, is perfectly malleable, and being only $11\frac{3}{4}$ specific gravity, has an advantage over platina in its introduction". (4)

In September 1836 another letter appeared in *The Mining Journal* (5), pointing out an error in the previous week's issue where it had been stated that palladium had so far only been obtained from crude platinum and had never found any applications. The writer went on to emphasise that this metal had been extensively procured from Brazilian gold "and is being used for many valuable purposes". His letter continued:

"Mr. Percival Johnson of 79, Hatton-garden, lately delivered a lecture on metallurgy, at Hampstead, and thus describes the metal:

"Its resisting the atmosphere and most of the weak acids, render it useful for dental purposes, for the graduated scales for mathematical instruments, and as an alloy for the tips of pencil-cases, in lieu of steel; having the same elasticity, without being liable to corrosion. Its oxide gives a hair-brown in enamel painting."

It may be added, that this metal is capable of a very high polish, and might be most beneficially used for reflectors to lighthouses, and for surgical instruments, particularly for foreign use in climates where the atmospheric damp is so prejudicial to steel. It is partially used for vaccine points, and here its superiority over steel must be obvious, since it has not unfrequently happened that patients have been vaccinated in distant counties from the metropolis with a rusty lancet, and the inflammation caused by the rust has been mistaken for the disease; and hence, in some cases, the occurrence of small-pox after supposed vaccination. It is also particularly useful for fine experiment balances: the Americans are so convinced of this fact, that several assay balances are now being made for the United States Mint, in Philadelphia, of this metal. I have also seen two in use in England.

If you think this notice worthy a place in your valuable publication, I shall feel obliged by its insertion, and beg to subscribe myself your obedient servant,

W. M. POUSSETT.

Chapter-house, St. Paul's, Aug. 29, 1836.

P.S—I am able to state, from my own knowledge, that several hundred ounces have been used for these purposes.   W. M. P.

The writer of this letter, William Poussett, had married yet another of Mrs. Johnson's sisters and in 1823 had been taken into the office at Hatton Garden by Johnson as clerk and salesman! The address he gave was that of their father-in-law, Thomas Smith, who held the post of Receiver-General to the Dean and Chapter of St. Paul's.

Some years earlier Johnson had made a lengthy visit to Germany and had spent several weeks at the Mining Academy at Freiberg, establishing a friendship with the professor of chemistry and metallurgy there, Wilhelm August Lampadius (1772–1842). In 1836 he sent over to Lampadius samples of the Brazilian ore, of the double salt of palladium as precipitated and of the metal produced by its reduction on heating. These were all closely examined and formed the subject of a paper in the following year (6) which the author hoped would:

"come opportunely as a contribution to the chemical history of palladium with grateful recognition of the appreciation which Mr. Johnson has won through this new method of treatment of palladium-gold".

A further promotional exercise occurred in 1845 when Johnson presented to the Geological Society, of which he had been a member since 1824, a quantity of palladium sufficient to provide their Wollaston medals for some years to come.

The process devised by Thomas Cock had taken no account of the other metals of the platinum group discovered by Wollaston and Tennant between

200

One of the Wollaston Medals awarded annually by the Geological Society for outstanding research. For some years this was struck in palladium from a quantity presented to the Society by Percival Norton Johnson who had been a member since 1824

1803 and 1805, but Johnson, with Cock's cooperation, was successfully extracting and refining these very shortly after establishing his business. One of the early researches carried out by Michael Faraday at the Royal Institution, in collaboration with a much older man, James Stodart (1760–1823), a well known maker of cutlery and surgical instruments, was on the effects of additions to steel of platinum, palladium, iridium, osmium and rhodium, as well as of silver and gold. The object of the investigation, which continued for some five years from its beginning in 1819, was to ascertain whether any such alloy steels would yield better cutting edges or would prove less susceptible to corrosion. A first account was presented to the Royal Institution in 1820 (7), with a longer paper to the Royal Society two years later (8). Although this research was too far ahead of the time when alloy steels would be understood and put into service, some considerable success was achieved, particularly with the platinum and rhodium additions. A platinum-steel specimen, for example, "after lying exposed for many months had not a spot on its surface" while in a letter to his friend Professor Gaspard de la Rive in Geneva, Faraday wrote:

> "Perhaps the best alloy we have yet made is that with Rhodium. Dr. Wollaston furnished us with the metal so that you will have no doubts of its purity and identity. One and a half per cent of it was added to steel and the button worked. It was very malleable but much harder than common steel and made excellent instruments. Razors made from the alloy cut admirably".

No appreciable results followed from these laborious experiments of Faraday and Stodart, reviewed in detail by Sir Robert Hadfield in 1931 (10), but at least

201

Michael Faraday
1791–1867

Almost an exact contemporary of Johnson, Faraday drew his supplies of platinum and its associated metals from him for the researches he carried out at the Royal Institution on electrochemistry, on the magnetic properties of metals and compounds and on the melting of optical glass in platinum. He was among those sponsoring Johnson for election to the Royal Society

one steel and cutlery manufacturer was sufficiently interested to pursue the matter. This was a Sheffield firm, Green Pickslay and Co., who wrote to Faraday in April 1824 to the effect that they were proposing to "make experiments in the large way". Later in the same year they wrote again:

"Green, Pickslay & Co., have great pleasure in informing M. Ferrady that they have made a number of experiments with the alloys recommended by him and find the steel greatly improved by them; they send a specimen alloyed with silver, iridium and rhodium which they consider the best they have produced, these alloys with some valuable practical hints have been furnished by Mr. Johnson, No. 79 Hatton Garden; the report of the forgers is that the steel works better under the hammer than any they have before used, and likewise hardens in a much superior manner. Green Pickslay & Co. beg Mr. Ferrady's acceptance of a pair of rasors made from this steel. They will have great pleasure in sending other specimens of cutlery etc., as they continue their experiments" (11).

Thus by 1824 at latest Johnson was able to produce and supply the other metals associated with platinum. A later testimony to this and to his growing reputation concerns the development of iridium for the tipping of the early gold nibs. Following Wollaston's use of a rhodium-tin alloy for this purpose, an English engineer, John Isaac Hawkins (1775–1865), found that iridium gave a better performance and this he obtained for a time from Johnson. But by 1835

202

Hawkins had selected all the suitable particles of iridium suitable for his purpose from Johnson's stock. He then records:

> "I therefore went to the British Association for the Advancement of Science which met at Dublin on the 10th of August, 1835, to inquire of the great Chemists of the time, expected to be there assembled, where I could be supplied with the precious material. On asking Dr. Dalton of Manchester, Dr. Thomas of Glasgow, Dr. Daubeny of Oxford, and many other eminent Chemists present at the meeting where I could procure the substance each, without communicating with any of the others, answered that I could obtain it of Mr. Johnson, Hatton Garden, London" (12).

Melting Optical Glass in Platinum

After Stodart's death in 1823 Faraday discontinued his investigations on steels and in the following year he was commissioned by the Royal Society to undertake a quite different study on the improvement of glass for optical purposes. A distinguished committee including the President, Humphry Davy, Wollaston, Warburton, Hatchett, Brande and Dolland was appointed to discuss the subject with the members of the Board of Longitude in an endeavour to improve the quality of telescopes while the leading glass makers Apsley Pellat and James Green of the Falcon Glass Works in Southwark were requested to build a suitable furnace. Little progress was made for some time, and then in 1827 a furnace was built at the Royal Institution so that Faraday could be freed from travelling to the works. Melting in the traditional clay crucibles still produced many defects in the glass, and then in April 1828 Faraday turned to the use of platinum and recorded his first comment:

> "Find that in small quantities in platina foil a clear glass may be made containing as much as 70 litharge to 10 of silica" (13).

He pursued this new method for a further six months, encountering many difficulties caused by the reduction of the lead and its attack on the platinum and from the iron plate on which stood the platinum vessel. But by substituting a platinum foil beneath the latter, by January 1829 he could record:

> "The platinum vessel has stood well but a very little glass had crept over on to the foil beneath but there was no appreciable loss" (14).

The results of this long and tedious research were reported to the Royal Society in the Bakerian Lecture in November and December 1829 "On the Manufacture of Glass for Optical Purposes", a long paper requiring three evenings for its presentation (15). In the course of this Faraday recommended that the glass be melted first in crucibles of pure porcelain and then poured from a platinum ladle into a platinum dish. The rough glass so obtained was then to be re-melted in a platinum tray about 10 inches square that must have come from "a good ingot that had been rolled very gradually and carefully".

The much improved samples of glass were prepared for use by the optician George Dolland and sent for examination to the astronomer Sir John Herschel. A report by the Committee for the Improvement in Optical Glass in 1831 stated:

Some of the pieces of optical glass made by Faraday. He first proposed the melting of glass for telescopes in platinum equipment during a long investigation carried out for the Royal Society in cooperation with the Board of Longitude but it was very many years before this became standard practice in the glass industry

"The telescope made with Mr. Faraday's glass has been examined by Captain Kater and Mr. Pond. It bears as great a power as can reasonably be expected, and is very achromatic. The Committee therefore recommend that Mr. Faraday be requested to make a perfect piece of glass of the largest size that his present apparatus will admit, and also to teach some person to manufacture the glass for general sale".

Faraday begged to decline any further involvement in this work but once again he was well ahead of his time and it was very many years before the melting of optical glass in platinum became standard practice.

William John Cock joins Johnson

Johnson developed a number of interests in the then booming lead, copper, silver and tin mines of Devon and Cornwall, first as assayer and then as the owner of several mines, this side of his activities involving him in long and frequent absences from London. Feeling the need for assistance, in 1826 he engaged an assayer named George Stokes whom he took into partnership in 1832, but this ended with the death of Stokes only three years later. He then turned for advice

William John Cock
1813–1892

The son of Thomas Cock, Johnson's brother-in-law who had devised a method of producing malleable platinum while working with William Allen at Plough Court in 1805, W. J. Cock was a partner of Johnson's in the development of the platinum business from 1837 until 1845 when he retired on account of his poor health. Both he and his uncle were founder members of the Chemical Society and he contributed a paper to them on the refining and alloying of palladium in 1843

to his old friend, brother-in-law and collaborator Thomas Cock, who proposed the name of his second son William John, now approaching 21. He had been articled to a solicitor in 1828 but like his father had a bent for the physical sciences and on receiving an offer from Johnson abandoned this profession and entered the business, becoming a partner in 1837, the firm then changing its name to Johnson and Cock.

The younger Cock quickly acquired a knowledge of chemistry and metallurgy and began to increase the size of the equipment and thus of the platinum ingots from a mere six ounces to about sixty, introducing a sectional mould and replacing an old screw press with an hydraulic one. There is however, no evidence of any attempt to profit by the work of Wollaston and to apply his methods for maintaining the surface energy of the metal in the most active possible state. In particular, the platinum sponge was always pressed dry as recommended in Cock's original specification, and never wet as prescribed in Wollaston's elutriation process. There can be little doubt that the metal was not so good as Wollaston's and suffered from blisters and porosity, but it was apparently good enough to satisfy the demands of the period.

As time went on the ingots and the sheets produced from them gradually increased further in size. Following Wollaston's example, Johnson and Cock entered the business of making sulphuric acid boilers, and in the course of a series of lectures on mineralogy and metals given to the City of London Literary Institute in 1843 Johnson stated that he had made such a vessel holding 63 gallons at a cost of £750 (16). The fabrication of this vessel was most probably the work of the Kepp firm who had made most of Wollaston's boilers and with whom Johnson had close business relations.

Another application for Johnson's platinum, following the French precedent, was for standard weights. In 1829 the German scientist Christian Heinrich Schumacher (1780–1850), Professor of Astronomy at Copenhagen, who had carried out experimental work on legal units of weight for the Danish government, ordered a copy in platinum of the Imperial Standard Troy Pound that had been made in brass in 1758 from Thomas Charles Robinson, the famous maker of balances, for comparison with the Danish weight. This copy he considered "must not be made of a metal liable to oxydation, but of platina. A platina pound was therefore ordered of Mr. Robinson". To carry out the comparison he sent over to London one of his assistants, Captain Nehus, who also took the opportunity to calibrate this new weight against another platinum troy pound that had been made some years earlier by William Cary at the request of the Royal Society (17).

After the destruction of the Imperial Standard Weights in the fire that destroyed the Houses of Parliament in 1834 it was decided to have new standards made in platinum. The new standard pound, together with four copies and a number of smaller weights, were made in 1844 by Henry Barrow from a hundred ounces of platinum provided by Johnson and Cock. The standard pound shown here was formerly held by the Exchequer but is now in the care of the National Physical Laboratory.

206

To celebrate the coronation of Queen Victoria in 1838 a number of medals were struck by the Royal Mint in platinum in addition to those made in silver and bronze. Designed by the royal medallist Benedetto Pistrucci, these showed the crowned head of Victoria, with on the reverse the Queen holding her sceptre and orb with three helmeted figures representing England, Scotland and Ireland presenting her with a crown.

Then in 1834 a disastrous fire burnt down the Houses of Parliament and among many other valuable objects the Imperial Standard Weights were either totally lost or rendered useless. In 1838 a Commission was appointed to supervise the restoration of the standards and three years later made their report recommending that the avoirdupois pound of 7000 grains be adopted instead of the old troy pound of 5760 grains and that the new standard, together with four copies of it, should be made in platinum. These together with a number of smaller auxiliary weights, were duly made in 1844 by Henry Barrow (1801–1870) who had taken over Robinson's business on his death in 1841, some 101 ounces of platinum being provided by Johnson and Cock. The calibration of these standards, in which Professor Schumacher gave some assistance, was described in great detail by Professor W. H. Miller of Cambridge in a paper to the Royal Society (18) and the standards were solemnly legalised by Act of Parliament, the principal weight being deposited in the Office of the Exchequer at Westminster, one copy going to the Royal Observatory at Greenwich, one to the Royal Society and one to the Royal Mint, while the fourth was immured in a wall of the rebuilt Houses of Parliament.

The possible use of platinum for medals was also not entirely overlooked in this country. On the occasion of Queen Victoria's coronation in 1838 a number of platinum medals were struck to the design of Benedetto Pistrucci, the Royal Medallist, in addition to the many silver and bronze medals (19).

Both Johnson and Cock became founder members of the Chemical Society in 1841 and the latter contributed a paper on the extraction of palladium to the first volume of their proceedings, also presenting the Society with a specimen of this metal (20). Johnson was the Society's first Honorary Auditor and served on its Council from 1842 to 1844, later presenting a quantity of palladium to provide the Society's first ten Faraday Medals. As already mentioned, Johnson was elected a Fellow of the Royal Society in 1846, his sponsors including Michael Faraday, W. H. Pepys and Charles Wheatstone. In the previous year he had been closely in touch with Faraday – with whom he was almost exactly contemporary – when he was studying the magnetic properties of a wide range of metals and other substances and discovering the phenomenon of diamagnetism. Faraday's paper to the Royal Society on this subject (21) includes the two following paragraphs:

> "Platinum – I have, as yet, found no wrought specimens of this metal free from magnetism, not even those prepared by Dr. Wollaston himself and left with the Royal Society. Specimens of the purest platinum obtained from Mr. Johnson were also found to be slightly magnetic.
> Palladium – All the palladium in the possession of the Royal Society prepared by Dr. Wollaston amounting to ten ingots and rolled plates, is magnetic. Specimens of the metal from Mr. Johnson, considered as pure were also slightly magnetic".

Faraday's diary for the period also mentions his obtaining from Johnson specimens of rhodium, iridium and osmium as well as various compounds of these metals for the same investigation (22).

The Beginnings of Electrochemistry

Faraday's famous series of quantitative researches on electrochemical decomposition, beginning in 1832 and following up Davy's earlier studies, put the new subject of electrochemistry firmly upon its foundations. With the advice of the Reverend William Whewell of Trinity College Cambridge he also proposed the terms we still use today, anode and cathode, electrolysis and electrolyte. Most of his brilliant experimental work was carried out with platinum plates as electrodes in the ingenious apparatus he designed, as he appreciated "the great advantage of the opportunity afforded amongst the metals of selecting a substance for the pole which shall not be acted upon by the elements to be evolved".

The enormous batteries devised by Pepys, Wollaston and others now began to give place to simple primary cells as a source of current, although the earlier designs were found to be subject to deterioration on standing and to polarisation caused by bubbles of gas accumulating on the surfaces of the electrodes. Not until 1836 was this problem solved when Professor J. F. Daniell (1790–1845) of King's College, London, a great friend and admirer of Faraday, devised his self-depolarising cell or "constant battery" (23). This was followed in 1839 by the cell proposed by W. R. Grove, the outstanding feature of this being the use of a

208

William Robert Grove
1811–1896

Born in Swansea, and educated at Brasenose College Oxford and called to the Bar in 1835, Grove spent several years at home on researches in electrochemistry, devising a primary cell using a platinum electrode and also inventing the now familiar fuel cell. In later years he had a distinguished legal career, becoming a Q.C. in 1853, a judge in 1871 and being knighted a year later

platinum electrode immersed in strong nitric acid contained in a porous pot which separated it from the zinc element immersed in weak sulphuric acid (24). The idea of the porous diaphragm was due to Antoine Cesar Becquerel (1788–1878), professor of physics at the Muséum d'Histoire Naturelle, and the two collaborated in the design of batteries. At the meeting of the British Association for the Advancement of Science in Birmingham in September 1839 Grove read a paper "On a small Voltaic Battery of Extraordinary Energy", describing a battery of his construction that had been presented to the Académie des Sciences by Becquerel in the previous April (25). This consisted of "seven liqueur glasses containing the bowls of common tobacco pipes, the metals of zinc and platinum" that produced a current equal to the most powerful batteries of the old type. A diagram of this construction is given over page. By assembling a number of such units in a wooden box Grove was able to provide a most useful source of constant and continuous current, and the first such battery is still preserved in the Science Museum in London (Inventory 1876–82). They were used successfully for many years, particularly in the early days of telegraphy in the United States and also for the first Atlantic cable.

Grove was also responsible for the origin of the fuel cell, generally regarded as a mid-twentieth century development. In a postscript dated January 1839 to a letter to *The Philosophical Magazine* about his battery he described an

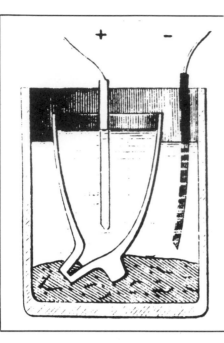

Grove's Original Primary Cell

The primary battery devised by W. R. Grove in 1839 consisted of a number of small cells contained in glass vessels, the electrolytes being separated by means of the broken-off bowls of clay tobacco pipes. The positive pole was of zinc and the negative pole of platinum. These he demonstrated to both the British Association for the Advancement of Science and the Académie des Sciences in Paris. Later he employed a more sophisticated design and the batteries were used for many years, particularly in telegraphy

experiment (26) on "an important illustration of the combination of gases by platinum" in which a galvanometer was permanently deflected when connected with two strips of platinum covered by tubes containing oxygen and hydrogen. By 1840 he had been elected F.R.S., and had been appointed Professor of Experimental Philosophy at the London Institution in Finsbury Circus and it was from there that he addressed a further letter, dated October 29, 1842, to *The Philosophical Magazine*, "On a Gaseous Voltaic Battery" (27). This paper, as well as a similar private letter to Faraday, describes the first practical fuel cell, employing platinum foil coated with spongy platinum produced by electrolysis of the chloride. A series of fifty pairs, constructed as shown here in Grove's original diagram, with dilute sulphuric acid as the electrolyte, was found "to whirl round" the needle of a galvanometer, to give a painful shock to five persons joining hands, to give a brilliant spark between charcoal points, and to decompose hydrochloric acid, potassium iodide and acidulated water.

Electroplating with Platinum and Palladium

The introduction of the electrodeposition of gold and silver in the early 1840s, mainly by the Elkingtons of Birmingham and their associate John Wright, naturally prompted attempts to electroplate with platinum and palladium.

It was at first assumed that platinum, having so many similar properties to those of gold, would be equally amenable to electrodeposition, but unfortunately

The First Fuel Cell

In 1842 W. R. Grove devised the first fuel cell, employing platinum foil coated with spongy platinum as the electrodes and sulphuric acid as the electrolyte. This is his letter to Faraday written a week before he wrote of his invention to the editor of *The Philosophical Magazine.* His diagram shows the construction, *ox* denoting the tube supplying oxygen, *hy* that supplying hydrogen, and the hatched lines the platinised platinum electrodes

By courtesy of the Royal Institution

London Institution
Saturday Oct 22
1842

My dear Sir

I have just completed a curious voltaic pile which I think you would like to see, it is composed of alternate tubes of oxygen & hydrogen through each of which passes platina foil so as to dip into separate vessels of water acidulated with sulphuric acid the liquid just touching the extremities of the foil as in the rough figure below. the

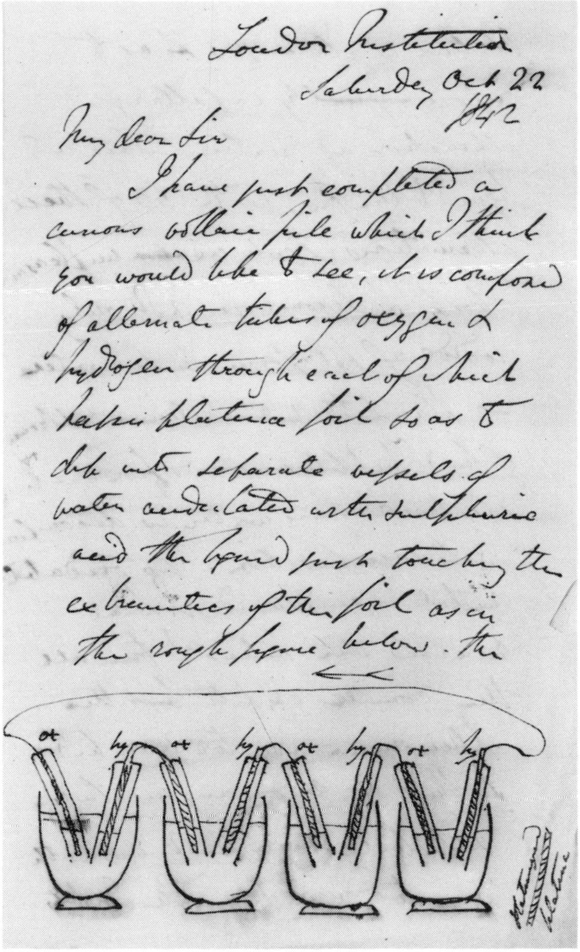

its insolubility as an anode and the complex chemistry of its salts presented severe technical problems. These were nevertheless tackled with enthusiasm by a number of scientists in England, France and Germany and attempts to develop a reliable process continued throughout most of the nineteenth century until satisfactory standards of quality were achieved.

But it was one of Daniell's old students in the chemistry department at King's College who tackled the problems most successfully. This was Alfred Smee whose grandfather and father had both been employed by the Bank of England, the father becoming Chief Accountant in 1831. At that time the Chief Accountant lived with his family in an official residence in the Bank, the necessary security confining them at home in the evening hours. On leaving college young Alfred therefore set up a laboratory in a room leading out of the

Alfred Smee
1818–1877

Born in Camberwell, educated at St. Paul's School and King's College, London, and then trained as a surgeon at St. Bartholomew's Hospital, Smee carried out all his researches in electroplating in a room set aside for him in the Bank of England where his father lived as Chief Accountant. He was elected a Fellow of the Royal Society in 1841, the year after he published his "Elements of Electrometallurgy, or the Art of Working in Metals by the Galvanic Fluid" at the early age of 22. The distinguished surgeon Sir Astley Cooper interested himself in the career of the young Alfred Smee, and it was upon his emphasising to the then Governor of the Bank of England that "you don't know what a treasure you have got in that young man" that Smee was given an appointment there as surgeon. Sir Astley was related by marriage to both Thomas Cock and Percival Norton Johnson and Smee's interest in platinum and palladium was no doubt encouraged and supported by them

family drawing room and next to the ledger office and here, working alone with elementary equipment, some lent by Daniell, he carried out a remarkable series of experiments in electrochemistry.

His first self-imposed task was to devise a battery more suitable for work in electrodeposition, and on February 28th, 1840 he was able to read a paper to the Royal Society, "On the Galvanic Properties of the Metallic Elementary Bodies with a description of a new Chemico-Mechanical Battery" (29).

Smee's battery relied upon his observation that an electrode having a roughened surface caused the hydrogen formed during the reaction to disperse, preventing the build-up of a film of bubbles. His cathodes were either of silver etched with acid or of platinum abraded with sandpaper, both being placed in a cell containing "nitro-muriate of platinum" as the electrolyte. This produced a thin layer of platinum in the form of a black powder, yielding an electrode simply and cheaply and one unaffected by the strength of acid in any cell. Banks of cells could be employed varying "from the size of a tumbler to a 10 to 12 gallon vessel". Its success lay, however, in its simplicity, and it aroused great interest in the art of depositing one metal upon another.

212

By the end of 1840 Smee, still only 22 years old, had compiled and published a remarkable text book, "Elements of Electrometallurgy", a term he himself coined. In this he described his processes for both platinum and palladium plating, writing that:

"Hitherto the reduction of these metals, in any other state than that of the black powder, has been always considered impossible."

He claimed that processes for "platinating and palladiating" rested upon the authority of his book and went on:

"Platinating metals by the galvanic current is a new feature in science. The process is similar in all respects of gilding but is more difficult. The solution of the nitro-muriate of platinum must be very weak, and the battery must be charged with dilute acid. The object to be coated must be very smooth, and thoroughly cleansed by potash, before the process is commenced. Having proceeded thus far, and the solution of platinum being ready, a very fine platinum wire, in connection with the silver of the battery, must be placed so as to dip into the solution, but must not be immersed beyond a very short distance. The object to be platinated is now ready for connection with the zinc of the battery, after which is effected, it is to be dipped in the solution. Immediately, oxygen gas will be given off from the platinum wire, in connection with the silver. From the copper or other metal to be platinated, no gas will be evolved, provided too much electricity be not generated. In a few minutes the object will be coated with platinum. It is needless to say that it has a beautiful appearance. It would

A sketch of Alfred Smee's laboratory made by his brother-in-law William Hutchinson and autographed by the sole occupant. Using apparatus lent by Professor Daniell and batteries of his own design employing platinised silver cathodes, he carried out here a long series of investigations on the electrodeposition of metals without assistance of any kind. The laboratory lay between the family drawing room and the ledger room of the Bank of England, and the occasional dropping of a ledger by a clerk would be sufficient to ruin one of Smee's experiments by breaking the contact with his battery

My working room
in the Bank of England
Alfred Smee

be of great value as a coating for telescopes, microscopes, quadrants, and a hundred other articles which must be exposed to the action of the weather." (30)

For palladium plating Smee used a similar electrolyte, nitro-muriate of palladium with a palladium anode:

"This metal is whiter than platinum, but not so bright as silver. It might be used in the same cases, and with the same advantages as platinum; and we have, besides, twice the bulk of metal in the same weight."

Smee's reputation as an electrochemist was rapidly established. He was elected a fellow of the Royal Society in June 1841, while earlier in the same year a special post, somewhat honorary in its duties, was created for him in the Bank of England. This appointment was made largely upon the recommendation of Sir Astley Cooper, the distinguished surgeon, and brother-in-law of Thomas Cock, who was a friend of Sir John Rae Reid, the Governor, and who had sometimes visited Smee's laboratory to see his experiments. He considered that the Bank should "turn Smee's scientific genius to good account", and accordingly the young man was appointed Surgeon to the Bank of England on January 1st, 1841. Astley Cooper's relationship with Percival Norton Johnson leaves little doubt that Smee's interest in platinum and palladium plating would have been encouraged by him, and that he supplied the necessary metals and salts. That they were well known to each other is established by Smee's acting as a steward at a dinner given in 1844 by the Society for Teaching the Blind, an organisation just established on the initiative of Johnson and his wife.

214

Smee's electrolyte could not have been very successful, and it was only one of a number put forward during the course of the next twenty years or so. His remark that platinum plating was "similar to gilding but more difficult" proved to be well founded. One worker who realised that because of the insolubility of the anode the solution must be periodically supplied with additions of a suitable compound was Henry Beaumont Leeson (1803–1872), a lecturer in chemistry at St. Thomas's Hospital in London, who filed a massive patent in 1842 (31) covering the electrodeposition of a whole range of metals and including the concept of agitating either the article to be plated or the plating solution in order to obtain smooth deposits at high current densities. He realised the cause of Smee's difficulties with the plating of platinum and palladium and pointed out that:

"The solution must be supplied with a fresh portion of the metal by adding to or placing within such solution or electrolytic fluid a further supply of some suitable salt to be dissolved or taken up from time to time as the fluid becomes exhausted."

In 1846 a further patent was filed by one George Howell of London and this contained the first reference to a solution that became known as the stabilised platinum electrolyte, made by dissolving platinum chloride in caustic soda and adding oxalic, citric, tartaric or acetic acid followed by caustic potash (32).

Lastly, so far as the period at present under discussion is concerned, came Thomas Hetherington Henry (1816–1859) who was for many years from 1837 employed as chemist in the brewery of Truman Hanbury and Buxton in Spitalfields, London, later setting up as a consulting analytical chemist in Lincoln's Inn. He was another founder member of the Chemical Society and was elected F.R.S. in 1846. There is no record of the process he employed but preserved among the John Percy collection of specimens at the Royal School of Mines is a thin sheet of copper plated on both sides with palladium and accompanied by a note in Professor Percy's hand stating that he was given it by the late T. H. Henry. The beams of two Oertling balances shown at the Great Exhibition of 1851 were also plated by Henry, one with platinum and the other with palladium (33).

All these investigations, together with similar researches carried out in France and Germany and the later and more successful developments in the field have recently been described in detail by Dr. Peta Buchanan (34).

Conclusion

Thus the three decades from the time when Wollaston had to abandon his production of malleable platinum until about 1850 were dominated on the one hand by the scientific genius of Michael Faraday, later supported by several of his disciples, and on the other by the metallurgical skill and enterprise of Percival Norton Johnson. In 1845, however, the younger Cock's health was giving him great trouble and he retired from his partnership, the firm again becoming P. N. Johnson and Company. He was succeeded in the platinum work by a young man named George Matthey who had entered the business as an

215

To celebrate the production of a particularly large ingot of platinum in 1850 Percival Norton Johnson had the lid of a snuff-box made from part of the rolled sheet. The body of the box is in silver bearing the Birmingham hallmark for that year, while the lid carries a relief showing the Roman Consul Lucius Junius Brutus condemning his sons to death for conspiracy. It has now been passed to the Victoria and Albert Museum.

apprentice in 1838 at the age of thirteen. His extraordinarily long career and his great achievements in building up the platinum industry in England will be related in a later chapter, but one last reference to Johnson should be included here. By 1850 the size of platinum ingots had increased considerably and had exceeded the capabilities of the London metal workers to roll them. In that year, to celebrate the production of his largest ingot so far, Johnson took it to Birmingham, in company with his brother-in-law Smith (who had now taken his mother's maiden name of Sellon) to have it rolled. From a piece of the resulting sheet he had made the lid of a snuff-box showing in relief the Roman Consul Lucius Junius Brutus condemning his two sons to death. The snuff box, the body of which is made in silver, is now preserved in the Victoria and Albert Museum in London.

References for Chapter 11

1   D. McDonald, Percival Norton Johnson, London, 1951
2   Letter from P. N. Johnson dated March 11th, 1846, Archives of the Royal Society; abridged in *Phil. Mag.*, 1846, **29**, 130
3   Letter from Sir Astley Cooper to Sir William Knighton, June 25th, 1826, Royal Archives, Windsor Castle, RA/26296
4   P. N. Johnson, *Mining J.*, 1835, **1**, 83
5   W. M. Poussett, *Mining J.*, 1836, **3**, 75
6   W. A. Lampadius and P. N. Johnson, *J. Prakt. Chem.*, 1837, **11**, 309–315
7   J. Stodart and M. Faraday, *Quarterly J. Science*, 1820, **9**, 319–329
8   J. Stodart and M. Faraday, *Phil. Trans. Roy. Soc.*, 1822, **112**, 253–270
9   Letter from Faraday to de la Rive, June 26, 1820, quoted in L. Pearce Williams, The Selected Correspondence of Michael Faraday, Cambridge, 1971, **1**, 117–120
10 Sir Robert Hadfield, Faraday and his Metallurgical Researches, London, 1931
11 Letter from Green Pickslay & Co., to Faraday, Archives of the Royal Institution
12 J. Foley, History of the Invention and Making of Foley's Diamond Pointed Gold Pens, New York, 1875, 58–60
13 M. Faraday, Glass Furnace Notebook, *Roy. Soc. MSS*, October 8, 1828, 234
14 *ibid*, January 16, 1829, 287
15 M. Faraday, *Phil. Trans. Roy. Soc.*, 1830, **120**, 1–58
16 P. N. Johnson, *Mining J.*, 1843, **13**, 182
17 C. H. Schumacher, *Phil. Trans. Roy. Soc.*, 1836, **126**, 457–495
18 W. H. Miller, *Phil. Trans. Roy. Soc.*, 1856, **146**, 753–945; *Phil. Mag.*, 1856, **12**, 540–552
19 Public Record Office, Mint 4/48, 49
20 W. J. Cock, *Proc. Chem. Soc.*, 1843, **1**, 161–164
21 M. Faraday, *Phil. Trans. Roy. Soc.*, 1846, **136**, 41–62
22 M. Faraday, Royal Institution Archives, December 12–19, 1845
23 J. F. Daniell, *Phil. Trans. Roy. Soc.*, 1836, 107–124
24 W. R. Grove, *Phil. Mag.*, 1839, **14**, 388–390
25 W. R. Grove, British Assocn. Report, 1839, 36–38
26 W. R. Grove, *Phil. Mag.*, 1839, **14**, 129–130
27 W. R. Grove, *Phil. Mag.*, 1842, **21**, 417–420
28 E. M. Smee (Mrs. Odling), Memoirs of Alfred Smee, London, 1878
29 A. Smee, *Phil. Mag.*, 1840, **16**, 315–321
30 A. Smee, Elements of Electrometallurgy, London, 1840, 94
31 H. B. Leeson, British Patent 9374, 1842
32 G. Howell, British Patent 11,065, 1846
33 Report of the Juries, Exhibition of 1851, 258
34 P. D. Buchanan, *Platinum Metals Rev.*, 1981, **25**, 32–41

Johann Wolfgang Döbereiner
1780–1849

Born the son of a coachman in Bavaria, Döbereiner first practised as an apothecary and then, to his great surprise, was invited to become Professor of Chemistry at Jena where he also acted as chemical adviser to Goethe. His discovery of the power of finely divided platinum to promote the oxidation of alcohol and to synthesise water from hydrogen and oxygen caused a great sensation in the chemical world and founded the study of catalysis

From a portrait in the possession of
Goethe-Nationalmuseum der Nationale Forschungs
und Gedenkstätten, Weimar

12

The Discovery and Early History of Catalysis

"I have tried to produce these phenomena with various metals but I have succeeded only with platinum and palladium."

HUMPHRY DAVY

The point had now been reached when the more important physical properties of platinum and some of its associated metals had become reasonably well understood and had formed the basis of several applications. One major property, and one that was later to make a most significant contribution to chemical industry, was, however, yet to be discovered.

As is so often the case, this discovery came incidentally from a quite different investigation. A disastrous explosion in a coal mine in the North of England in 1812, shortly followed by a number of similar explosions, prompted an appeal for advice to Humphry Davy at the Royal Institution and resulted in his well-known researches on flame and his invention of the miner's safety lamp. In a paper read to the Royal Society on January 23rd 1817 Davy described his experiments on the increase in the limits of combustibility of mixtures of coal gas and air with increasing temperature:

"For this purpose, I introduced a small wire-gauze safe-lamp with some fine wire of platinum fixed above the flame, into a combustible mixture containing the maximum of coal gas, and when the inflammation had taken place in the wire-gauze cylinder, I threw in more coal gas, expecting that the heat acquired by the mixed gas in passing through the wire-gauze would prevent the excess from extinguishing the flame. The flame continued for two or three seconds after the coal gas was introduced; and when it was extinguished, that part of the wire of platinum which had been hottest remained ignited, and continued so for many minutes.

It was immediately obvious that this was the result which I had hoped to attain by other methods, and that the oxygen and coal gas in contact with the wire combined without flame, and yet produced heat enough to preserve the wire ignited, and to keep up their own combustion. I proved the truth of this conclusion by making a similar mixture, heating a fine wire of platinum and introducing it into the mixture. It immediately became ignited nearly to whiteness, as if it had been itself in actual combustion, and continued glowing for a long while, and when it was extinguished, the inflammability of the mixture was found entirely destroyed.

219

The opening page of the manuscript of Sir Humphry Davy's paper read to the Royal Society on January 23rd, 1817, "Some new experiments and observations on the combination of gaseous mixtures, with an account of a method for keeping a continued light in mixtures of inflammable gases and air without flame". The paper goes on: "I had intended to expose fine platinum wires to oxygen and olefiant gas and to oxygen and hydrogen during their slow combination under different circumstances, when I was accidentally led to the discovery of a new and curious series of phenomena." Davy had discovered heterogeneous catalytic oxidation but unfortunately he did not appreciate the significance of his findings

By courtesy of the Royal Institution

I have tried to produce these phenomena with various metals; but I have succeeded only with platinum and palladium; with copper, silver, iron, gold and zinc, the effect is not produced." (1)

Thus Davy discovered the phenomenon of heterogeneous catalytic oxidation and although he obtained the same results with ether and alcohol, obtaining an acidic product that Faraday and later Daniell identified as a mixture of acetaldehyde and acetic acid, he did not carry further the study of this profoundly important effect. His mind was fully occupied with problems of combustion, he was by now fully satisfied with the safety-lamp and he tended to regard the phenomenon he had discovered, as he wrote to one of his colliery manager friends, as "more like magic than anything I have seen ... it depends upon a perfectly new principle in combustion" (2).

The Researches of Edmund Davy

Despite this Davy's paper aroused great interest; it was translated into German and French and it also encouraged his cousin Edmund Davy to take up the matter. Edmund, seven years younger than Humphry, had been engaged as the

Edmund Davy
1785–1857

A younger cousin of Humphry Davy, Edmund first assisted the latter at the Royal Institution and then held the chair of chemistry at the Royal Cork Institution. Here he prepared finely divided platinum and found that it would oxidise alcohol vapour at room temperature, sufficient heat being generated to raise the metal to a white heat. Again, Edmund Davy failed to grasp the significance of his discovery, but it led immediately to the much more effective researches of Döbereiner

From a portrait in the possession of the Royal Dublin Society

latter's assistant in 1804 and then in 1831 had secured an appointment as Professor of Chemistry at the Royal Cork Institution. While still at the Royal Institution in London he had made several studies in the chemistry of platinum and published long papers in *The Philosophical Magazine* (3). In 1817 he carried these researches further and presented a paper to the Royal Society on a new fulminating compound of platinum (4). He stood much in the shadow of Humphry, and when in 1820 he published a paper describing the preparation of finely divided platinum by reducing a solution of platinum sulphate with alcohol and its great activity at room temperature in the oxidation of a further quantity of alcohol, he merely wrote:

> "In this case the acid first noticed by Sir H. Davy (in his beautiful experiment of the ignited platinum wire, and since more fully examined by Mr. Daniell) is produced ... This mode of igniting a metal seems to be quite a new fact in the history of chemistry, but the means of keeping it in a state of ignition is only another illustration of the facts previously pointed out by Sir H. Davy, in his late valuable researches." (5)

Johann Wolfgang Döbereiner

Edmund Davy's paper was reproduced in German in Schweigger's *Journal für Chemie* in the spring of the following year and attracted the attention of Johann

221

Wolfgang Döbereiner who was to open up a much more vigorous study of the phenomenon and to cause quite a sensation among his scientific contemporaries. Döbereiner, born the son of a coachman in 1780, first served an apprenticeship to an apothecary and then spent five years as an apothecary's assistant in Karlsruhe and Strasburg, attending lectures on chemistry and mineralogy in his free time. Opening his own pharmaceutical manufacturing business, he began to contribute papers to Gehlen's *Neues Allgemeine Journal für Chemie* but his business failed and he was left impoverished.

In 1810, to his great surprise, Döbereiner was appointed to the Professorship of Chemistry in the University of Jena, the chair created by Carl August, Duke of Saxe-Weimar-Eisenach in 1789 and now left vacant by the death of its first holder, Johann Göttling. The appointment was made by the Duke, an enlightened patron of the arts and sciences, together with his Minister of State Goethe on the recommendation of Gehlen, and Döbereiner remained for the rest of his life most grateful for this opportunity, refusing several offers of chairs in other universities. An extraordinary friendship grew between Döbereiner, the Grand Duke (as he became in 1815) and Goethe, both interested in chemistry.

By the August of 1821 he had not only repeated Edmund Davy's experiments but had fully appreciated their significance, rightly regarding the important discovery as the activity of the platinum rather than, as had both the Davys, as the action upon it of the alcohol. He believed at the time that Edmund Davy's product was a sub-oxide of platinum and he wrote:

> "The platinum sub-oxide moreover, does not undergo any change during this transformation of the alcohol and can immediately be used again to acidify fresh, perhaps limitless, quantities of alcohol . . . a circumstance that permits its use for the large scale preparation of acetic acid." (7)

In July 1823 he turned his attention to the metal itself, and prepared platinum in powder form by heating ammonium chloroplatinate. This he found would ignite a mixture of hydrogen and either air or oxygen even at room temperature or below:

> "There now followed in a few moments that strange reaction; the volume of the gases diminished and after ten minutes all the oxygen in the admitted air had condensed with the hydrogen to water." (8)

He also gave some thought to the most suitable form to adopt for the platinum powder and made use of small moulded pellets of potter's clay impregnated with platinum – the first example of a supported catalyst.

Döbereiner was quick to make his findings known. Two days after this experiment – which he repeated "at least thirty times that day and always with the same result" – he wrote to Goethe:

> "Permit me, your Excellency, to give you news of a discovery that seems to be important in the highest degree from the points of view of both physics and electrochemistry." (9)

He also wrote similarly to Lorenz Oken, the editor of a scientific journal *Isis*,

222

Döbereiner began his career as Professor of Chemistry in rooms in the palace of his patron, Duke Carl August of Saxe-Weimar-Eisenach, but these soon proved inadequate and in 1816 a large house was acquired, shown on the right of this illustration, for use as both a laboratory and a home. Later, in 1833, a new laboratory, visible behind the trees, was built for him to plans drawn up by the Duke's Minister of State, Goethe

published in Jena, and sent detailed accounts to his friend J. S. C. Schweigger in Bayreuth and to L. W. Gilbert in Leipzig who both published his letters in their respective journals (10), while he quickly produced a small book, "Über neu Entdeckte höchst Merkwürdige Eigenschaften des Platins," published in Jena, also in 1823.

On August 3rd Döbereiner made an even more striking experiment by merely directing a stream of hydrogen at the platinum powder so that it was mixed with air before reaching its target. The platinum immediately became white hot and ignited the hydrogen. This, reported at once to the editors already mentioned, and published by them as appendices to his first announcement (10), caused even more excitement. An unnamed friend in Paris wrote to Döbereiner to say that his news had aroused "a great sensation here and excited the liveliest interest", while his patron, the Grand Duke Carl August wrote to him:

"I am delighted that your splendid discovery excites the attention of foreign countries. I return to you the letter from Paris and thank you for the published paper." (11)

At the second annual meeting, held in Halle in September, of the Gesellschaft deutscher Naturforscher und Ärtze, Döbereiner presented a paper and, to the surprise and delight of the members, gave a practical demonstration. Among his more significant comments was:

"Most likely a new natural principle is operative here that will become apparent through further investigation" (12).

An incidental result of Döbereiner's discoveries was his invention of the first lighter. This employed hydrogen, generated from zinc and sulphuric acid, passing over finely divided platinum which then glowed sufficiently to ignite the gas. A number of different types produced in Germany and in England became very popular and many thousands of them were in use over a long period of time. Döbereiner refused to file a patent for his lighter with the comment " I love science more than money". These are but three of the many designs that were produced

By courtesy of the Science Museum

The reaction in France was also swift. On Sunday August 24th a brief notice of Döbereiner's work was published in a daily newspaper, the *Journal des Debats*, including the comment that

"This beautiful discovery is going to open up a new field of research in physics and chemistry." (13)

This also quickly attracted attention, while Professor Karl Kastner of the University of Erlangen wrote to his former student Liebig, then studying under Gay-Lussac in Paris, to give an account of Döbereiner's work. Liebig immediately passed on the news to the Académie des Sciences through Louis Jacques Thenard, Professor of Chemistry at the École Polytechnique, who had earlier been interesting himself in reactions such as the decomposition of

224

Louis Jacques Thenard
1777–1857

Born in the village of La Louptiere, now known in his honour as La Louptiere-Thenard, near Nogent-sur-Seine, Thenard came to Paris at the age of seventeen during the Reign of Terror to study pharmacy. His keenness brought him to the notice of Vauquelin and Fourcroy, both of whom advanced his career. In 1804 he replaced Vauquelin in the chair of chemistry at the College de France while in 1810 he was appointed Professor at the École Polytechnique and succeeded Fourcroy in the Académie des Sciences. Immediately the news of Döbereiner's discovery reached Paris he and his colleague Dulong began to investigate the effects of platinum, palladium, rhodium and other metals on the combustion of hydrogen and oxygen

ammonia over heated metals. Thenard was most excited by this development and, together with his younger collaborator Pierre Louis Dulong, he immediately began an investigation into the new phenomenon by studying the effects of heated solid metals on inflammable mixtures of gases. They obtained samples of palladium and rhodium from Wollaston, and they demonstrated that platinum, palladium, rhodium, cobalt, nickel, gold and silver required increasingly high temperatures, in that order, to bring about the combination of hydrogen and oxygen. As promptly as September 15th they read a first report to the Académie and through the influence of their friend Gay-Lussac, the editor, this was squeezed into the September issue (in small type) of the *Annales de Chimie* (14) before Döbereiner's paper had appeared in French translation in the same journal.

Shortly afterwards Thenard and Dulong produced palladium, rhodium and iridium in powder form and found that they were effective at room temperature, this result being reported to the Académie only a week after their original paper was read (15).

During the winter of 1823 further research was carried out by Adolf Pleischl the Professor of Chemistry in the University of Prague, who was anxious to discover why it was that the experiments sometimes failed. He concluded quite rightly that the important factors were the degree of fineness and the porosity of

225

Adolph Martin Pleischl
1787–1867

A native of Bohemia and Professor of Chemistry at the University of Prague from 1815 until 1833 when he moved to Vienna, Pleischl quickly followed up Döbereiner's work and in November 1823 he established that the activity of the platinum powder depended upon its fineness. He found that platinum powder that had previously been effective failed if pressed hard together, while an especially fine powder worked extremely well at a lower temperature than usual

his platinum powder, a finding that drew an appreciative letter from Döbereiner, published as an appendix to the paper. A few weeks later he established that palladium also behaves in a similar way to platinum but with lower activity, and he added that unfortunately he had no iridium, rhodium or osmium available to him although he felt that they might well exhibit the same property (17).

In his well known annual reports on the progress of chemistry Berzelius gave the highest praise of which he was capable to Döbereiner for the year 1823:

> "From any point of view the most important and, if I may use the expression, the most brilliant discovery of last year is, without doubt, that fine platinum powder has the ability to unite oxygen and hydrogen even at low temperatures." (18)

One by-product of Döbereiner's work was the invention of the lighter that still bears his name, the Döbereiner Feuerzeug. In this hydrogen, generated from zinc and sulphuric acid, passed through a nozzle over a small amount of finely divided platinum held on thin platinum wires and became ignited, the flame then being passed to a candle. Many thousands of these lighters were produced in Germany and in England but Döbereiner refused to accept any financial reward for his invention.

226

For a more detailed account of Döbereiner's life and work a most valuable source is to be found in an unpublished thesis by Dr. Peter Collins and in a shorter version of this published in 1976 (19) upon which the present writer has drawn heavily and gratefully in the foregoing section.

Further Investigations in Great Britain

On the day following the reading of Dulong and Thenard's first paper to the Académie des Sciences the French physicist J. N. P. Hachette (1769–1834) wrote one of his many letters to Michael Faraday (20). The concluding paragraph reads in translation:

"I saw demonstrated yesterday the beautiful experiment of Döbereiner, a German scientist (of Stuttgart, I think). You are no doubt familiar with it. It consists in directing a current of hydrogen on to platinum powder obtained from the solution of this metal in aqua regia by precipitation with the ammonium salt. The hydrogen gas inflames, by simple contact. In your hands this fact will not be the last of its kind".

Faraday was of course highly interested in this piece of news, and by September 27th he had prepared platinum in powder form by heating ammonium chloroplatinate and had repeated Döbereiner's experiment to his entire satisfaction (21). He at once wrote a short note for the October issue of the *Quarterly Journal of Science*, edited from the Royal Institution by Professor William Brande, Davy's successor. In this he wrote:

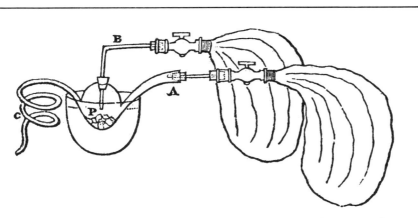

Concerned at the lack of detail in Döbereiner's account of his experiment on the combination of oxygen and hydrogen, William Herapath, an analytical chemist who later became Professor of Chemistry and Toxicology at the Bristol School of Medicine, repeated the experiments in October 1823, using the apparatus shown here, and read a paper to the newly-formed Bristol Philosophical Society of Inquirers very soon after the publication of Faraday's brief note. Hydrogen and oxygen were supplied from the two bladders and passed over the spongy platinum in the vessel P held in a capsule

227

"A most extraordinary experiment has been made by M. Döbereiner. It was communicated to me by M. Hachette; and having verified it I think every chemist will be glad to hear its nature" (22).

Later in the same year Faraday prepared a review of the work so far carried out, "On the Action of Platina on Mixtures of Oxygen, Hydrogen and other Gases" (23). In this, after describing Döbereiner's experiments, he reviewed Dulong and Thenard's papers most favourably and went on to mention a paper by William Herapath of Bristol who had repeated Döbereiner's experiments in October 1823 and confirmed that the platinum underwent no chemical change and that other metals did not give the same effect (24), and to one Alexander Garden, a chemist of Oxford Street, London, who had shown that the black residue left when crude platinum is dissolved in aqua regia was also active in the same way (25). However, Faraday turned to other matters and did not return to this subject for a further ten years, as will appear later in this chapter.

In 1825 Humphry Davy made his only reference to the phenomenon since his original discovery, and in a rather obscure place. In 1818 he had published a book "On the Safety Lamp . . . with Some Researches on Flame", but in 1825 he was moved to publish a second edition because his experience, as he wrote in the preface

"having shewn that the precautions which it was intended to describe either are not known or are not attended to, I have thought it might assist the cause of humanity to advertise the book a second time" (26).

This edition contained several appendices, and in Appendix No. 2 he returned to the subject and to the later work of Döbereiner with finely divided platinum such as Edmund Davy had used, and of Dulong and Thenard, who had shown that

"various metals in a finely divided state have the same property of hastening or producing combinations at a lower temperature than those at which they usually occur".

Davy continued:

"It is probable that the rationale of all these processes is of the same kind. Whenever any chemical operation is produced by an increase of temperature, whatever occasions an accumulation of heat, must tend to give greater facility to the process; a very thick wire of platinum does not act upon a mixture of oxygen and hydrogen, at a heat below redness; but if beat into thin laminae, it occasions its combustion at the heat of boiling mercury, and, when in the form of the thinnest foil, at usual temperatures. I cooled the spongy platinum to 3° of Fahr., and still it inflamed hydrogen nearly of the same temperature, issuing from a tube cooled by salt and ice.

It may be supposed that the spongy platinum absorbs hydrogen, or that it contains oxygen; but neither of these hypotheses will apply to the fact that I first observed, of the ignition of fine wires in different mixtures of inflammable gases and air, at temperatures so far below ignition."

Immediately after the publication of this second edition Davy became ill and exhausted, and after four years of wandering in Europe he died in Geneva at the early age of 50 on May 29, 1829.

228

In the meantime William Henry (1775–1836), the Manchester chemist and manufacturer, had carried out an investigation on the use of Döbereiner's reaction in the analyses of gases, using the latter's pellets of china clay and platinum sponge (27) while Edward Turner (1796–1857), later to become the first professor of chemistry at University College, London, but then an extra-mural lecturer in Edinburgh, read a paper to the Royal Society of that city on the applications of Döbereiner's discovery to eudiometry (28). Again he made use of small pellets of pipe-clay mixed with spongy platinum, dried and then fired in a spirit lamp. He found that the action of platinum "affords a neat and expeditious method of ascertaining the purity of hydrogen or oxygen", but could not agree with Henry that platinum could be helpful in gas analysis.

The Contact Process for Sulphuric Acid

As early as 1831 the industrial possibilities offered by heterogeneous catalytic oxidation were grasped by a young man engaged, with his father and the latter's partner, in the manufacture of vinegar in Bristol, an enterprise started in 1824

A.D. 1831 . . . . . . Nº 6096.

Manufacture of Sulphuric Acid.

PHILLIPS' SPECIFICATION.

TO ALL TO WHOM THESE PRESENTS SHALL COME, I, PEREGRINE PHILLIPS, Junior, of Bristol, Vinegar Maker, send greeting.

WHEREAS His present most Excellent Majesty King William the Fourth, by His Letters Patent under the Great Seal of Great Britain, bearing date
5 at Westminster, the Twenty-first day of March, in the first year of His reign, did, for Himself, His heirs and successors, give and grant unto me, the said Perigrine Phillips, His especial licence, sole privilege and authority, that I, the said Peregrine Phillips, my exors, admors, and assigns, or such others as I, the said Peregrine Phillips, my exors, admors, or assigns, should
10 at any time agree with, and no others, from time to time and at all times during the term of years therein mentioned, should and lawfully might make, use, exercise, and vend, within England, Wales, and the Town of Berwick-upon-Tweed, my Invention of " CERTAIN IMPROVEMENTS IN MANUFACTURING SULPHURIC ACID COMMONLY CALLED OIL OF VITRIOL;" in which said Letters Patent
15 is contained a proviso that I, the said Peregrine Phillips, shall cause a particular description of the nature of my said Invention, and in what manner the same is to be performed, to be inrolled in His said Majesty's High Court of Chancery within six calendar months next and immediately after the date of the said in part recited Letters Patent, as in and by the same,
20 reference being thereunto had, will more fully and at large appear.

NOW KNOW YE, that in compliance with the said proviso, I, the said Peregrine Phillips, junior, do hereby declare that the nature of my said improvements are herein set forth and explained; but for the better understanding

The patent specification filed by the mysterious Peregrine Phillips in 1831 for the manufacture of sulphuric acid by passing sulphur dioxide and air through a platinum tube containing finely divided platinum. Apart from the fact that he was the son of a vinegar manufacturer in Bristol nothing is known of him or of how he acquired his knowledge of chemistry. A year after the patent was granted the vinegar business closed and Peregrine Phillips, although still a young man, was never heard of again

229

and trading as Phillips, Thorne and Co. Father and son were both named Peregrine Phillips, and the younger one withdrew from the partnership in 1831. Unfortunately nothing is known about this young man despite a meticulous research carried out in the local records in 1926 by Sir Ernest Cook, a distinguished chemist who became Lord Mayor of Bristol, except that his father had been a tailor from 1803 until the establishment of the vinegar firm and that only a year after the son's withdrawal the business was abandoned and the premises sold at auction. (29)

None the less Peregrine junior must have acquired an adequate knowledge of chemistry to become one of the relatively few men to devise an entirely new chemical process, in his case the manufacture of sulphuric acid. His patent No. 6096 covers "Certain Improvements in Manufacturing Sulphuric Acid commonly called Oil of Vitriol." The specification states:

> "The first improvement then, namely, the instantaneous union of sulphurous acid with the oxygen of the atmosphere, I effect by drawing them in proper proportions by the action of an air pump or other mechanical means thro' an ignited tube or tubes of platina, porcelain, or any other material not acted on when heated by the sulphurous acid gas. In the said tube or tubes I place fine platina wire or platina in any finely-divided state, and I heat them to a strong yellow heat, and by preference in the chamber of a reverberatory furnace; and I do affirm that sulphurous acid gas being made to pass with a sufficient supply of atmospheric air through tubes as described, properly heated and managed, will be instantly converted into sulphuric acid gas, which will be rapidly absorbed as soon as it comes into contact with water."

In the following year Professor Gustav Magnus of the University of Berlin published a brief review of Phillips' proposed process, confirming his findings but drawing attention to the fact that the platinum powder was effective only at high temperatures (30). It was, however, to be another seventy years before the process was adopted in industry after some further research into the conditions required for commercial success; details of this are set out in Chapter 21.

Faraday's Return to the Subject

Ten years after his review of the work of Döbereiner and Thenard, Faraday had occasion to take up again the curious phenomenon he had then discussed. In the middle of his classic researches on electrochemistry, believing that he had established the constant chemical action of a given quantity of electricity and busily constructing his new voltameter, he was mystified by the peculiar reaction of the platinum electrodes on the gases produced in electrolysis:

> "I was occasionally surprised at observing a deficiency of the gases resulting from the decomposition of water and at last an actual disappearance of portions which had been evolved, collected and measured."

His immediate reaction was that this arose from a completely new phenomenon that required thorough investigation, and in the long paper he read to the Royal Society in January 1834 (31) he wrote:

"These experiments reduced the phenomenon to the consequence of a power possessed by the platina, after it had been the positive pole of a voltaic pile, of causing the combination of oxygen and hydrogen at common, or even at low, temperatures. This effect is, as far as I am aware, altogether new."

He went on to emphasise that this property of inducing combination was the same as that discovered by Döbereiner "to belong in so eminent a degree to spongy platina", and then turned to the discussion of a theory advanced some years earlier by a physician in Florence, Ambrogio Fusinieri (1773–1853). In his paper (32) Fusinieri contended that during the reactions first observed by Humphry Davy "concrete laminae" of the combustible substance could be seen on the platinum surface, these then disappearing by burning. He proposed the concept of "native caloric" (calorico nativo) as an explanation.

Faraday had learnt Italian during his European tour with Sir Humphry and Lady Davy in the years 1813 to 1815, but he was puzzled by the rather archaic language in which Fusinieri's paper was written and confessed that he could not form "a distinct idea of the power to which he refers the phenomena". (Faraday's difficulties in this connection have been discussed more recently by Professor A. J. B. Robertson and two of his colleagues (33)).

The theory then advanced by Faraday involved the simultaneous adsorption of both reactants on the platinum surface, the only essential condition being a perfectly clean and metallic surface:

"The effect is evidently produced by most, if not all bodies, weakly perhaps by many of them, but rising to a high degree in platina . . . The platina is not considered as causing the combination of any particles with itself, but only associating them closely around it."

Although a century and a half later this phenomenon remains something of a mystery to chemists, Faraday was indeed close to the truth of the matter.

The Enterprise of Frédéric Kuhlmann

Important investigations into the production of both nitric and sulphuric acids by passing the appropriate gases over heated platinum were carried out in France by an energetic and enterprising chemist and manufacturer, Frédéric Kuhlmann (1803–1881). Combining the Professorship of Industrial Chemistry at the University of Lille with industrial projects after studying under Vauquelin, he had studied the work of Humphry and Edmund Davy, Döbereiner and Thenard. He was also greatly interested in the part played by nitrogen products in agriculture, and in fact in later years carried out investigations on the use of ammonium salts as fertilisers on his own estate.

With this as his background, Kuhlmann went to work vigorously on the nitric acid problem, and found that by passing a mixture of ammonia and air over platinum sponge heated to about 300°C in a glass tube he obtained nitric acid. He filed a patent application for this invention in December 1838. (34)

At this time saltpetre was readily and cheaply available, and the new process could not offer any commercial advantage, but in the course of a paper given to the Académie in the same year, Kuhlmann made these prophetic remarks:

> "If in fact the transformation of ammonia to nitric acid in the presence of platinum and air is not economical, the time may come when this process will constitute a profitable industry, and it may be said with assurance that the facts presented here should serve to allay completely any fears felt by the government on the difficulty of obtaining saltpetre in sufficient quantities in the event of war." (35)

On the same day as his patent for ammonia oxidation was granted Kuhlmann secured a second patent for the manufacture of sulphuric acid by the oxidation of sulphur dioxide over finely divided platinum supported on glass or other material not attacked by the acid (36). He was probably aware of Peregrine Phillips' patent of 1831, filed only in England, but unlike Phillips he was able to put his idea into practice in the chemical firm that he founded, Etablissements Kuhlmann at Loos, although it was again very many years before it became a full scale commercial process.

Berzelius Coins the Term Catalysis

It may have been noticed that throughout this chapter, other than in its heading, the word "catalysis" has hardly been used. It was not in fact ever employed by those whose researches have been described, but was coined by J. J. Berzelius only in 1836. In his annual report to the Swedish Academy of Sciences for the previous year he reviewed some of the results of these workers and realising that there must be a common cause for this effect, he coined a name for it, writing:

> "This is a new power to produce chemical activity belonging to both inorganic and organic nature, which is surely more widespread than we have hitherto believed and the nature of which is still concealed from us. When I call it a new power, I do not mean to imply that it is a capacity independent of the electrochemical properties of the substance. On the contrary, I am unable to suppose that this is anything other than a kind of special manifestation of these, but as long as we are unable to discover their mutual relationship, it will simplify our researches to regard it as a separate power for the time being. It will also make it easier for us to refer to it if it possesses a name of its own. I shall therefore, using a derivation well-known in chemistry, call it the ***catalytic power*** of the substances, and decomposition by means of this power ***catalysis***, just as we use the word analysis to denote the separation of the component parts of bodies by means of ordinary chemical forces. Catalytic power actually means that substances are able to awaken affinities which are asleep at this temperature by their mere presence and not by their own affinity" (37).

Thus we owe to Berzelius, to his perception and his gift for enriching chemistry with appropriate terminology, a term that has come to signify a phenomenon of vital and far-reaching importance in modern chemical industry and one associated pre-eminently with the platinum metals.

References for Chapter 12

1   H. Davy, *Phil. Trans. Roy. Soc.*, 1817, **107**, 77–85

2   Letter from Davy to John Buddle, British Museum Add. MSS 33963 f 114

3   E. Davy, *Phil. Mag.*, 1812, **40**, 209–220; 263–278; 350–365

4   E. Davy, *Phil. Trans. Roy. Soc.*, 1817, **107**, 136–157

5   E. Davy, *Phil. Trans. Roy. Soc.*, 1820, **110**, 108–125

6   E. Davy, *J. für Chem. (Schweigger)*, 1821, **31**, 340–356

7   J. W. Döbereiner, *Ann. Phys. (Gilbert)*, 1822, **72**, 193–198

8   J. W. Döbereiner, *J. für Chem. (Schweigger)*, 1823, **38**, 321–326

9   J. Schiff, Briefwechsel zwischen Goethe und Johann Wolfgang Döbereiner, Weimar, 1914, letter 94, July 29, 1823

10 *J. für Chem. (Schweigger)*, 1823, **38**, 321–326; *Ann. Phys. (Gilbert)*, 1823, **74**, 269–273

11 O. Schade, Briefe des Grosherzogs Carl August und Goethe an Döbereiner, Weimar, 1856, letter 21, August 9, 1823

12 J. W. Döbereiner, *J. für Chem. (Schweigger)*, 1823, **39**, 3–4

13 *J. des Debats*, 1823, August 24, 4

14 P. L. Dulong and L. J. Thenard, *Ann. Chim.*, 1823, **23**, 440–443

15 P. L. Dulong and L. J. Thenard, *Ann. Chim.*, 1823, **24**, 380–387

16 A. Pleischl, *J. für Chem. (Schweigger)*, 1823, **39**, 142–159

17 A. Pleischl, *J. für Chem. (Schweigger)*, 1823, **39**, 201–205; 351–356

18 J. J. Berzelius, Jahres-Bericht, 1825, **4**, 60–61

19 P. Collins, The Development of Heterogeneous Catalysis, 1817–1823, Thesis submitted in June 1974, University of Oxford: *Ambix*, 1976, **23**, 96–115

20 Letter from J. N. P. Hachette to Faraday, September 16, 1823, I.E.E. Archives

21 M. Faraday, Royal Institution Laboratory Notebook No. 8, 81

22 M. Faraday, *Quart. J. Science*, 1823, **16**, 179

23 M. Faraday, *Quart. J. Science*, 1823, **16**, 375–377

24 W. Herapath, *Phil. Mag.*, 1823, **62**, 286–289

25 A. Garden, *Ann. Phil.*, 1823, **22**, 466–467

26 H. Davy, On the Safety Lamp for Preventing Explosives in Mines, Houses Lighted by Gas, Spirit Warehouses, or Magazines in Ships etc., with Some Researches on Flame, 1825, Appendix 2, pp 148–151

27 W. Henry, *Phil. Trans. Roy. Soc.*, 1824, **114**, 266–289

28 E. Turner, *Edin. Phil. J.*, 1824, **11**, 99–113; 311–318

29 Sir Ernest Cook, *Nature*, 1926, **117**, 419–421

30 H. G. Magnus, *Ann. Phys., (Poggendorff)*, 1832, **24**, 610–612

31 M. Faraday, *Phil. Trans. Roy. Soc.*, 1834, **124**, 55–122

32 A. Fusinieri, *Giornali di Fisica*, 1824, **7**, 371–376

33 M. Farinelli, A. L. B. Gale and A. J. B. Robertson, *Ann. Sci.*, 1974, **31**, 19–20

34 F. Kuhlmann, French Patent 11331

35 F. Kuhlmann, *Compte Rendus*, 1838, **7**, 1107–1110

36 F. Kuhlmann, French Patent 11332

37 J. J. Berzelius, *Jahres-Bericht*, 1836, **15**, 243; *Ann. Chim.*, 1836, **61**, 146–151

Peter Grigorievich Sobolevsky
1781–1841

The son of a surgeon, Sobolevsky served in the army until 1804
when he took up an administrative and technical career. In 1826
he was appointed Director of the large new chemical laboratory
just erected for the Mining Cadet Corps in St. Petersburg and
here he devised a successful process for the extraction and
refining of Russian platinum. His bust in the library of the Mining
Institute commemorates this achievement

13

The Foundation of the Russian Platinum Industry

*"Anxious to find a sure market for the new
precious metal whose production promised to
be rich and considerable, the Finance Minister
Count Kankrin planned to employ it as
coinage."*

ERNST KHRISTIANOVICH FRITSMAN

Some years before its actual or official discovery rumours had been circulating that platinum was to be found in Russia. In 1806, for example, Vauquelin wrote:

"A rumour has spread for some years that platinum has been discovered in Siberia but this has not yet been confirmed" (1).

Again, Guyton de Morveau, writing on platinum in 1810, mentioned that:

"it has also been let drop, without denial, and no doubt prematurely, that this metal exists in Siberia" (2).

It was not until 1819, however, that small pieces of a heavy white metal found in the gold fields on the eastern slopes of the Urals south of Ekaterinburg (now Sverdlovsk) were brought to the attention of anyone in authority. The Russian establishment was much more favourably placed than the Spanish authorities in Columbia had been to deal with such a discovery. Since about 1745 gold mining had been carried on in this area, and there was a mining laboratory in Ekaterinburg to which samples of the new white metal were sent for examination. In 1822 the Director of the laboratory, Ignatyevitch Varvinsky (1797–1832) began a study of these samples and finding that they contained platinum sent a further specimen to St. Petersburg for examination in the chemical laboratory of the Mining Cadet Corps that had been established by Count Mussin-Pushkin in 1804. The analysis was carried out by Vasily Vasilyevitch Lyubarsky (1795–1852), who showed it to be osmiridium, identical with that found in Colombia. By 1824 Lyubarsky was able to complete a quantitative analysis showing: iridium 60 per cent, osmium 30 per cent, platinum 2 per cent, gold 0.7 per cent and iron 5 per cent.

In August of the same year the first native platinum was found in gold placers to the north of Ekaterinburg and further similar discoveries quickly followed in the neighbourhood of Goroblagodat where there was an iron works managed by Nikolai Rodionovich Mamyshev, a man who not only located these deposits but who was to play an important part in the early refining of Russian platinum.

In the year 1825 eleven poods, or 180 kilograms, of native platinum had been collected and sent to St. Petersburg, where it was analysed and reported to contain 87 per cent of platinum with gold, silver and the other platinum metals.

The earliest analyses of the Russian deposits made in Western Europe were reported by André Laugier (1770–1832) Fourcroy's cousin and his successor at the Muséum d'Histoire Naturelle. These were carried out on two quite small and different samples that the famous German naturalist and traveller extraordinary Alexander von Humboldt had managed to secure "after two years of fruitless attempts" (3). Laugier showed that one sample from the gold bearing sands gave 67 per cent of platinum, while the other, from Ekaterinburg, showed roughly 25 per cent of osmiridium, only 20 per cent of platinum, 50 per cent iron with small amounts of copper, and rhodium, but no palladium.

In June 1825 even richer deposits running at about 83 per cent of platinum, began to be located to the north-west of Ekaterinburg, in the district of Nizhny-Tagil on land belonging to the ancient and noble family of the Demidovs who owned profitable gold, silver and copper mines in the area. While little development work was taking place in the Goroblagodat field, the Demidovs were highly energetic at Nizhny-Tagil and their output increased rapidly, the head of the family, Count Nikolai Nikitich Demidov (1774–1828) sending four young men to be trained at the Mining Academy in Freiberg according to a letter from Professor August Breithaupt, the Professor of Mineralogy there (4), who had already made a study of the Russian deposits (5).

Platinum Declared a State Monopoly

The importance of the discoveries of 1824 and 1825 was not lost upon the Imperial Russian Government who at once declared platinum to be a State Monopoly, in which no dealings could take place except under licence. While private individuals were allowed to conduct the mining operations, all refining was to take place under the care of the St. Petersburg Mint and no export of native metal was allowed. A tax of 10 per cent was levied on private output, but the owners were allowed to dispose of it, subject to licence and priority of State demand. The administration of this order came under the Department of Mining and Salt Industry. From the beginning, however, the conditions of State Monopoly exercised in a remote part of the country were very favourable to the existence of flourishing smuggling activities and these duly made their appearance, growing to such an extent as to vitiate all the published statistics for the Russian output of platinum even long after the State Monopoly had ceased.

Early Refining Processes

It is to Mamyshev that credit must be given for the initial attempts at refining and fabrication. These took place in the Urals under his energetic leadership. In 1825 there happened to be in his works, for the moment without employment, a mining engineer named Alexander Nikolaevich Arkhipov, who had already shown an interest in platinum by working out methods for separating it from gold and demonstrating that the native mineral "excelled that of American origin in purity, containing up to 75 per cent of platinum". Mamyshev therefore turned to him with the request to examine the refining of the metal and its fabrication. He gave him two assistants, one a fellow mining engineer named G. A. Jossa and the other a works mechanic, V. Sysoev, and the work began in the laboratory of the Kushvinsky works at Goroblagodat. The process used was that of Janety, by which platinum was rendered malleable by means of arsenic; even in that remote neighbourhood knowledge of this process existed and was employed with a quite considerable degree of success. Following Mamyshev's suggestions, they began by making a platinum ring, followed by a teaspoon, and then Sysoev made an ink-stand from all three metals produced in the Goroblagodat area, cast iron, gold and platinum. Then came a little chain, some priming-pans for pistols and finally an altar shrine or tabernacle of considerable size. Next Arkhipov made up alloys of copper and platinum (20:80 and 33:67) for which he claimed valuable properties in colour, malleability and tarnish resistance, and he also experimented in the decoration of glass and porcelain by means of the new metal. Last of all, also in 1825, Mamyshev suggested that it should be applied to coinage (6). That all this work should have been so successfully conducted in a works laboratory in a distant part of the Urals, when transport was slow or non-existent, is an enormous tribute to the energy and knowledge of Mamyshev and his assistants. Several of the articles they made are still preserved in the Mining Institute at Leningrad.

The Work of Sobolevsky

While this work was taking place in the field, other researches were being actively pursued in St. Petersburg. These went on in the combined laboratory of the Mining Cadet Corps and the Department of Mining and Salt Industry, to which the Government had released 20 pounds of Goroblagodat platinum for experimental purposes in 1826. The work was put under a very able and highly qualified chemist, Peter Grigorievich Sobolevsky (1781–1841), the outstanding figure in this period of the history of platinum in Russia, who had as his chief assistant the very capable chemist Lyubarsky.

Sobolevsky's first action on taking up the work was to send to the Urals for Sysoev, in order that he might repeat the work of Arkhipov and determine whether the arsenic process was satisfactory for larger-scale work. He soon decided that it was not, since it took several days to prepare a single pound of platinum, it was dangerous to the workers and it could be applied only to very

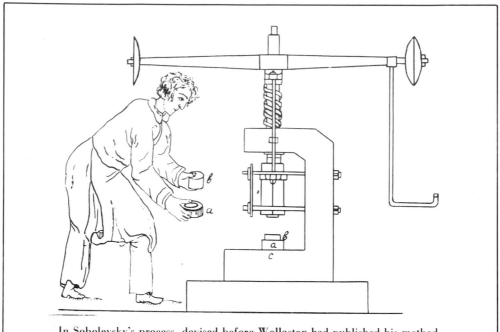

In Sobolevsky's process, devised before Wollaston had published his method, the platinum powder was packed into the mould *a*, covered with the steel cylinder *b* and put under the screw press *c*. The resulting discs were then brought to a white heat while still in the mould and again compressed. They were then sufficiently compact to be **forged and rolled** into sheet

small pieces. So he and Lyubarsky set to work to find a new process, first examining all the work that had been carried out in Western Europe as far as it was recorded in accessible literature.

By 1827 Sobolevsky was able to read a paper to the Annual Meeting of the Scientific Committee of the Mining Corps and the Department of Mining and Salt Industries. He prefaced his account of their process with a very complete summary of the results of earlier workers, and therefore it is possible to see how some modifications and improvements were introduced in the Russian work. The native metal was charged into a retort connected with a couple of receivers containing water, and boiled with four times its weight of aqua regia. After boiling down to a thick consistency, the liquor was poured off from the black insoluble residue and evaporated to dryness, while a fresh portion of ore was treated in the retort. The evaporated material was taken up in hot water, filtered and treated with a cold saturated solution of sal-ammoniac.

The preparation of malleable metal from the calcined chloroplatinate followed upon an observation made by Sobolevsky when trying to melt the platinum sponge in a carbon-lined crucible. He noticed the agglomeration of the

238

metal and, applying compression in a screw press, obtained a fragile metallic cake. This he heated to a white heat and again applied compression to give a satisfactory malleable metal. His large-scale working involved pressing the spongy metal while cold:

> "We pack cold refined platinum in the form of sponge very tightly in a thick iron circular mould of a chosen size (this mould can equally be in any other form; the round appeared to us to be the more convenient); we press it with the strong compression of a screw press, and having extracted it from the mould, we have a dense round tablet with a metallic sheen. In this state the platinum disc has as yet no malleability and the cohesion of the particles among themselves does not resist a strong blow; it fractures and crumbles up. To convert such discs into malleable platinum, it is only necessary to heat them to a white heat and at this temperature to subject them to the pressure of the same press. From a single compression the disc of platinum entirely alters its appearance; its granular structure becomes dense and it becomes completely malleable. The size of the tablets does not make any difference in this affair; large and small discs at a single compression become equally malleable and ductile. After this pressing the discs are forged into strips or rods of any desired form."

Sobolevsky and his Russian followers claim that he evolved this process himself, but the first part of the paper of 1827 in which he describes it shows that he was well acquainted with the work of most of the men who had worked on platinum in Western Europe down to that time. He mentions the work of Baruel but not that of Cock; he knew about Bréant's work but not the nature of his process; in 1827 Wollaston's work had not yet been published. The screw press is reminiscent of Cock, but the simplicity of Sobolevsky's process does him credit. In justification of his claim, he himself points out these differences: in the final form of his process the sponge is first pressed cold, the first cake is heated to whiteness separate from the mould and it is pressed only once afterwards, with such hammering as is necessary for fabrication. In all these respects his procedure differs from Cock's, as Sobolevsky himself says in a footnote to his paper.

The platinum was now available in pieces of any desired size, there was practically no loss in refining, and it gave reliable results. Sobolevsky first demonstrated his process in January 1827 after producing 20 pounds of refined metal, and on the following March 21st he made it fully public to an assembly of the Scientific Committee of the Mining Corps and Department of Mining and Salt Industry, a full account being published in the Russian *Mining Journal* for 1827 (7). When reading the paper he displayed a number of objects made of platinum, including some medals, wires, dishes, crucibles and several ingots, one of which weighed 6 pounds.

> "Platinum prepared by us was used for the stamping of medals and counters; not a small part of it was also used by goldsmiths who showed that articles could be made out of it, and these showed that this platinum was not inferior in this respect to that prepared by the French."

One series of medals was struck in platinum to commemorate the coronation of the Tsar Nicholas I in 1826. A specimen has survived in the possession of Johnson Matthey and is illustrated over page.

239

To commemorate the coronation of the Tsar Nicholas I in 1826 medals were struck in platinum from the recently discovered deposits in the Urals. The obverse shows the head of the emperor with the inscription:

His Majesty Nicholas I, Emperor and Absolute Ruler of All Russia.
The reverse reads:

A Pledge of Happiness for One and All.
Crowned in Moscow 1826.
The medals were 5 centimetres in diameter and weighed 131 grams

While the work of Sobolevsky and Lyubarsky was going on, the energetic Minister of Finance and ex-officio head of the Mining Department, Count Egor Frantsevich Kankrin was sending out samples of native platinum to foreign chemists and scientific societies, asking them to experiment with it and report their results to him together with suggestions for its application. He sent a pound to the Royal Society in London and half-a-pound to Wollaston; a pound each to the Institut and the Société d'Encouragement in France; half-a-pound to Berzelius in Sweden and four pounds to Osann in Dorpat. The enquiry also found its way, perhaps via the Royal Society, to P. N. Johnson in London, and George Matthey records in a letter dated June 9th, 1851, of which the copy is still preserved by Johnson Matthey, that:

"Mr Johnson twenty-four years ago supplied the Russian Government at the request of Chevalier Benkhausen (the Russian Consul in London) with every information respecting the method of preparing platina, the intrinsic value, uses, and adaptation to coinage as a currency or circulating medium."

But the most spectacular results came from Berzelius in the form of the publication of his classic work on the chemical properties of the metals of the platinum group and the methods to be used for the analysis of substances containing them and their separation one from another; these important researches will be discussed in the next chapter.

240

**Count Egor Frantsevich Kankrin
1775–1845**

As Minister of Finance to Tsar
Nicholas I Kankrin was also head of
the Department of Mining and he it
was who conceived the idea of employing platinum as coinage despite warn-
ings from Alexander von Humboldt

The Introduction of Platinum Coinage

The growing output from the mines, coupled with the success of Sobolevsky and
Lyubarsky's work on refining and fabrication, urged Count Kankrin to still
more energetic exploration of the fields in which the new metal might be
exploited as a means of strengthening the national economy and bringing some
much needed money into the Treasury. Unfortunately there was not in Russia
the same scope as had been available in Western Europe. There was not so
much scientific work going on and therefore there were fewer laboratories to be
equipped; there was no chemical industry to speak of, and as for jewellery the
flamboyant tastes of the wealthy Russians were not attracted by the dull white
colour of the new metal. It was necessary therefore that new and previously
untried uses should be sought, if the Government and the country were to profit
by the new discovery, so Kankrin turned back to Mamyshev's suggestion of 1824
that platinum should become a coinage metal. Accordingly specimen coins were
made by the Mint and in August 1827 the Emperor approved the design, but
"expressed the wish that the opinion of competent people should be sought".
Therefore Kankrin wrote to Alexander von Humboldt, who was then travelling
in Colombia inspecting the platinum mines, asking his advice and sending some
sample coins. After three months spent in collecting information, von Humboldt
replied from Berlin expressing considerable misgivings about the proposal. The

Many hundreds of thousands of platinumn coins of three, six and twelve roubles in value were struck in the Mint at St. Petersburg, beginning in 1828 and continuing until Kankrin's death in 1845. The three rouble piece weighed 10.55 grams, the six rouble 20.7 grams and the twelve rouble 41.4 grams

chief grounds for this were that platinum was available in South America, and that it would therefore be difficult to control the price sufficiently to prevent depreciation and counterfeiting of the proposed currency. Kankrin's enthusiasm was not damped, however, and he replied maintaining his arguments and asking for further consideration. But he was too impatient to wait for the answer, and on March 8th, 1828, he ordered the preparation of dies for a 3-rouble coin. On April 24th issue of the coins was authorised, and he wrote to von Humboldt sending him a specimen. Meanwhile the work proceeded on 6-rouble and 12-rouble coins and in due course issue to the public began, a large quantity of platinum having been purchased from Count Demidov. But in his anxiety to avoid the troubles predicted by Humboldt he seems to have pitched too high the value at which the platinum was coined, as will be seen a little later.

The exciting news from Russia prompted several visits from overseas by geologists and others. In 1829, at the invitation of Nicholas I, and at his expense, Alexander von Humboldt undertook an eight-month long tour of Siberia accompanied by Gustav Rose (1798–1873), Professor of Mineralogy in Berlin, and his colleague Christian Gottfried Ehrenberg (1795–1876). Their journey was too hurried – they covered over 9,000 miles – for any intensive study of the

242

platinum deposits, although Humboldt pointed out how remarkable it was that platinum was found in abundance only on the western European slopes of the Urals while the rich gold placers were on the Asiatic side. Some time after their return Rose contributed a paper on the geology of the platinum deposits (8).

Then in 1831 an American mineralogist, James Dickson, who had had experience of the platinum workings in South America, paid a visit to Russia. Count Nikolai Demidov had died three years earlier and had been succeeded by his son Analoti (1812–1870) who was away travelling in Italy but Dickson was given every assistance by his agent and relative Danilov. He was astonished at the collection he was shown of large and beautiful specimens of native platinum, some weighing seven or eight pounds, and most of them destined as presents for the crowned heads of Europe. He went on in his report:

"The Demidovs, Davidovs and many other Russian families are acquiring princely revenues ... more especially since all the platina is now coined at the Imperial Mint and established as part of the current coin of the realm. Though many hundred pounds weight of platina are coined monthly they disappear rapidly from circulation ... I consider their price much above the London price of malleable platinum, which is at present about twenty-five shillings per ounce: considering that the crude platina is the product of the country, the Russian price for malleable platina, which is about twenty-eight shillings, is too extravagant" (9).

A rather different point of view was expressed a little later by an English visitor, William Marshall (1799–1844), also a geologist and the Vice-President and Honorary Curator of Mineralogy to the Yorkshire Philosophical Society, to which he made a full report on his return, presenting them with a large specimen of native platinum (10). Marshall was at some pains to criticise Wollaston's procedure for the production of malleable platinum by comparison with the much simpler method adopted by Sobolevsky which he regarded as producing the metal in a sufficiently pure state for practical purposes.

At the meeting of the Gesellschaft deutscher Naturforsche held in Stuttgart in September 1834 Sobolevsky presented a paper giving details of the discovery of platinum in the Urals, of his refining procedure, and now comparing it with the Wollaston process made known in 1828, concluding:

"From all this it can be seen on what a great scale platinum is being worked in Russia today and to what degree of simplicity this process has been brought which formerly gave so much trouble to metallurgists" (11).

In fact Sobolevsky had greatly increased the speed of his operations from their painfully slow beginnings. At first only 3 poods (1,575 ounces troy) could be refined in a week, yielding only 1,050 ounces of pure platinum; this gave 2,000 coin blanks and 440 ounces of scrap but by 1834 the refinery was able to cope with one pood (525 troy ounces) of native platinum per day and the coiners had increased their capacity accordingly. In increasing the speed of refining, however, Sobolevsky had by no means improved his process. Instead of evaporating his aqua regia to dryness, taking up with water, and then precipitating incompletely with sal-ammoniac he had now abandoned the evaporation to dryness in seeking to save time and was adding sal-ammoniac directly to the original aqua regia solution, relying on an excess of aqua regia to retard precipitation of the iridium, and on long washing of the precipitate with water to remove such chloroiridate as nevertheless appeared.

The evaporation to dryness tended to decompose the iridium chloride into the soluble iridous form, and the incomplete precipitation tended to leave any undecomposed iridic salt in solution. In the new process the presence of excess aqua regia kept the iridium salt oxidised and therefore easily precipitated, the acidity of the solution did nothing to prevent this precipitation, and the excessive washing removed considerable quantities of platinum in large volumes of washwater which had to be avaporated for its recovery. The procedure was bad and after Sobolevsky's death in 1841 it gave place to an improved process.

Döbereiner's Refining Process

A close connection between the Russian authorities and Johann Wolfgang Döbereiner and his work on platinum in Jena, described in the last chapter, developed through Maria Pavlovna, a daughter of Tsar Paul I and a sister of his successors Nicholas I and Alexander I. In 1804 she had married Carl Friedrich, the son and heir of Duke Carl August of Weimar and of course had settled there.

The Grand Duchess Maria Pavlovna 1786–1859

A daughter of Tsar Paul I who had witnessed Lavoisier's melting of platinum in Paris in 1782 and a sister of his successors Nicholas I and Alexander I, the Grand Duchess married Carl Friedrich, the son and heir of Döbereiner's patron, Duke Carl August of Weimar. As well as being interested in chemistry she was in close touch with her brother's Minister of Finance, Count Kankrin, and was able to secure ample supplies of native platinum for Döbereiner's researches. Later she was instrumental in securing his services to improve the methods of refining Russian platinum

244

She took a great interest in chemistry and in Döbereiner's activities and occasionally helped to support him financially. She was also in close touch with her brother's Minister of Finance, Count Kankrin who was also Head of the Mining Department in St. Petersburg. Neither Kankrin nor Sobolevsky was satisfied with the platinum refining processes in use and the Grand Duchess drew their attention to the possibility that Döbereiner's skill and experience might be of value. Kankrin at once adopted her suggestion, but Döbereiner was unwilling to leave Jena. Instead his son Franz travelled to Russia and accommodation was arranged for him in Osann's laboratory in the University of Dorpat (now known as Tartu), and a Dr. Friedrich Weiss of the University Chemical Department was assigned to assist him, with Döbereiner himself acting as consultant. Unlimited supplies of native platinum were of course available to them, as well as access to the refining operations in St. Petersburg. In their approach to the problem they proposed to free the solution in aqua regia from iron and copper by the addition of alkalies, a procedure that had been tried by Chabaneau in Spain in the 1790s without success. In the meantime the English astronomer and chemist Sir John F. W. Herschel (1792–1871) had made a remarkable discovery about platinum in the course of some work he was doing on the influence of light on chemical action, work which has placed his name among the list of the inventors of photography. In a paper read to the British Association at its Oxford meeting in 1832 (12) he showed that if a solution of the metal in aqua regia is neutralised with lime, and after filtering some more lime is added, nothing happens as long as the mixture is kept in the dark but that if it is exposed to sunlight there is quickly formed a copious white precipitate. This happened less rapidly in cloudy daylight and did not take place at all in red or even yellow light. Herschel found the precipitate to be "a combination of the oxide of platinum with lime" in which the oxide seems to perform the part of an acid.

Further exploration was carried out in Dorpat by Weiss and the younger Döbereiner (13). They showed that from solutions of native platinum, the lime appeared to precipitate at once the iridium, rhodium, copper, iron, lead, titanium, etc., present in the solution, as oxides containing no calcium, while the platinum remained dissolved as long as the liquid was not warmed or exposed to light. In the latter case the white precipitate described by Herschel was formed; in the former, if the mixture is boiled for some time the platinum is converted into a form in which it is not precipitated by ammonium chloride. The palladium behaved in an intermediate and uncertain manner, slowly yielding even in the dark a white precipitate containing lime.

This, however, did not prevent an attempt to apply the new knowledge to the large-scale refining of platinum from the Urals at St. Petersburg, and in 1841 a new process based upon it was introduced and refining operations were transferred to the St. Petersburg Mint. In the new treatment, described in some detail in a paper by a Colonel Kovanko in the *Russian Mining Journal* in 1843 (14), the aqua regia solution had milk of lime added to it until it remained only slightly acid. The precipitate was filtered, the solution evaporated to dryness, and the residue calcined at a bright red heat until no further evolution of chlorine took place. This decomposed the calcium chloroplatinate into calcium chloride and metallic platinum. The former was leached out with water and the platinum pressed and forged. This latter part of the process had also been altered, and not for the better. The pressing of the cold sponge was carried out as before, but the subsequent heating was effected by placing a pile of the pressed discs in a furnace where the firing of porcelain was going on at the same time,

> "so that their heating continues for a day and a half. Given the conditions of first-class refining of the platinum and, especially, good washing, a very malleable and serviceable platinum is produced by this firing".

But the results were not very successful and the reason for this was without doubt the erratic behaviour of the various metals of the platinum group under refinery conditions. Not only was the finished metal of poor quality, but piles of residues began to accumulate which were even larger and more complex than those produced by the earlier processes.

Before, however, passing on from the work of the Döbereiners it will probably be appropriate to describe in modern terms what the fundamental occurrences were in the course of their process. The addition of lime to the solution of platinum in aqua regia in the dark did not permit any appreciable precipitation of the metal, but under the influence of actinic light a process of hydrolysis set in, leading to precipitation of the platinum in the form of successive members of a series of hydroxy-chloroplatinates, culminating in hexahydroxyplatinate, all of calcium. This resulted in the solution being robbed of some or all of its platinum according to the intensity of the light and of the length of the exposure.

The process was too long and clumsy, and it was finally abandoned, but not before Döbereiner senior had published a monograph, "Zur Chemie des Platins", dedicated to Count Kankrin that was remarkably comprehensive for its time.

The Cessation of Coining

The issue of the rouble coins went on satisfactorily under Kankrin's watchful guidance and under the stimulus of this demand the output of the mines increased rapidly to some 60,000 ounces in 1830, reaching a record of about 120,000 ounces in 1843 but then declining rapidly. This was occasioned by a serious fall in the demand for platinum outside Russia, causing a fall in price in the western markets and a drop in production in Colombia. In 1843 the output there was only 1,600 ounces and in due course the price fell to a level lower than the exchange value of the Russian platinum coins. Kankrin, alarmed at the prospect of extensive counterfeiting, put a severe limitation on the amounts coined and strengthened his customs regulations at the frontiers. But shortly after 1844 he retired from office and his successor Vroncenko was a timid man with none of his strength, experience and convictions. Under him something like

246

panic supervened, and in 1846 he ordered the cessation of coining and the withdrawal of the whole platinum currency. So von Humboldt's worst fears of 1827 were fully realised and the public returned to the Treasury 3,263,292 roubles, or about three-quarters of the amount that had been issued, the remainder being either hoarded or smuggled abroad. The coins had been in issue for eighteen years and the total number of them was: 3-roubles, 1,373,691; 6-roubles, 14,847; and 12-roubles 3,474; with a total face value of 4,251,843 roubles. They weighed 0.333, 0.666 and 1.332 ounces troy each respectively. The total amount of platinum used was 485,505 ounces troy (15).

Karl Klaus and the Discovery of Ruthenium

Before concluding this survey of the Russian platinum industry in its early years there has to be recorded the discovery of the sixth and last member of the platinum group of metals.

It will be remembered that very soon after its discovery Count Kankrin sent samples of native platinum to a number of scientists. Among these was Gottfried Wilhelm Osann (1797–1866) who had been one of Döbereiner's students in Jena and was now Professor of Physics and Chemistry in Dorpat, becoming Professor at Wurzburg in 1828. By means of blowpipe tests he came to the conclusion that he had found indications of the presence of three new metals to be called pluran, ruthen and polin (16). He sent samples of his preparations to Berzelius just at the time when the latter had attained a great deal of confidence in his knowledge of the behaviour of the platinum metals. Berzelius reported that he could not recognise in pluran anything already known to him, and that it might be the oxide of a new metal since it had some peculiar properties. It was a heavy white powder, volatile, without odour, and giving a solution coloured grey by sulphuretted hydrogen. The polin and ruthen were simply a mixture of silica, titania, zirconia and oxides of iridium. Moreover, Osann's analyses of native platinum and the residues did not compare in accuracy with those of the Swedish chemist. Eventually he published a notice revoking the discovery of any new metal (17). Nevertheless this painstaking work of Osann's showed that there was still a great deal about the platinum metals remaining unexplained, and that much had still to be done.

Some years later this was taken up by Karl Karlovich Klaus, or Carl Ernst Claus in the German version of his name, a pharmacist from Dorpat who was appointed to a new chair of chemistry in the University of Kazan in 1838. He worked chiefly on analytical methods, but his main efforts were directed to processes to be used on work on an industrial scale and especially to the working up of the many difficult residues produced in such operations. In 1844 in the course of the work he discovered and confirmed the sixth member of the platinum group metals, and named it ruthenium in honour of his adopted country and to perpetuate the memory of Osann.

In 1840 he had turned his attention to the residues which were being

**Karl Karlovich Klaus
1796–1864**

Born in Dorpat, now Tartu in Estonia, Klaus first practised as a pharmacist there, becoming an assistant in the chemical department of the university. In 1836 he was appointed Professor of Chemistry in the University of Kazan and a few years later began work on the insoluble residues from the refining of platinum. Besides discovering the sixth member of the group, ruthenium, he carried out extensive researches on iridium and rhodium and finally assembled the results of more than twenty years of research in a monograph that was not published until some years after his death

produced by Sobolevsky in the St. Petersburg platinum refinery in order to see if they threw any light on the problem of Osann's new metals. He had been interested for some time in the composition of native platinum, and these residues seemed to present a concentrated opportunity.

He asked Sobolevsky for, and obtained, 2 pounds of them and proceeded to study these, first by repeating carefully the work of his predecessor Osann, and then launching out for himself. In addition to the four other platinum metals, he was surprised to find a content of 10 per cent of platinum itself, and this "richness of the residues, lying uselessly in considerable quantities in the laboratory of the Mint at St. Petersburg, appeared to me to be so important that I reported to the mining authorities and in 1842 set off for the capital", where he saw Kankrin. The latter fully approved of his desire to examine a larger quantity of the material and he returned to Kazan with half a pood. This, however, turned out to be much poorer in platinum than the first and "the hope of my method for the profitable extraction of this metal from them disappeared; there remained only investigations of interest to science".

It was in the course of two years of arduous work that he discovered the new metal. This he did by taking a portion of the insoluble osmiridium residue and fusing it with potash and nitre in a silver crucible. After keeping the mixture molten for an hour and a half at bright redness he poured out the contents into an iron capsule. He then dissolved the cake in water to yield an orange-coloured solution and treated this with nitric acid, producing a black precipitate consist-

248

ing of the oxides of osmium and ruthenium. He distilled this with aqua regia and condensed the osmium tetroxide. The residue consisted mainly of the two chlorides of ruthenium and on adding ammonium chloride, ammonium chlororuthenate was precipitated. This when calcined yielded ruthenium sponge to the quantity of 6 grams. Klaus retained the name "ruthenium" both for patriotic reasons, since ruthenium is a latinised name for Russia, and also in honour of the earlier work of Osann, whose "ruthen" did contain a small quantity of the new metal although most of it had been lost in the hydrochloric acid solutions which Osann did not examine.

Klaus sent a sample of his potassium chlororuthenate to Berzelius who at first believed it to be a salt of iridium, but after some further examination he withdrew this verdict and expressed the opinion that the salt contained a metal which was unknown to him.

Klaus published his work as it was being done in a series of short articles in the Bulletin of the Russian Academy of Sciences and a consolidation of the whole appeared in 1844 in the Scientific Records of the University of Kazan. It ran to 188 pages and was reprinted as a booklet in 1845 under the title "Chemical Examination of the Residues from Uralian Platinum Ore and of the Metal Ruthenium". This appears to have been a private publication and did not reach the outside world. A summary, under the heading of "Discovery of a new Metal", appeared in Poggendorff's *Annalen* for 1845 and was reproduced in *The Philisophical Magazine* (18) and again by Berzelius in his *Jahresbericht* for 1841 (19), but the first full account of the whole work given to Western Europe seems to be that in Liebig's *Annalen* for 1847 as "Beiträge zur Chemie der Platinmetalle" (20). This last also was published as a pamphlet in German and under the above title, at Dorpat in 1854 (21).

After this Klaus continued to take a close interest in the refining operations going on at St. Petersburg and in the properties of platinum. At one time he gave particular attention to methods for keeping the chloroplatinate solution free from iridium and, more importantly, free from silica, a matter which had bedevilled many workers including Wollaston:

"This silica remains in the platinum sponge and is capable later during the heating of the platinum of uniting intimately with the platinum, indeed of alloying with it, and therefore making it brittle and unsuitable for use."

Klaus continued to carry out research on the platinum metals while at Kazan and compiled the results of nearly twenty years of research on ruthenium, osmium and rhodium in a small book published there in 1859 to commemorate the fiftieth anniversary of his university.

Meanwhile, in 1852 Klaus had returned to the University of his birthplace, Dorpat, where he had served as an assistant between 1831 and 1837, to occupy a newly-founded Chair of Pharmacy and there he formed a desire to publish a monograph on the platinum metals and their alloys, incorporating the results of his twenty years' research in this field, as well as the history of the metals, references to the work of other scientists and a description of their technical and industrial applications. To help him to this end, the Russian Government sent him to France and Germany to visit the laboratories and refineries in which platinum was being handled and to study the history of the metals in the libraries of Berlin and Paris. He left Dorpat in May 1863 and his journey was something of a triumphal progress. In Berlin he met Heinrich and Gustav Rose, Poggendorff, Magnus and others, and the Prussian Academy of Sciences made him a Corresponding Member. Menschutkin in a paper on Klaus (22) goes on to say:

"Before him doors opened to which entrance was strongly forbidden to strangers, and he succeeded in making a thorough acquaintance with the organisation of the business in platinum in the works of Heraeus at Hanau, Desmoutis Quennessen in Paris, Chapuis & Co in Paris, Matthey in London".

Early in 1864 he returned to Dorpat and set about the collation of his material, but an important meeting of the Pharmaceutical Society called him to St. Petersburg, and on his way home he took a chill, became ill and died on March 12th. The manuscript of the first three chapters of his book was found among his papers and prepared for publication in 1865 by his colleagues, the Academicians Fritzsche and Jakobi, but circumstances and then the death of the two sponsors intervened and nothing was achieved. Fortunately, however, the script was rescued from oblivion eighteen years later by Klaus's former pupil and his successor in the Chair at Kazan, Alexander Mikhailovich Butlerov (1828–1886) and this time publication took place at the hands of the Russian Academy of Sciences under the title "Fragment einer Monographie des Platins und der Platinmetalle" and as a token of respect to Klaus's memory. The little book is of the greatest value to the historian of platinum.

Conclusion

With the cessation of the demand for coinage there ended what the Russian writers call "The First Period" of the Russian platinum industry. There was no other application available to support the market, and price and output fell to such an extent that over the next eight years, 1845 to 1852, the whole of the Urals produced only 113 poods (60,000 ounces). After that, however, a revival took place, stimulated by the activity in the West by the firms of Johnson and Matthey in London and of Desmoutis Quennessen in Paris, and later on by the scientific and technical work of Deville and Debray to which reference will be made in Chapter 15.

But to sum up the very important results of this early work: first, it effectively explored and opened up the important new source of native platinum that was destined to supply practically the whole world's needs of what gradually became a vital raw material from the eighteen-sixties until at least 1917. After that, its importance declined as other sources were found and exploited, but even then it was able to supply all Russian needs and still to make occasional appearances in the western markets. Secondly, it was the means by which were made the first intensive examination of the chemical and physical properties of the whole group of six platinum metals. The other great feature of the chapter is the lifetime's work of Klaus. The result, as we have seen, was the discovery of the sixth platinum metal, ruthenium, and a satisfying elucidation of its properties as well as those of iridium, osmium and rhodium. A lot was left for the future, but the foundations were well and truly laid.

References for Chapter 13

1 N. L. Vauquelin, *Ann. Chim.*, 1806, **60**, 317–322

2 L. B. Guyton de Morveau, *Ann. Chim.*, 1810, **73**, 334–335

3 A. Laugier, *Ann. Chim.*, 1825, **29**, 289–295; *Phil. Mag.*, 1825, **66**, 285–288

4 *Edin. New Phil. J.*, 1828–29, **6**, 197–198

5 A. Breithaupt, *Ann. Phys. (Poggendorff)*, 1826, **8**, 500–505

6 E. K. Fritsman, *Ann. Inst. Platine (Leningrad)*, 1927, **5**, 23–74

7 P. G. Sobolevsky, *Mining J. (Russia)*, 1827, **2**, 84: reproduced in *Ann. Inst. Platine (Leningrad)*, 1927, **5**, 206–219

8 G. Rose, *Ann. Phys. (Poggendorff)*, 1834, **31**, 673–676

9 J. Dickson, *Monthly Am. J. Geology (Featherstonehaugh)*, 1831, Sept., 118–124

10 W. Marshall, *Phil. Mag.*, 1832, **11**, 321–323; *J. Tech. Chem.*, 1832, **14**, 319–322

11 P. G. Sobolevsky, *Ann. Phys. (Poggendorff)*, 1834, **33**, 99–109; *Ann. Chem. (Liebig)*, 1835, **13**, 42–52

12 J. F. W. Herschel, *Phil. Mag.*, 1832, **1**, 58–60

13 F. Weiss and F. Döbereiner, *Ann. Chem. (Liebig)*, 1835, **14**, 15–21

14 Kovanko, *Mining J. (Russian)*, 1843, 447; reproduced in *Ann. Inst. Platine*, 1927, **5**, 219–225

15 E. K. Fritsman, *Ann. Inst. Platine*, 1927, **5**, 23–74

16 G. W. Osann, *Ann. Phys. (Poggendorff)*, 1827, **11**, 311–322; 1828, **14**, 329–357

17 G. W. Osann, *Ann. Phys. (Poggendorff)*, 1829, **15**, 158

18 C. Claus, *Ann. Phys. (Poggendorff)*, 1845, **64**, 192–197; *Phil. Mag.*, 1845, **27**, 230–231

19 J. J. Berzelius, *Jahres-Bericht*, 1847, **26**, 181–184

20 C. Claus, *Ann. Chem. (Liebig)*, 1847, **63**, 337–360

21 C. Claus, Neue Beiträge zur Chemie der Platinmetalle, Dorpat, 1863. This is based on Claus's papers in *J. prakt. Chem.*, 1860, **79**, 28–59; 1860, **80**, 282–317; 1861, **85**, 129–161; 1863, **90**, 65–105

22 B. N. Menschutkin, *Ann. Inst. Platine*, 1928, **6**, 1–10

Jöns Jacob Berzelius
1779–1848

After studying at the University of Uppsala Berzelius was appointed
assistant professor of botany and chemistry in the University of
Stockholm, becoming professor in 1807 and later, in 1818, secretary of
the Royal Swedish Academy of Sciences. His enormous output of work
included a brilliant investigation of the platinum metals, their separation, methods of analysis and their atomic weights

From a portrait by Johan Way, by courtesy of
the Royal Swedish Academy of Sciences

14

The Platinum Metals in Early Nineteenth Century Chemistry

"Only one who himself has worked with the chemistry of the platinum metals can fully understand and evaluate the difficulties Berzelius had to overcome. He not only determined the atomic weights of platinum, palladium, rhodium, iridium and osmium but also investigated a large number of the most important compounds of these metals."

HENRIK GUSTAV SÖDERBAUM

The first quarter of the nineteenth century saw the transformation of chemistry into an exact science. Following the great contributions of Lavoisier, his appreciation of the role of oxygen, his use of the balance and his introduction of the new nomenclature, a second major advance was made with the establishment of the quantitative basis of chemical combination. It is possible here only to summarise very briefly the development of this numerical approach and to set the scene for the great change brought about in inorganic chemistry and its impact on the study of the platinum metals.

This began with the rather obscure and somewhat neglected writings of Jeremias Benjamin Richter (1762–1807), the originator of the term "stoichiometry" and of the concept of equivalent weights of bases and acids. Richter studied under the great German philosopher Immanuel Kant at Königsberg and his choice of subject for his doctoral thesis, "The Use of Mathematics in Chemistry", was most probably inspired by Kant. His book, published in 1792, was prefaced with a quotation in Greek from the Apocrypha:

"Thou hast ordered all things in measure and number and weight." (1)

In 1799 Joseph Louis Proust, whose work on platinum in the Royal Laboratory in Madrid has been described in Chapter 6, put forward his Law of Constant Composition, a concept that he later summarised in the phrase:

253

John Dalton
1766–1844

The concept of atomic weight introduced by Dalton together with his laws of constant and multiple proportions in chemical combination gave a great impetus to inorganic and analytical chemistry and prompted Berzelius to embark on his systematic investigations

"A compound is a substance to which Nature assigns fixed proportions; it is in a word a being which she never creates even in the hands of man, except with the aid of a balance." (2)

This brought Proust into a long controversy with Claude Louis Berthollet (1748–1822), who maintained that fixed proportions were exceptional rather than a general rule, but the atomic theory propounded by John Dalton, incorporating his doctrine that the elements were composed of atoms of constant weight and that compounds were formed by their union in simple numerical proportions, put an end to this argument. On October 21st, 1803 Dalton read a paper to the Manchester Literary and Philosophical Society – published only later in 1805 – in which he concluded:

"An enquiry into the relative weights of the ultimate particles of bodies is a subject, as far as I know, entirely new: I have lately been prosecuting this enquiry with remarkable success." (3)

Dalton began his determination of atomic weights with the gases, but by 1804 he had turned to the metals and the calculation of their atomic weights from the published analyses of other workers, selecting the values which disagreed least with the often inconsistent analyses of their oxides and adding new results as further literature appeared.

His views on platinum, about whose atomic weight he remained uncertain, may be seen in the reproduction here of a passage from "A New System of Chemical Philosophy" published between 1808 and 1810. Dalton also introduced his "arbitrary marks" or symbols to represent the elements.

254

Dalton began publishing his 'New System of Chemical Philosophy' in 1808, the second part following two years later. This extract from Part II shows his thoughts on platinum at that time. He also devised a number of symbols for the elements, three of them illustrated here

Silver

Gold

Platina

The weight of the ultimate particle of platina cannot be ascertained from the data we have at present: from its combination with oxygen, it should seem to be about 100; but, judging from its great specific gravity, one would be inclined to think it must be more. Indeed the proportion of oxygen in the oxides of platina cannot be considered as ascertained.

Platina is chiefly used for chemical purposes; in consequence of its infusibility, and the difficulty of oxidizing it, crucibles and other utensils are made of it, in preference to every other metal. Platina wires are extremely useful in electric and galvanic researches, for like reasons.

At first Dalton's views made very little impression, but in 1807 Thomas Thomson, Professor of Chemistry in the University of Edinburgh and the founder of the *Annals of Philosophy*, published the third edition of his book, A System of Chemistry, and included in this an account of the atomic theory based upon a conversation he had had with Dalton in Manchester in 1804. A year after this Thomson read a paper to the Royal Society in which he confirmed the law of multiple proportions by the analysis of some neutral compounds (4). Immediately following this a paper from W. H. Wollaston not only gave further support to Dalton but disclosed that he had himself been turning his mind in the same direction:

> "I thought it not unlikely that this law might obtain generally . . . and it was my design to have pursued the subject with the hope of discovering the cause, to which so singular a relation might be ascribed. But since the publication of Mr. Dalton's theory of chemical combustion the enquiry which I had designed appears to be superfluous as all the facts that I had observed are but particular instances of the more general observations of Mr. Dalton that in all cases the simple elements of bodies are disposed to unite atom to atom singly or if either is in excess, it exceeds by a ratio to be expressed by some simple multiple of the number of its atoms." (5)

The Massive Researches of Berzelius

It was this paper of Wollaston's, reproduced in Nicholson's *Journal* in November 1808, that first gave Berzelius any indication of Dalton's theory as the Napoleonic blockage had prevented his receiving earlier publications from England. He was in the course of preparing a text book of chemistry in Swedish and had been impressed by his reading of Richter's work on stoichiometry when Wollaston's paper reached him and he realised at once the enormous

255

significance of Dalton's generalisation. Immediately Berzerlius set himself the task of determining "the definite and simple proportions in which the constituent parts of inorganic substances are united with each other", proposing to establish the atomic weights of all the elements then known and to this end to improve the infant techniques of quantitative analysis.

By 1814 he was able to publish in Thomson's *Annals of Philosophy* (translated much later into German) a tentative table of atomic weights, based on oxygen taken as 100, including those for rhodium, platinum and palladium, with all the evidence on which he had based his calculations, while at the same time, becoming wearied with the need to refer repeatedly to the elements by their full names, he proposed the use of our now familiar symbols based upon the initial letter of their Latin names (or upon this followed by a second letter from the body of the name) to replace the rather cumbersome circular symbols devised by Dalton. Thus we acquired the symbol **Pt** for platinum, while his first suggestion for palladium **Pl**, was later revised to **Pa** to avoid confusion with platinum and finally to **Pd**. Similarly he proposed **I** for iridium later revised to **Ir**, and **R**, changed to **Rh**, for rhodium. (6)

In 1818 a fuller version of the atomic weights of almost all the elements appeared but still excluding iridium and osmium (7). These were again all based upon the atomic weight of oxygen taken as 100 and were reasonably close to modern values except that several were in error, being twice their real value because of his assumption that most oxides had a simple formula, consisting of one atom of metal and two of oxygen, an error that he later corrected.

This amazing achievement was surpassed some ten years later in so far as one of his favourite pursuits, the chemistry of the platinum metals, was concerned. Reference was made on page 240 to Count Kankrin sending to Berzelius two small samples of native platinum from the Urals together with a personal note asking him to analyse them and to be good enough to report his results. He had already noticed that the early results obtained in the Russian mining laboratory and in Paris by Laugier showed marked differences and he realised that there was a great deal still to be established about the means of separating the five metals of the platinum group so far known. Berzelius therefore embarked upon a thorough examination of these specimens. He began with rhodium and quickly found that his earlier work (6) carried out on a sample presented to him by Wollaston, contained errors in his calculation of its atomic weight. This he proceeded to correct, and he then went on to determine the atomic weights of palladium, iridium, platinum and osmium. From all this work he put together a rather complicated scheme for the separation of the constituents of native platinum on an analytical scale and so for the first time provided means for a reasonably accurate analysis for this material and the products obtained from it (7). He found that the Nizhny Tagil mineral could be divided into magnetic and non-magnetic portions and he analysed these separately. On the other hand, the mineral from Goroblagodat was completely non-magnetic, but from another point of view was noteworthy in being almost completely free from iridium

256

(apart from osmiridium). For the sake of comparison he conducted at the same time analyses of some native platinum from Colombia. His results are given in his paper as follows:

	Nizhny Tagil Non-Magnetic	Nizhny Tagil Magnetic	Goroblagodat	Colombia
Platinum	78.91	73.58	86.50	81.30
Iridium	4.97	2.35	—	1.46
Rhodium	0.86	1.15	1.15	3.46
Palladium	0.28	0.30	1.10	1.06
Iron	11.01	12.98	8.32	5.31
Copper	0.70	5.20	0.45	0.74
Osmiridium	1.96	—	1.40	1.03 (Os)
Insoluble	—	2.30	—	—

This massive and masterly account of the chemistry of the entire group of platinum metals then known to exist, including a study of their salts and double salts, was presented to the Royal Swedish Academy of Sciences, in 1828 and reproduced in German in Poggendorff's *Annalen* in the same year and in French a year later (8). Klaus, in his monograph referred to in the last chapter, wrote that "this treatise became a governing factor for all the subsequent work on this subject".

A further ten years passed before Berzelius came to express his atomic weights on the basis of hydrogen as unity. His revised table of 1838 (9) includes these values, alongside which have been set the accepted figures today.

	O = 100	H = 1	Modern Value
Platinum	1215.026	194.753	195.09
Rhodium	750.680	120.305	102.9055
Palladium	714.618	114.526	106.4

And so Berzelius, among his many other activities, put the analysis of the platinum group of metals on a sound and proper basis and established most of their atomic weights.

The Contributions of Friedrich Wöhler

A number of German students spent periods of time with Berzelius in Stockholm before taking up important careers in chemistry in their own country. One of the first of these was Friedrich Wöhler, who made the journey in 1823 on the advice of his teacher at Heidelberg, Leopold Gmelin, spending a year undergoing a rigorous training in chemical analysis and forming a life-long friendship with Berzelius. Before leaving Heidelberg, however, Wöhler in collaboration with Gmelin had published a paper on the discovery of the double potassium cyanides of both platinum and palladium. (10)

The discovery of new elements was a prominent feature of inorganic chemistry during this period, and after Oersted's discovery of aluminium Wöhler, on his return to Berlin, set himself the task of its isolation by heating its

257

chloride with potassium but found that the reaction was so violent that his glass apparatus was immediately shattered. He therefore turned to the use of a platinum crucible with a cover, held on by a piece of platinum wire. He succeeded in obtaining molten aluminium in 1827 recording that:

> "At the moment of reduction the crucible became intensely red hot, both within and without, but the metal of the crucible was not sensibly acted upon."

A year later he successfully undertook the preparation of metallic beryllium (then known as glucinium) by exactly the same procedure. (11)

Some years later, when Professor of Chemistry at Cassel, and following up an earlier observation by Berzelius, Wöhler investigated the separation of the platinum metals and devised a method for the refining of iridium and osmium by heating the insoluble residues with sodium chloride in a stream of chlorine. This converted the two metals to soluble chlorides that could then be separated in solution (12). This process was used by platinum refiners for many years, while Wöhler's interest in the subject also remained alive for a long period. The great American historian of chemistry, Dr. Edgar Fahs Smith (1854–1928), who had spent two years in the 1870s under Wöhler, now Professor of Chemistry at Göttingen University, once reminisced towards the end of his life about this formative

258

period and recalled that one of Wöhler's favourite questions to a candidate for the Ph.D. degree was:

"Will you tell me how you would separate the platinum metals from each other." (13)

At least two of Wöhler's students were to play leading parts in the development of the platinum industry in Germany, as will be seen later.

Platinum in the Development of Analytical Chemistry

The general acceptance of the atomic theory and of the concept of definite numerical proportions in chemical composition gave a great impetus to analytical chemistry. In the time of the earlier analysts, Bergman and Vauquelin for example, the composition of very few minerals was known with any degree of certainty. Klaproth was probably the first to determine the reasonably exact constitution of a great number of minerals, but it was Berzelius who laid the foundations of modern gravimetric analysis, improving its accuracy and devising a number of procedures that are still in use today. An important part in this progress was brought about by the introduction of crucibles made in platinum to replace the clay crucibles formerly used for ignitions and fusions. Early in his career he lodged with the wealthy iron master and minerologist Wilhelm Hisinger (1766–1852) and in 1808 Berzelius could record that:

"Only one single platinum crucible was to be found in Sweden. It was owned by Hisinger, who was kind enough to put it at my disposal, but it was too heavy for the balance." (14)

Two years later, however, he was reporting the use of a platinum crucible as a matter of routine (15), while much later on, in a paper dealing with analytical procedures and apparatus, he wrote:

"Since platinum crucibles have been available they have become quite indispensable in chemical research and so much has become possible that was formerly impracticable." (16)

The development of mining and of the engineering industry in the nineteenth century necessitated even greater attention being paid to methods of analysis and this statement from Berzelius was later echoed by Justus von Liebig (1803–1873). Writing from his famous research and teaching laboratory in Giessen in 1844 he commented:

"Without the use of platinum the composition of most of the mineral species would still be unknown." (17)

The Co-ordination Compounds of Platinum

Early in the nineteenth century the study of the compounds of platinum began to interest a number of chemists. The existence of ammonium chloroplatinate was of course well known as the source of the pure metal on heating, and other similar triple salts were soon recognised. One of the first to undertake an investigation of a wide range of compounds was Edmund Davy, the young cousin of Humphry Davy, whose initial discoveries in the field of catalysis were described in the previous chapter. In 1812, while still an assistant to his cousin at the Royal Institution in London, he published a long paper in *The Philosophical Magazine* (18), "On Some Combinations of Platina", in which he reported his findings on its reactions with sulphur, phosphorus, oxygen, chlorine and ammonia, concluding his paper with a rather prophetic comment:

> "Platina appears to be characterised no less by its valuable properties than by its disposition to form peculiar triple compounds; and there can be little doubt that a more extensive acquaintance with the combinations of this metal will add considerably to the number of such substances. It would be an interesting inquiry, whether the same laws which seem to govern the formation of binary compounds extend likewise to those of ternary compounds. For an investigation of this kind no metal seems so well adapted as platina."

In the following year Vauquelin reported the discovery of a compound of palladium with ammonia and chlorine, "a salt of a very agreeable rose colour" that we now know as $[Pd(NH_3)_4]$ $[PdCl_4]$ or as Vauquelin's salt. (19)

Two chlorides of platinum were known at an early stage, platinic chloride, $PtCl_4$, which was investigated by Berzelius who employed its double salt with potassium chloride in his first determination of the atomic weight of platinum, and platinous chloride, $PtCl_2$. It was while working with this latter compound during his stay with Berzelius in 1827 that Heinrich Gustav Magnus discovered the corresponding platinum salt, $[Pt(NH_3)_4]$ $[PtCl_4]$. (20)

The constitution and structure of these first co-ordination compounds remained unknown for many years and they were customarily named after their discoverers. This, his new compound – the first platinum ammine to be prepared – was known then, and still is, as Magnus' Green Salt; it has been the subject of a great deal of study ever since and has played an important part in the development of co-ordination chemistry because of the great stability of this and other platinum compounds. A review of this pioneering work by Magnus and its significance was contributed to *Platinum Metals Review* a few years ago by Professor George Kauffman. (21)

Ten years later a twenty-one year old student from Alsace named James Gros, working in Liebig's laboratory in Giessen, prepared several chlorinated derivatives of Magnus' salt which he described in Liebig's *Annalen* in a paper "On a New Class of Platinum Salts" (22), while in 1840 Jules Reiset (1818–1896), working in the private laboratory of Professor Pelouze in Paris, prepared the base $Pt(NH_3)_4(OH)_2$, of which he regarded the salts of both Magnus and Gros as compounds. (23)

The subject was taken up by an Irish chemist, Robert John Kane (1809–1890) – afterwards Sir Robert – the Professor of Natural Philosophy to the Royal Dublin Society. In 1836 Kane had also spent some time in Liebig's laboratory and in a contribution to *The Philosophical Magazine* in 1841 he argued that Reiset's salt was not at all constituted as the latter had described it (24).

Heinrich Gustav Magnus
1802–1870

After taking his doctoral degree in the University of Berlin in 1827 Magnus, advised by Mitscherlich and Wöhler, went to Stockholm for further study under Berzelius. It was here that he discovered the first complex compound of platinum and ammonia, known always as Magnus' Green Salt, and the forerunner of a great number of co-ordination compounds. He was later appointed Professor of Technology and Physics in the University of Berlin

Reiset pursued the matter, and later obtained a second series of compounds containing only half the amount of ammonia as in the first series from which the so-called Reiset's Second Chloride was prepared (25).

Simultaneously, and quite independently, an Italian, Michele Peyrone (1814–1883), again in Liebig's laboratory, prepared a yellow chloride having the same composition as this second chloride of Reiset's but differing entirely in its properties (26), a finding that was treated with incredibility by many chemists until the concept of isomerism, another term coined by Berzelius, was accepted for inorganic compounds. Peyrone had studied medicine at the University of Turin, but at the age of twenty-five abandoned this profession to take up chemistry, studying first under Dumas in Paris and then going to Giessen. His paper was abstracted by Reiset in the journal he had just founded, *Annuaire de Chimie,* but the editor claimed the discovery as his own (27). Peyrone responded smartly to the effect that his own discovery had been made more than ten months before its publication and that two of Reiset's friends and colleagues had been observers of his work in Giessen during 1843. (28)

In the succeeding volume of his new journal Reiset remained unimpressed. He duly abstracted Peyrone's second paper, but added:

> "The study of these isomers is of great interest, but the conditions surrounding these phenomena must be subjected to a rigorous definition which M. Peyrone has not always managed to obtain." (29)

261

Peyrone was not, however, deserving of these strictures on his ability. He was in fact the first to identify the true formulation of Magnus' Green Salt as $[Pt(NH_3)_4]$ $[PtCl_4]$.

Peyrone later returned to Italy as a Professor, first in Genoa and then in the University of Turin, taking a leading part in agricultural chemistry. He continued to work on isomers of Magnus' Salt, reporting his results to the Royal Academy of Science of Turin in 1847, his paper being published by Liebig with whom he maintained a life-long correspondence (30). Peyrone has some claim to distinction in that his first compound, still known as Peyrone's chloride but now identified as *cis*-dichlorodiammine-platinum (II), has been found in recent years to be a most effective means of treating a number of types of cancer.

Further investigations were carried out by a Russian chemist named Rajewski, working under Pelouze at the College de France in Paris and known there as Raewsky, who reacted Magnus' salt with nitric acid and obtained yet another series of compounds differing from those prepared by Gros (31). Much discussion about the constitution of these compounds was now going on and the debate was entered by the two famous but unfortunate collaborators Charles Gerhardt and Auguste Laurent. In two papers published in the journal they had themselves established (32) and in a further report to the Académie des Sciences by Gerhardt alone (33) they ascribed formulae in the manner of the time to all these preparations and gave a plausible explanation of their structures and of the relationship of one series to another.

A few years later interest was still increasing, not only in platinum but in its allied metals, and by 1854 Karl Klaus could open a paper with these comments:

"The discovery of Gros' Salt led the way to a series of curious compounds of platinum which so attracted the attention of chemists that very shortly thereafter Reiset, Peyrone, Raewsky, Laurent and Gerhardt occupied themselves with the subject and considerably increased their number. Just recently Skoblikoff [in St. Petersburg] and Hugo Müller [in Göttingen] have made known two new such compounds of iridium and palladium and I can now add two others based upon rhodium and a second of iridium, to join them, while at the same time I am in the course of preparing similar compounds of osmium and ruthenium." (34)

One of the rhodium complexes mentioned, chloropentammine rhodium chloride, was inevitably known from then onwards as Klaus' Salt.

Much later a major study of these ammoniacal platinum compounds was undertaken by the Swedish chemist Per Teodor Cleve (1840–1905) working in the laboratory of the Academy of Sciences in Stockholm after a short period of study under Professor Wurtz in Paris. He first examined the compounds obtained by Gros, Reiset and Raewsky, publishing his findings only in Swedish, and then discovered an entirely new series of salts containing two atoms of platinum by the action of ammonia on the iodine derivatives of Gros' base, while he also found that tetravalent platinum yielded a series of ammonium compounds. This work occupied him for six years, and he finally assembled all his results in a memoir in English detailing the systematic arrangement of the many compounds then known (35).

Alfred Werner
1866–1919

For many years the structure of the many complex compounds of platinum and other metals remained uncertain. In 1893 Werner, just appointed Professor of Chemistry at Zürich, after carrying out research on the metal-ammonia compounds, put forward his co-ordination theory and explained the spatial arrangements of the components of these complexes. In 1913 he was awarded the Nobel Prize in Chemistry for this achievement

Great difficulty was still experienced in the assigning of structures to the ever growing number of complex compounds of this type, but in 1893 the well known co-ordination theory put forward by Alfred Werner, Professor of Chemistry at Zürich at the age of only twenty-six, not only yielded a satisfactory explanation but also enabled new compounds with predictable properties to be synthesised (36). He followed with a paper jointly with his friend and collaborator Arturo Miolati (1869–1956), an Italian who had been a fellow student of Werner's at the Technical High School in Zürich. Together they·measured the molecular conductivities of many complex compounds including those of platinum, finding that the number of ions always agreed with the co-ordination theory (37). In 1966, the centenary of Werner's birth, a review of his work on the structure of platinum complexes was contributed to *Platinum Metals Review*, by Dr. W. A. Smeaton, while a more detailed account has been given by Professor George B. Kauffman of Werner's controversy with yet another Dane, Professor Sophus Mads Jørgensen (1837–1914) of the University of Copenhagen, who had also prepared a number of co-ordination compounds of platinum and rhodium. (39)

263

The First Organometallic Compounds

The study of organometallic compounds of the platinum metals has become intensive only during the last two or three decades, while their great usefulness in homogeneous catalysis has been recognised and is now being adopted in the chemical industry. The origin of this major development goes back very many years, however, and once again to Scandinavia. The first platinum compound of this type – in fact the first such compound of any transition metal – was a platinum ethylene complex prepared in 1830 by William Christopher Zeise in Copenhagen and known inevitably from then onwards as Zeise's Salt. The son of an apothecary, Zeise became a lecture assistant to Hans Christian Oersted and then in 1822 was appointed Professor of Chemistry in the University of Copenhagen after spending some time under Professor Stromeyer at Göttingen. In that year he discovered the xanthates and in 1825 he prepared a compound of platinum, carbon and oxygen which he found would ignite alcohol just as would Edmund Davy's preparation mentioned earlier in Chapter 12 (40). Following this line of research five years later he investigated the reaction between alcohol and platinous chloride and prepared the potassium salt $K[PtCl_3(C_2H_4)]$ in the form of "beautiful yellow crystals". This basic discovery was actually first reported in Latin to the Royal Danish Academy of Sciences in November 1830 and was reproduced in the following year in most of the German chemical

William Christopher Zeise
1789–1847

Beginning his career as a pharmacist, Zeise became an assistant to Oersted in the University of Copenhagen and then, after a period with Stromeyer in Göttingen, returned there as Professor of Chemistry. He prepared the first organometallic compounds of platinum in 1830, so laying the foundations of the much later investigations of the great number of such compounds of all the platinum metals and their usefulness in homogeneous catalysis

264

journals (41). Zeise gave full details of his analysis of the new compound, but he unfortunately became the recipient of some harsh criticism from Justus von Liebig, already in deep controversy with Dumas about the constitution of alcohol and ether, who claimed that Zeise's analysis was in error (42). Objecting strongly to Liebig's unpleasant comments, Zeise repeated his analyses, completely verifying the composition of the new salt (43), but Liebig, reproducing the paper in his own *Annalen*, followed it immediately with another long polemical contribution "On the Ether Theory with Special Reference to the Preceding Communication of Zeise" (44). The correctness of Zeise's results was established, and he went on to prepare further organometallic compounds of platinum with acetone (45), as well as some ammonium and potassium ethyl platinum complexes that he described in a text book in Danish that he published during the last year of his life. (46)

This work was followed up by Charles Adolphe Wurtz (1817–1884), another student of Liebig's, who discovered methylamine and ethylamine in 1849 just after he had succeeded Dumas as lecturer in organic chemistry at the École de Médecine in Paris. In the course of this investigation he prepared golden-yellow crystalline complexes of platinous chloride with both these new compounds. (47)

The thirty-year old controversy between Zeise and Liebig was revived in London by two assistants of August Wilhelm Hofmann (1818–1892), yet another Liebig student who had been persuaded by the Prince Consort to accept the post of Professor of Chemistry in the newly founded Royal College of Chemistry in 1845. These were Johann Peter Griess (1829–1888) who had studied first at Jena and then under Kolbe at Marburg and who had been invited to London in 1858 by Hofmann, and Carl Alexander Martius (1838–1920) whose inaugural dissertation at Göttingen before coming to London had been on the cyanides of the platinum metals. They first confirmed Zeise's formula and demonstrated that ethylene was liberated when his compound was decomposed, going on to prepare analogous salts containing some of the new organic molecules discovered by Hofmann such as diphenyl amine, ethylene diammine and aniline (48). Griess is better known for his discovery of the diazo compounds and for his later long career as chemist to Allsopps' brewery in Burton-on-Trent, later becoming a Fellow of the Royal Society and one of the founders of the Institute of Chemistry, while Martius was the discoverer of naphthalene dyes and the founder of the AGFA concern in Berlin.

The study of these organometallic compounds of platinum was also taken up by Karl Birnbaum (1839–1887), Professor of Chemistry in the Karlsruhe Polytechnic, who in 1867 undertook their synthesis by a different route, starting with ethylene itself. He succeeded in obtaining not only the original Zeise salt but its homologues with propylene and amylene. (49)

No further work seems to have been carried out on these complexes for a very long time, in fact not until 1907 when W. J. Pope (later Sir William Jackson Pope) and his colleague Stanley John Peachy at the University of Manchester announced to a meeting of the Chemical Society that all the six platinum metals had been found to react with magnesium methyl iodide. Two years later they described in detail the preparation of a number of alkyl compounds of platinum (50). The chairman of the earlier meeting, Sir Henry Roscoe, congratulated these two authors on

"having opened out an entirely new branch of investigation which might indeed be said to be a wonderful find".

There is now, of course, a vast literature on the organo-metallic compounds of all six platinum metals, on their constitution and on their actual and potential applications in chemical industry.

The Absorption of Hydrogen by Palladium

One further nineteenth century discovery that has become of considerable industrial importance was the ability of palladium, far in excess of that of any other metal, to absorb hydrogen. This stemmed from the researches of Thomas Graham (1805–1869) who succeeded Andrew Ure as Lecturer in chemistry at Anderson's College in Glasgow and then in 1837 followed Edward Turner as Professor of Chemistry at University College, London. In a paper read to the Royal Society of Edinburgh in 1831 (51), Graham succeeded in establishing the relationship between the rate of diffusion of a gas and its density – the well known Graham's Law – and for the remainder of his life he continued to interest himself in problems of diffusion.

In 1854 he resigned from University College on his appointment as Master of the Royal Mint in succession to Sir John Herschel and his energies were for some years entirely devoted to the duties of his office, including the supervision of the major change from copper to bronze coinage. Later he was able to continue his work on diffusion, and in 1866 he presented a paper to the Royal Society, "On the Adsorption and Dialytic Separation of Gases by Colloid Septa" (52). He had first studied the behaviour of hydrogen exposed to red-hot platinum and had found that not only was it absorbed but that it could be retained for an indefinite time, and that no other gases showed the same effect.

Turning from platinum to palladium, Graham found that it could absorb five or six hundred times its own volume of hydrogen and that when exposed to coal gas only the hydrogen penetrated the metal. No such effect could be obtained with iridium and osmium.

A year later he showed that hydrogen could be occluded by palladium electrolytically when immersed in dilute sulphuric acid and in contact with a piece of zinc (53). His last major contribution, published only a few months before his death, was on the relation of hydrogen to palladium (54). In this he considered that they formed an alloy or compound and that hydrogen was "the vapour of a highly volatile metal" which he named Hydrogenium. He also studied the occlusion of hydrogen by alloys of palladium containing varying percentages of silver made for him by George Matthey, and found them to give the same effect provided that the alloying element did not exceed 50 per cent.

266

In 1869, only shortly before his death, Graham had a number of medallions made at the Mint in what he conceived to be an alloy of palladium and "hydrogenium", the name he coined for the form in which hydrogen existed when absorbed. These he distributed to his many friends to demonstrate the nature of this unusual combination of a gas and a metal

In all this work Graham had the co-operation of his personal assistant, a young graduate from the Royal School of Mines named W. C. Roberts. On Graham's death Roberts was appointed Chemist to the Mint and became famous later as the metallurgist Sir William Chandler Roberts-Austin whose work with platinum will be referred to later.

Exercising his influence at the Mint, Graham had a number of small medallions struck in his "palladium-hydrogenium" alloy and distributed these to his many friends. In presenting one to the then Chancellor of the Exchequer, Robert Lowe, he wrote rather quaintly:

"The little medallion is composed of about 9 parts of palladium (a rare metal) and 1 part of hydrogenium by *bulk*. If the latter took the form of gas, it would measure 8 or 9 cubic inches, or 3 port wine glasses full, to be very plain."

Some of the curious effects of this unusual phenomenon, particularly the dimensional changes that occur when palladium is alternatively heated and cooled in hydrogen, were investigated in 1893 by Copius Hoitsema in the University of Leyden, working under Professor Bakhuis Roozeboom who had done much to interpret alloy systems in terms of Willard Gibbs' Phase Rule. He concluded that two distinct hydrogen-palladium phases were formed, changing as hydrogen is occluded or released (55). (Curiously Hoitsema also left the chemical profession to become Master of the Mint in Utrecht in 1909). This study was but the first of literally hundreds of investigations that have been carried out over the past century and are still continuing. But it was Graham who laid the foundations, and the modern outcome of his work is to be seen in the design and operation of equipment to generate high purity hydrogen from a variety of intake gases that are now relied upon by many types of industrial users. Hydrogen generators, built by Johnson Matthey, incorporating diffusion tubes in a silver-palladium alloy, are now in service in applications as diverse as the hydrogenation of edible oils, the manufacture of semi-conductors, the annealing of stainless steel and the cooling of power station alternators. (56)

References for Chapter 14

1 J. B. Richter, Anfangsgründe der Stöchyometrie oder Messkunst chymische Elemente, Breslau and Hirschberg, 1792

2 J. L. Proust, *J. de Phys.*, 1806, **63**, 367–369

3 J. Dalton, *Mem. Manchester Lit. and Phil. Soc.*, 1805, **1**, 271–287

4 T. Thomson, *Phil. Trans.*, 1808, **98**, 63–95

5 W. H. Wollaston, *Phil. Trans.*, 1808, **98**, 96–102, *J. Nat. Phil.*, (*Nicholson*), 1808, **21**, 164–169

6 J. J. Berzelius, *Ann. Phil. (Thomson)*, 1814, **3**, 252–257; 353–364; *J. Chem. (Schweigger)*, 1817/18, **21**, 307–341; **22**, 51–77; 317–343

7 J. J. Berzelius, Essai sur la Théorie des Proportions Chimiques, Paris, 1819

8 J. J. Berzelius, *Kongl. Vetensk. Akad. Handl.*, 1828, **16**, 25–116; *Ann. Phys. (Poggendorff)*, 1828, **13**, 435–488; 527–565; *Ann. Chim.*, 1829, **40**, 51–82; 138–165; 257–285; 337–350

9 J. J. Berzelius, *Ann. Chim.*, 1838, **28**, 426–427

10 L. Gmelin and F. Wöhler, *J. Chem. (Schweigger)*, 1822, **36**, 230–231

11 F. Wöhler, *Ann. Phys. (Poggendorff)*, 1827, **11**, 146–161

12 F. Wöhler, *Ann. Chim.*, 1828, **39**, 77–84

13 E. F. Smith, *J. Chem. Ed.*, 1928. **5**, 1554–1557

14 J. J. Berzelius, Autobiographical Notes, trs. O. Larsell, Baltimore, 1934, 63

15 J. J. Berzelius, *Afhandl. Fysik. Kemie Mineralogen*, Stockholm, 1810; *Ann. Phys. (Gilbert)*, 1811, **37**, 249–339

16 J. J. Berzelius, *J. tech. Chem.*, 1832, **13**, 358

17 J. Liebig, quoted by H. F. Keller, *J. Franklin Inst.*, 1912, **174**, 541

18 E. Davy, *Phil. Mag.*, 1812, **40**, 209–220; 263–278; 350–365

19 N. L. Vauquelin, *Ann. Chim.*, 1813, **88**, 167–186

20 H. G. Magnus, *Ann. Phys. (Poggendorff)*, 1828, **14**, 239–242

21 G. B. Kauffman, *Platinum Metals Rev.*, 1976, **20**, 21–24

22 J. Gros, *Ann. Chem. (Liebig)*, 1838, **27**, 241–256

23 J. Reiset, *Comptes Rendus*, 1840, **10**, 870–872

24 R. J. Kane, *Phil. Mag.*, 1841, **18**, 293–296

25 J. Reiset, *Ann. Chim.*, 1844, **11**, 417–433

26 M. Peyrone, *Ann. Chem. (Liebig)*, 1844, **51**, 1–29; *Ann. Chim.*, 1844, **12**, 193–211

27 J. Reiset, *Annuaire de Chim.*, 1845, **1**, 166

28 M. Peyrone, *Ann. Chem. (Liebig)*, 1845, **55**, 205–213

29 J. Reiset, *Annuaire de Chim.*, 1846, 224–225

30 M. Peyrone, *Ann Chem. (Liebig)*, 1847, **61**, 178–181

31 M. Raewsky, *Comptes rendus*, 1846, **23**, 353–354; 1847, **24**, 1151–1154; **25**, 794–797

32 A. Laurent and C. Gerhardt, *Compt. rend. Trav. Chim.*, 1849, **5**, 113–115; 1850, **6**, - 273–304

33 C. Gerhardt, *Comptes rendus*, 1850, **31**, 241–244

34 C. Claus, *J. prakt. Chem.*, 1854, **63**, 99–108

35 P. T. Cleve, On Ammoniacal Platinum Bases, Stockholm, 1872

36 A. Werner, *Z. anorg. Chem.*, 1893, **3**, 267–330

37 A. Werner and A. Miolati, *Z. Phys. Chem.*, 1893, **12**, 35–55

38 W. A. Smeaton, *Platinum Metals Rev.*, 1966, **10**, 140–144

39 G. B. Kauffman, *Chymia*, 1960, **6**, 180–224

40 W. C. Zeise, *Oversegt Kongl. Dansk. Videnskab Forhandl.*, 1825–6, 13; *Ann. Phys. (Poggendorff)*, 1827, **9**, 632

41 W. C. Zeise, *Ann. Phys. (Poggendorff)*, 1831, **21**, 497–541; *J. Chem., (Schweigger)*, 1831, **62**, 393–441; **63**, 121–135

42 J. Liebig, *Ann. Chem. (Liebig)*, 1834, **9**, 1–39

43 W. C. Zeise, *Ann. Chim.*, 1836, **63**, 411–431; *Ann. Chem. (Liebig)*, 1837, **23**, 1–11

44 J. Liebig, *Ann. Chem. (Liebig)*, 1837, **23**, 12–42

45 W. C. Zeise, *Ann. Phys. Poggendorff*, 1838, **45**, 332–336: 1839, **47**, 478–480; *Phil. Mag.*, 1839, **14**, 84–87

46 W. C. Zeise, *Haandbog de Organiske Stoffers almindelige Chemie*, Copenhagen, 1847

47 C. A. Wurtz, *Ann. Chim.*, 1850, **30**, 443–507

48 J. P. Griess and C. A. Martius, *Ann. Chem. (Liebig)*, 1861, **120**, 324–327; *Comptes rendus*, 1861, **53**, 922–925

49 K. Birnbaum, *Ann. Chem. (Liebig)*, 1868, **145**, 67–77

50 W. J. Pope and S. J. Peachey, *Proc. Chem. Soc.*, 1907, **23**, 86; *J. Chem. Soc.*, 1909, **95**, 571–576

51 T. Graham, *Phil. Mag.*, 1833, **2**, 175–190; 269–276; 351–358

52 T. Graham, *Phil. Trans.*, 1866, **156**, 399–439

53 T. Graham, *Proc. Roy. Soc.*, 1867/8, **16**, 422–427

54 T. Graham, *Proc. Roy. Soc.*, 1868/9, **17**, 212–220; 500–516

55 C. Hoitsema, *Z. phys. chem.*, 1895, **17**, 1–42

56 D. A. Stiles and P. H. Wells, *Platinum Metals Rev.*, 1972, **16**, 124–128; M. J. Cole, *Platinum Metals Rev.*, 1981, **25**, 12–13

Robert Hare
1781–1858

Born in Philadelphia the son of a brewer who had emigrated from
England. Hare studied chemistry under James Woodhouse and carried
out research on the production of high temperatures in his free time. By
means of his oxy-hydrogen blowpipe he became the first scientist to
succeed in melting platinum in substantial quantity

From a portrait in the possession of the University of Pennsylvania

15

The Melting of Platinum and the New Metallurgy of Deville and Debray

"It is therefore necessary to have a method of treatment more expeditious and more practical than that now adopted. It is for this reason that we are proposing an entirely new metallurgy for platinum."

HENRI SAINTE-CLAIRE DEVILLE
AND JULES HENRI DEBRAY

All the platinum so far described as having been worked into useful articles had been prepared from the powder form by hot forging, and all attempts at its melting had succeeded only on a most minute scale. The time was approaching, however, when melted platinum was to become available to industry. The first effective procedure was devised by Robert Hare, the young son of a brewer of the same name in Philadelphia who had emigrated from England in 1773 and who later became Speaker of the State Senate of Pennsylvania. While attending the lectures on chemistry given by James Woodhouse at the University of Pennsylvania and also helping his father in the brewery business the younger Robert Hare undertook research in his spare time in the cellar of the house in which he lodged on the corner of Dock Street and Walnut Street (1). He was deeply interested in the possibility of obtaining very high temperatures by means of an oxy-hydrogen blow-pipe and he was also aware of Lavoisier's melting of a very small amount of platinum by the use of oxygen alone, described in Chapter 4. He felt that Lavoisier's apparatus was both too unwieldy and too expensive, while he reasoned that an even higher temperature would be obtained by the combustion of hydrogen and oxygen together. He therefore designed and built the apparatus illustrated on the following page, based upon a barrel from the brewery and not noticeably less unwieldy or less complicated than Lavoisier's equipment! By supplying a lamp or a candle with a continuous stream of oxygen and hydrogen kept in storage vessels and expelled by hydrostatic pressure through a common orifice, Hare was able to melt a number of substances, among them platinum, that had hitherto been regarded as extremely refractory. He had in fact achieved the highest temperature as yet attained.

271

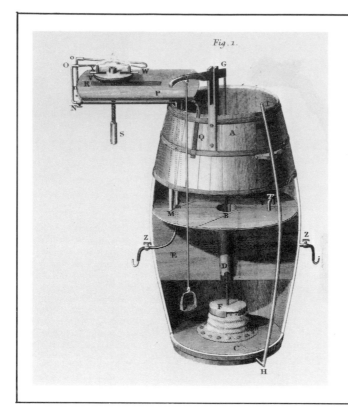

Fig. 1.

The apparatus designed and built by Robert Hare in 1801 for the melting of platinum and other refractory substances. Based upon a barrel from his father's brewery, it comprised two separate storage vessels for hydrogen and oxygen, the gases being driven through a common orifice into the flame of a lamp, shown at the top left, mounted on an adjustable table

Hare had become a member of the Chemical Society of Philadelphia, founded in 1792, of which James Woodhouse was President, and in December 1801 he demonstrated his apparatus before the society. His long and detailed account of the equipment and his experiments was immediately published in pamphlet form by the society and was reproduced in England in *The Philosophical Magazine* and in France in the *Annales de Chimie* (2).

Among his fellow boarders in the house was Benjamin Silliman (1779–1864) who had been invited to accept the chair of chemistry at Yale but having no knowledge of the subject had been sent to study at the University of Pennsylvania. He and Hare co-operated in improving the apparatus and in obtaining a higher purity oxygen by heating potassium chlorate in stone retorts. Silliman later recorded in his autobiography:

"The retorts were purchased by me at a dollar each and as they were usually broken in the experiment the research was rather costly, but my friend furnished experience and as I was daily acquiring it I was rewarded both for labour and expense by the brilliant results of our experiments." (3)

During the winter of 1802–1803 Hare had an opportunity of demonstrating his blow-pipe to the discoverer of oxygen, Joseph Priestley, now approaching seventy, who it will be remembered had been informed by Benjamin Franklin twenty years earlier of Lavoisier's experiment. Priestley acknowledged that Hare's experiments were "quite original". (4)

For a number of years Hare was occupied in the brewery business, and then in 1818 he became for a short time Professor of Natural Philosophy at the College of William and Mary in Williamsburg but was soon appointed to the post he had long earnestly desired, Professor of Chemistry in the Medical School of the University of Pennsylvania. Here he took up the study of electricity, collaborating with Joseph Henry and corresponding with Michael Faraday. Silliman, on the other hand, annually demonstrated the fusion of platinum and of other substances by means of Hare's blow-pipe to his students at Yale.

The Controversy with Professor Clarke

Although Hare did not pursue his blow-pipe researches any further for a number of years he was aroused to great indignation in 1819 by what appeared to him an appropriation of his invention. A letter from a Mr. H. I. Brooke of Keppel Street in London to Thomson's *Annals of Philosophy* in 1816 described the design of a new form of blow-pipe which the writer had asked the instrument maker John Newman to construct (5). This came to the notice of Edward Daniel

Edward Daniel Clarke
1769–1822

Professor of Mineralogy at Cambridge and a friend of Wollaston, Smithson Tennant and Humphry Davy, Clarke experimented with an improved form of blow-pipe from 1816 until his death, successfully melting small quantities of platinum. After describing his results in several papers, in 1819 he published a small book that provoked Robert Hare to an indignant response

Clarke, Professor of Mineralogy at Cambridge, who had been in the habit of submitting his mineral specimens to the ordinary blow-pipe. Clarke persuaded Newman to modify his apparatus to some extent and then proceeded to carry out a number of experiments with it, using hydrogen and oxygen in the carefully controlled proportions of two to one. He melted a range of substances, including platinum and palladium, and began to publish a series of papers, first in the Royal Institution's *Quarterly Journal of Science* and then in the *Annals of Philosophy* (6). In the first of these papers he wrote:

"Platinum was not only fused the instant it was brought into contact with the flame of the ignited gas, but the melted metal ran down into drops."

Wollaston then wrote to him suggesting that he tried to melt "ore of iridium" and this he claimed to have done successfully as well as melting a small quantity of iridium-osmium residue that had belonged to Smithson Tennant. After many further experiments, and after suffering a dangerous explosion of his apparatus, Clarke devised a wooden screen to protect himself or his man-servant, and again he melted platinum but only in small quantities. Then in 1819 he published a small book (7) describing the whole range of his work, the introduction including the sentence:

"The American Chemists lay claim to it as their invention in consequence of

Instead of using two separate containers for his oxygen and hydrogen Clarke employed only one, with the consequent danger of explosion. After one such experience he devised a wooden screen to protect himself or his man servant while melting platinum and other substances with his blow pipe.

274

experiments made in 1802 by Mr. Robert Hare, junior, Professor of Natural Philosophy in Philadelphia."

On receiving a copy of the book Hare immediately reacted angrily, addressing a long letter to his friend Silliman who had just recently founded the *American Journal of Science and Arts*, quoting some lines of Virgil written on hearing that the Roman dramatist Bathyllus had claimed some of his own verses (8). The opening page of this diatribe is reproduced here. One of his more interesting comments was:

"He (Clarke) would evidently wish the reader to adopt the false impression that the facility with which platinum may be fused is owing to 'the great improvements' made fourteen or fifteen years after I had devised and used them. Will Britons tolerate such conduct in their professors?"

The opening page of Hare's letter to the *American Journal of Science* in which he attacked Clarke for what he considered to be the appropriation of his invention. The Latin lines from Virgil read in translation:

"I indeed have made these verses but another has received the honour,
So your birds make nests, not for yourselves,
So your sheep bear fleeces, not for yourselves,
So your bees make honey, not for yourselves,
So your oxen do not bear the plough for yourselves."

CHEMISTRY, PHYSICS AND THE ARTS.

ART. XIII.—*Strictures on a publication, entitled Clark's Gas Blowpipe; by* ROBERT HARE, *M. D. Professor of Chemistry in the medical department of the University of Pennsylvania, and the real inventor of the compound or hydro-oxygen blowpipe, in that safe and efficient form by which the fusion of the most refractory earths, and the volatilization and combustion of Platinum was first accomplished.*

Hos ego versiculos feci, tulit alter honores,
Sic vos non vobis nidificatis aves,
Sic vos non vobis Vellera ferti oves,
Sic vos non vobis melificatis apes,
Sic vos non vobis fertis aratra Boves.
VIRGIL.

DR. CLARK has published a book on the Gas Blowpipe, in which he professes a "sincere desire to render every one his due." That it would be difficult for the conduct of any author to be more discordant with these professions, I pledge myself to prove in the following pages, to any reader whose love of justice may gain for them an attentive perusal.

In the year 1802, in a memoir republished in the 14th Vol. of Tilloch's Philosophical Magazine, London, and in the 45th Vol. of the Annales de Chimie, I had given the rationale of the heat produced by the combustion of the aeriform elements of water, and had devised a mode of igniting them free from the danger of explosion. I had also stated in the same memoir that the light and heat of the flame thus produced were so intense, that the eyes could scarcely sustain the one, nor the most refractory substances resist the other, and had likewise mentioned the fusion of the pure earths and volatilization of the perfect metals as among the results of the invention.

Subsequently in the first part of the 6th Vol. of American Philosophical Transactions, an account of the fusion of strontites, and the volatilization of Platinum, was published by me.

275

The letter also contained a series of drawings of the various forms of the compound blow-pipe as devised by Hare over the years. Clarke died in 1822 and there is no evidence that he ever responded to or acknowledged this rather harsh piece of criticism, or that he pursued his work on the melting of platinum any further.

On his part, Hare again took up the subject after a long interval. In 1836 he made a visit to England and attended the meeting of the British Association for the Advancement of Science, held that year in Bristol, meeting the leading scientists of the time including the seventy year old John Dalton, then a Vice-President of the Chemistry Section, and contributing several short papers on improvements in chemical apparatus. In August 1837 he wrote to Dalton to report his melting of a more substantial quantity of platinum:

> "I beg leave through you to communicate to the British Association for the Advancement of Sciences that by an improvement in the method of constructing and supplying the hydro-oxygen blow-pipe, originally invented by me in the year 1801, I have succeeded in fusing into a malleable mass more than three fourths of a pound of platina. In all I fused more than two pounds fourteen ounces into four masses, averaging of course nearly the weight above mentioned. I see no difficulty in succeeding with much larger weights."

This letter was duly read to the British Association by Dalton during its meeting in Liverpool a month later (9), but Hare was still not satisfied. In September 1838 he exhibited to the Chemical Society of Philadelphia an even larger specimen of platinum, writing to Silliman:

> "I have by improvements in my process for fusing platina succeeded in reducing twenty-five ounces of that metal to a state so liquid that, the containing cavity not being sufficiently capacious, about two ounces overflowed it, leaving a mass of twenty-three ounces. I repeat that I see no difficulty in extending the power of my apparatus to the fusion of much larger masses." (10)

After a further interval of some eight years Hare, towards the end of his occupation of the chair of chemistry, announced the melting of both iridium and rhodium, specimens of which he had obtained from Johnson and Cock. (11)

Hare took no commercial advantage of his long series of investigations, but his work did lead to two further developments in quite different directions. First, his oxy-hydrogen blow-pipe was adapted to good effect by the Scotsman Thomas Drummond (1797–1840) who served in his earlier years in the British Ordnance Survey where he devised a source of intense white light by directing the flame on to a block of lime. This was designed as an improvement to navigational safety, and his invention began to be installed in lighthouses in 1829, later being adopted for the theatre from where the expression "in the lime-light" originated.

Working with this means of illumination in Paris, Marc Antoine Gaudin (1804–1880), a former student of both Dumas and Ampére who was employed by the Bureau of Longitude, presented a paper to the Académie des Sciences in 1838 in which he described his method of preparing a crystalline form of lime from which he made crucibles (12). In these he was able to melt an alloy of 10 per cent iridium and 90 per cent platinum, almost certainly the first occasion on which a synthetic alloy of platinum had been produced. He commented on the lustre, the malleability and the extreme resistance to corrosion of his alloy.

The second and more significant outcome of Hare's work was due to his assistant in the University of Pennsylvania. This was Joaquim Bishop (1806–1886) who after working in the jewellery trade and then in a brass foundry became Hare's instrument maker in 1832 and took part in the experiments on platinum. Leaving Hare in 1839, he set up in business for himself and so founded the platinum works that bore his name for very many years and that will be dealt with more fully in a later chapter.

The Work of Deville and Debray

Hare noticed that if he maintained his molten platinum in an oxidising atmosphere it gradually freed itself from any base metals present by oxidation, and it was this observation that led to the next major step forward in the history of platinum, this taking place in Paris in the eighteen-fifties at the hands of Deville and Debray and constituting a revolution in the platinum industry.

Henri Sainte-Claire Deville was born in St. Thomas, one of the Virgin Islands, the son of a prosperous ship owner who had emigrated from France. He was sent to Paris to be educated, studying medicine but also attending the lectures of Thenard at the Sorbonne. In 1839 he set up a small private laboratory in which he began to make original investigations on organic substances including essential oils. On the recommendation of Thenard he was appointed Professor of Chemistry at Besançon in 1845 at the age of only twenty-six. When Balard was appointed Professor of Chemistry in the College de France in 1851 Deville succeeded him at the École Normale where he remained for the rest of his life, converting a most inadequate laboratory for the training of teachers into one of the outstanding centres of research in Europe. Here he found Jules Henri Debray, a native of Amiens who had been a student there since 1847 and whom he appointed as his assistant in 1855, the association developing into a great friendship and a most fruitful collaboration with Debray eventually succeeding him as professor.

The ten years following Deville's appointment were characterised by quite exceptional activity and achievement in the field of high temperature reactions. First he succeeded in producing aluminium by the reduction of its chloride with potassium, later with sodium, and his results being brought to the attention of Napoleon III by Dumas, he was given a government grant to establish a pilot plant at Javel. This was followed by the erection of a larger works at Nanterre, Deville and his colleagues Debray and Paul Morin subscribing capital to form the Société de l'Aluminium de Nanterre.

With production established, Deville returned to his laboratory and again gave his mind to the need for better methods of producing high temperatures. In

**Henri Sainte-Claire
Deville
1818–1881**

Professor of Chemistry at the
École Normale in Paris,
Deville followed up his researches on the production
of aluminium by studying the
metallurgy of the platinum
metals. In 1857 he and Jules
Debray devised the lime-
block furnace fired by a
mixture of oxygen and coal
gas which for the first time
made it possible to refine and
melt platinum and alloys on a
large scale. Their method
remained in use until the
development of the induction
furnace in the 1920s

1856 he had published a long study of the use of coal gas and oxygen in blow-
pipes and their application to both melting and welding in which he referred
to the earlier work of Gaudin (13). At first Deville and Debray used the equipment to produce manganese, chromium, nickel and cobalt in a state of purity,
using crucibles made of lime or magnesia.

They then turned to the platinum metals and for four years pursued their
researches with great activity. Already in the 1856 paper Deville had made the
point that his melted platinum had properties quite different from those of the
powder metallurgy product. He had enjoyed the co-operation of a Parisian goldsmith and manufacturer of "doublé" or rolled gold, Auguste François Savard,
and he commented:

> "Cast and refined platinum is a metal as soft as copper, as confirmed by the Paris
> Mint; it is whiter than ordinary platinum and does not have the porosity that has so
> far proved to be an obstacle in the manufacture of an impermeable doublé of
> platinum."

278

Savard, whose firm, founded in 1829, still exists in the Rue Saint Gilles, had put his furnaces and rolling mills at Deville's disposal and had made a very thin piece of platinum-clad copper that completely resisted the action of nitric acid.

Deville and Debray's final design of a furnace for the melting of platinum, illustrated here, comprised two cylindrical blocks of lime bound together with steel strip. A hollow was formed in the lower block to contain the metal, a pouring channel being provided to the edge. The upper block was also hollowed out to form a roof which was pierced to receive the coal gas and oxygen. The whole unit could be tilted for pouring the molten platinum. By this means platinum was not only melted in quantity for the first time but could also be refined to some extent by exposure to an oxidising atmosphere, while the lime served to absorb any slag formed by the oxidation of base metal impurities.

Following this, Deville and Debray went on to devise a process for refining native platinum. Their preliminary results were described to the Académie des Sciences in 1857 (14) and in the same year both French and British patents were filed, not in Deville's name but in that of Debray, and assigned for some reason to the Société de l'Aluminum de Nanterre (15). The British rights were at once

Jules Henri Debray
1827–1888

A native of Amiens, Debray studied at the École Normale and then became first an assistant and later a collaborator of Deville's, eventually succeeding him as Professor there in 1881. Their association was extremely close, and their joint work on the melting of platinum and its alloys extended over many years

279

François Auguste Savard
1803–1875

The founder of a firm producing rolled gold in Paris in 1829, Savard turned to the production of platinum-clad copper, brass and silver in 1854. During the researches of Deville and Debray he put at their disposal both his furnaces and his rolling mills so that they could investigate the physical properties of their alloys. He also produced extremely thin coatings of platinum on copper from their melted metal that displayed complete freedom from porosity. His company continues in operation to this day

acquired by George Matthey whose long career in the platinum industry will be described in the next chapter, while the French patent was shortly afterwards acquired by Desmoutis and Chapuis.

The response from Russia was even more positive. Just at this time the government in St. Petersburg was considering restarting the minting of a platinum coinage, and in June 1859 their representative, Academician Boris Semenovich Yacobi (Known in Western Europe as Moritz Hermann Jacobi), Industrial Adviser to the Ministry of Finance, visited Paris to study the techniques of Deville and Debray. His visit coincided with the publication of a long paper by them, On Platinum and the Metals that Accompany It (16), in which they detailed not only the properties of each of the six platinum metals but gave a complete scheme for their analysis and finally proposed a simple and economical process for yielding a ductile and industrially useful iridium-rhodium-platinum alloy directly from the Russian mineral. The process included the preliminary alloying of the native platinum with lead to separate osmiridium and earthy matter, the lead alloy then being cupelled to leave the platinum rich material ready for melting.

280

DE LA

MÉTALLURGIE DU PLATINE

ET DES

MÉTAUX QUI L'ACCOMPAGNENT,

PAR MM. H. SAINTE-CLAIRE DEVILLE ET H. DEBRAY.

Nous avons exposé, dans le LVIe volume de ces *Annales*, un projet de métallurgie nouvelle du platine fondée sur l'emploi des moyens de la voie sèche : notre travail était fini et publié, lorsque le Gouvernement russe, par l'intermédiaire de M. Jacobi, conseiller d'État et membre de l'Académie des Sciences de Saint-Pétersbourg, nous proposa d'étudier, sur une échelle relativement assez grande, toute la partie pratique de la question que le manque de matériaux nous avait obligés de négliger momentanément. Nous avons accepté cette mission avec empressement, en nous imposant à nous-mêmes la condition de diriger nos travaux de telle manière qu'ils pussent être adaptés le plus complétement possible aux besoins du Gouvernement russe. La Monnaie de Russie reçoit en effet chaque année des quantités de minerai de platine de l'Oural assez variables, qu'elle traite elle-même par un procédé qui sera bientôt

Deville and Debray demonstrated their process to Jacobi, who was so impressed that he asked his Minister of Finance to sponsor further research on melting and on the working up of the residues that had accumulated in great quantity in the refinery in St. Petersburg. An agreement was quickly concluded by which the Russian government financed the setting up of furnaces and of an oxygen plant in the École Normale and Deville and Debray were enabled to carry out their experiments on much larger quantities of material. On February 23rd they received from St. Petersburg some 56 kilograms of mineral from the Demidov workings at Nizhny Tagil, demonetised coinage and refinery residues. On June 15th "after three and a half months of incessant work day and night", they delivered to Jacobi 42 kilograms of ingots, rolled sheet and cast objects, together with an ingot of iridium weighing just over a kilogram. Their loss of platinum amounted to only 120 grams. This remarkable piece of work was described in detail in a further long paper (17), of which the opening page is reproduced here, with full particulars of the amounts of lead, oxygen and reagents consumed.

281

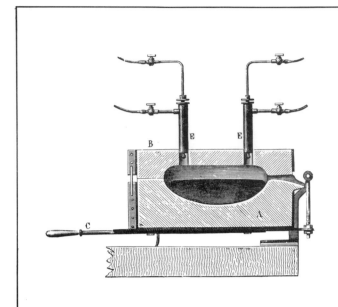

An original drawing of the lime-block furnace devised by Deville and Debray for the melting of platinum. The two cylinders A and B were hollowed out and pierced at E to admit the oxygen-coal gas burners. The molten platinum was poured through the channel on the right by tilting the furnace by means of the lever C. This type of furnace remained in use in the industry for many years until the development of the induction furnace

Jacobi also enjoyed the use of the melting furnaces and rolling mills of Savard as well as of the facilities of the Paris Mint through the co-operation of its Director, Professor Pelouze, and in the December of 1859 he was able to send to the Minister of Finance thirty-eight medals struck in several alloys of platinum containing 5, 10 and 20 per cent of iridium. These were first shown to the Académie des Sciences by Pelouze, together with an ingot of iridium weighing 267 grams which Jacobi presented to the Académie (18). Unfortunately, despite Jacobi's recommendation in a pamphlet published in 1860 (19), the Russian government abandoned all thoughts of a new platinum coinage.

Despite this disappointment, Deville and Debray continued their intense work, elaborating their methods of preparing platinum and the other metals of the group in commercial quantities and in a high state of purity. In 1860 they laid before the Académie two ingots of platinum weighing together 25 kilograms (20) while two years later they reported to them on the industrial progress so far made with their melting process (21). In this they referred to the casting of an ingot of platinum weighing 100 kilograms by George Matthey, "the mass becoming so liquid that it filled exactly with metal every part of the mould and reproduced all its imperfections with unexpected precision". In a footnote to one of their earlier papers (16) they had already recorded in referring to their two licensees that

"Today these procedures are functioning and are being perfected every day in the hands of the clever craftsmen to whom they have been confided."

282

Boris Semenovich Jacobi
1801–1874

Born in Potsdam in Germany and known in his early years as Moritz Hermann von Jacobi, he left to make his career in Russia and was appointed Professor at the University of Dorpat in 1834, later moving to St. Petersburg to take charge of the Physics laboratory of the Academy of Sciences. In 1859, by then serving also as Industrial Adviser to the Russian Ministry of Finance, he visited Paris and sponsored further investigations by Deville and Debray but failed to persuade his government to re-introduce a platinum coinage

The New Standard Metres

Jacobi was again destined to play a part in the work of Deville and Debray. In 1867 a Great International Exhibition was held in Paris, he was among the visiting scientists and he took the opportunity to join in discussions on standard weights and measures in which the French were deeply interested because of their long sustained efforts to spread the use of the metric system. With Jacobi's support a committee of delegates from a large number of countries was formed and strongly recommended the universal adoption of metric weights and measures, this resulting in the setting up of the International Metric Commission in 1869

> "for the construction and verification with the best appliances of modern science of new international standards of the metre and the kilogram".

The Franco-Prussian War delayed things, but in 1872 the first full meeting took place with twenty-nine countries represented, leading to the Convention du Metre signed in Paris in 1875. This finally achieved the support of all the

283

In 1873 the President of France, Louis Adolphe Thiers, together with a number of his ministers, paid a visit to Deville's laboratory in the École Normale to witness the melting of ten kilograms of iridium–platinum for the production of the new standard metre. Deville is standing in front of the door looking thoughtful; Debray is at the opposite end of the furnace, while their assistant Clement is tilting the lime-block to pour the alloy. The President is holding a protective glass in front of his eyes

important countries and the establishment of the still existing International Bureau of Weights and Measures with its headquarters in Paris (22).

An examination of the original standard metre, the Mètre des Archives, made by Janety in 1796 and described earlier in Chapter 10, showed that the ends were no longer plane, but the Convention decided that they would not go back to the original definition of the metre as one ten-millionth of a quarter of the world's meridian, as there was no agreement as to its exact value. They preferred to use the Mètre des Archives "in the state in which it is found" but to convert its length from an "end" standard to a marked "line" standard.

The next question to be decided was the most suitable material from which the new standard and its copies should be made, and in 1862 Deville and Debray put forward a suggestion that an alloy of platinum with 10 per cent of iridium offered the best possibilities. It has to recommend it high density, high melting point, great resistance to humidity and air, a fine grain, perfect polish, great hardness and full malleability. Deville and Debray claimed to have put in ten years of study, assisted by many other people, before they arrived at the possibility of preparing this material in a satisfactory form and were satisfied

284

that it was better than anything else available. One of their greatest troubles had been encountered in the preparation of large ingots of this alloy, completely free from blisters and cavities. The only method available in those days of detecting their presence was the undertaking of very careful density determinations. Deville himself carried out large numbers of these, working long hours into the night and exhausting his strength in the service of the posterity that he dreamed of but could not hope to see (23). In one way and another the attempts to produce a sufficiency of homogeneous metal occupied several years at the hands of Deville himself and his assistants, the Norwegian O. J. Broch, the Belgian J. S. Stas, and George Matthey (24). Time after time the metal failed and, even after attaining the necessary qualities at its original preparation, was found to have taken up iron during the subsequent mechanical fabrication.

A detailed account of these matters and of the major role played by George Matthey will be given in the next chapter.

During this work an immense number of analyses of the platinum metals and mixtures thereof had to be undertaken, and the conscientiousness of Deville and his collaborators was such that continual advances were made in the accuracy of the determinations. The final methods were set out by Deville and Stas in 1877–1878 and these remained the last word on the subject for a generation or more afterwards (25). All these labours connected with the metre filled up the last years of Deville's life, but he had the satisfaction of seeing them completed before he died in 1881.

Conclusion

The important invention of the lime furnace by Deville and Debray enabled platinum to be melted commercially on a large scale for the first time and remained the sole means of effecting this until the coming of the induction furnace a great many years later. But it must not be thought that this at once solved all the problems of the fabricator of platinum. It provided him with a means of preparing alloys but it was not the answer to all the questions raised in handling pure platinum. Here problems of purity gradually grew to be of paramount importance and it became evident that not only were the gases of the blowpipe capable of introducing impurities (and not only gaseous ones) into the melt, but that at the high temperature involved the refractory was capable of yielding them too. The former could introduce oxygen, iron dust, and carbon and sulphur products from the coal-gas; the latter calcium and silicon from the reduction of the oxides of these elements in the lime. So the forging of sponge continued for a long time, and was still used by refiners for many years.

In addition to the introduction of melting, Deville's work covered the commercial analysis, the refining, the alloying and the fabrication of all the six members of the platinum group metals. Over twenty-five years of careful experimental work in these fields by such an experienced scientist put at the disposal of the platinum industry a rationalised technique that formed the basis for the

For sixty years, until the development of the Ajax-Northrup high-frequency induction furnace in 1918, the melting of platinum and its alloys was carried out on an industrial scale in the lime block furnace introduced by Deville and Debray, with an oxy-hydrogen flame and the tilting of the furnace to pour the molten metal into an ingot mould

operations of all the firms working in it for at least the next thirty or forty years. In the hands of Deville, science and practice were brought together and an important industry made possible.

In July 1981 to commemorate the centenary of Deville's death, an appreciation of his life and work was compiled by Dr. J. C. Chaston (26) who concluded his contribution with these words:

"The platinum industry has, since then, expanded to an extent far beyond anything he could have foreseen, but among those who laid its foundations there are few who deserve commemoration more than the modest, hard-working research worker and teacher, platinum chemist and technologist, Henri Sainte-Claire Deville."

The successful adoption of Deville and Debray's method of melting platinum by leading refiners in England, France and Germany will be described in the following two chapters.

286

References for Chapter 15

1   E. F. Smith, The Life of Robert Hare, Philadelphia, 1917

2   R. Hare, Memoir on the Supply and Application of the Blow-Pipe, Chemical Society of Philadelphia, 1802; *Phil. Mag.*, 1802, **14**, 238–245; 298–308; *Ann. Chim.*, 1803, **45**, 113–138

3   N. Reingold, Science in Nineteenth Century America, London, 1966, 7–9

4   E. F. Smith, *loc. cit.*, 8

5   H. I. Brooke, *Ann. Phil.*, 1816, **7**, 367

6   E. D. Clarke, *Quart. J. Sci.*, 1817, **2**, 104–125; *Ann. Phil.*, 1817, **9**, 89–96

7   E. D. Clarke, The Gas Blow-pipe, or Art of Fusion by Burning the Gaseous Constituents of Water, London, 1819

8   R. Hare, *Am. J. Sci.*, 1820, **2**, 281–302

9   R. Hare, *Am. J. Sci.*, 1838, **33**, 195–196

10 R. Hare, *Am. J. Sci.*, 1839, **35**, 328–329; *Phil. Mag.*, 1839, **15**, 487–488

11 R. Hare, *Am. J. Sci.*, 1846, **2**, 365–369

12 M. A. Gaudin, *Comptes rendus*, 1838, **6**, 861–863

13 H. Sainte-Claire Deville, *Ann. Chim.*, 1856, **46**, 182–203

14 H. Sainte-Claire Deville and H. J. Debray, *Comptes rendus*, 1857, **44**, 1101–1104

15 French patent 18532, British patent 1947, of 1857

16 H. Sainte-Claire Deville and H. J. Debray, *Ann. Chim.*, 1859, **56**, 385–496. This was also published as a pamphlet

17 H. Sainte-Claire Deville and H. J. Debray, *Ann. Chim.*, 1861, **61**, 5–146

18 T. J. Pelouze, *Comptes rendus*, 1859, **49**, 896–897

19 B. S. Jacobi, About Platinum and its Use in Coinage, Imperial Academy of Science, St. Petersburg, 1860 (in Russian)

20 H. Sainte-Claire Deville and H. J. Debray, *Comptes rendus*, 1860, **50**, 1038–1039

21 H. Sainte-Claire Deville and H. J. Debray, *Comptes rendus*, 1862, **54**, 1139–1144; *Chem. News*, 1862, **6**, 150–151

22 A. Perard and C. Volet, Les Mètres prototype du Bureau International, Paris, 1945; B. Swindells, *Platinum Metals Rev.*, 1975, **19**, 110–113

23 H. Sainte-Claire Deville and H. J. Debray, *Comptes rendus*, 1875, **81**, 839

24 O. J. Broch, H. Sainte-Claire Deville and J. S. Stas, *Ann. Chim.*, 1881, **22**, 120–144

25 H. Sainte-Claire Deville and J. S. Stas, De l'Analyse du Platine iridie, Paris, Gauthier-Villars, 1878; Proces-Verbaux de Comité International des Poids et Mesures, 1877

26 J. C. Chaston, *Platinum Metals Rev.*, 1981, **25**, 121–128

George Matthey
1825–1913

Joining P. N. Johnson as an apprentice at the age of thirteen, he retired as Chairman of Johnson Matthey and Co. Limited when he was eighty-four. During this extraordinarily long career he developed the refining and fabrication of platinum from a laboratory scale to a successful industrial operation upon which his successors have been able to build. He was elected a Fellow of the Royal Society in 1879

From a portrait in the possession of Johnson Matthey

16

George Matthey and the Growth of the British Platinum Industry

"Many a physicist and many a chemist has gained distinction as the result of researches upon the properties of platinum, iridium and other metals which such metallurgists as Mr. Matthey have enabled them to obtain in the desired form and condition; and too often have they forgotten that the glory and honour should by right be shared with those who have laboured to produce the specimens upon which they have operated."

THE TIMES, 1913

The early history of the London firm of Johnson Matthey and Co. has already been outlined in Chapter 11. While always interested and active in the treatment and applications of platinum from its foundation in 1817, it did not begin to assume international importance in this field until after 1851. In the preceding years Johnson had developed a number of interests in the then booming lead, copper, silver and tin mines in Devon and Cornwall, and in 1838, with plans for the expansion of his then very modest business in mind he concluded an agreement with his great friend John Matthey, a wealthy stockbroker and foreign exchange dealer whose father Simon had left his home town of Le Locle in Switzerland in about 1790 to settle in London. By the terms of the agreement, in return for an injection of new capital, two of John Matthey's sons were to be apprenticed to Johnson. George, the third son, at once entered the business at the age of only thirteen to work in the assay laboratory, to be followed by his youngest brother Edward in 1850.

At the time of George's entry, and until 1845, the firm was known as Johnson and Cock, and besides benefiting from Johnson's own excellent training the young Matthey was fortunate in having W. J. Cock as his mentor in the very early days of platinum refining and fabrication. But in 1845 Cock's health failed

The young George Matthey, from a portrait painted in 1845 when he was twenty. It was in this year that he first took charge of the platinum laboratory after his training under Johnson and Cock and began to increase the size of the platinum ingot. Later he adopted the melting of platinum by the Deville and Debray method and successfully developed new applications

and he retired from the partnership, leaving George Matthey, still only twenty years old, in charge of the work.

He set about his new task with great determination, and there now began that remarkable period of persistent scientific endeavour combined with an acute business sense (Cock, although a great chemist, was no business man) in which George Matthey transformed a laboratory activity into an industrial enterprise and made platinum available for use throughout the world.

The work was hard, and the hours were long – from 7 in the morning until well after 6 in the evening, for six days a week. Cock had been given but a short time in which to begin re-organising platinum production, and George Matthey had to put in hand the preparation of larger, sounder and more malleable ingots as well as introducing more effective separation of the other platinum metals and their refining. By December of that year he was supplying Michael Faraday – for his studies on magnetism – with wire and foil in both platinum and palladium, with rhodium, iridium and osmium in metallic form and with a number of compounds of all the five metals then known.

In 1849 the Prince Consort, then President of the Royal Society of Arts, put forward a suggestion to hold a Great Exhibition to further the application of science to industry, but his proposal met with great opposition from many quarters. Johnson, like many other industrialists, was unwilling to support the exhibition, but young George Matthey realised the opportunity it presented and finally persuaded his employer to take part to the extent of a small glass case

290

containing platinum crucibles, capsules and a large basin, together with specimens of palladium and some of its alloys and specimens of iridium and rhodium. A prize medal was awarded for this exhibit, but any pride or pleasure that Matthey had taken in this success was very swiftly dissipated when he walked along to see the exhibit mounted by his French competitor, Quennessen of Paris. There he found a large platinum still, holding thirty gallons, made "in one piece without seam or solder", for the concentration of sulphuric acid as well as a complete apparatus in platinum for the distillation of hydrofluoric acid.

Russian Platinum Supplies and a Partnership

This at once stiffened his resolve to become pre-eminent in the platinum business, and he now also had a much more assured supply of raw material. Hitherto the only source of native platinum was the Chocó area of Colombia, and much of the meagre flow from there was smuggled out. The new source, discovered on the eastern slopes of the Ural mountains in Russia, was subjected to an Imperial monopoly by the Tsar's government which insisted that all refining should be undertaken by the Mint at St Petersburg. However, in 1850 George Matthey was successful in coming to an arrangement with one of the mine owners, Count Anatole Demidov, whereby he became the sole refiner and selling agent, platinum to be delivered in parcels of 1000 ounces at a time. This achievement, finalised in the October of 1851, prompted P. N. Johnson to take the young man into partnership, the name of the firm then being changed from P. N. Johnson & Co. to Johnson and Matthey. Thereafter Johnson, now aged 59, began to relinquish control of the business leaving Matthey in more or less full charge. Under his creative guidance and determination and his keen eye for new applications platinum refining and fabrication developed, as we shall see in this chapter, into a flourishing business.

The Commercial Melting of Platinum

At this time the only means of producing platinum in malleable form was by the powder metallurgy method of pressing the sponge in a mould and hot forging, introduced by Thomas Cock and Wollaston. While a number of scientists had succeeded in melting small samples, the resulting metal was nearly always brittle because of contamination by carbon or refractory materials. But a major development sprang from George Matthey's visit to the Paris Exhibition in 1855. Here he became friendly with Paul Francois Morin, one of Deville's collaborators in the work on aluminium at the École Normale and now works director of the Société de l'Aluminium, and as a consequence of this meeting a letter arrived from Jules Debray in August 1857 offering the British rights in the melting process described in the last chapter. Matthey immediately recognised the potential value of this development and at once replied expressing his interest and a wish to see the process in operation, his letter ending:

"You may not be aware that I am *the only* refiner and worker to any extent in Platina in this country; it will be therefore for your consideration which is most advantageous, to allow any one indiscriminately to participate in the patent, or to confine your negotiations entirely to me".

Before the end of September George Matthey was in Paris, bringing with him a quantity of platinum and residues for trial melting in the Deville and Debray furnace. The results were highly satisfactory, and the negotiations were finalised with M. Morin when a cheque for £500 was handed over for the assignment of the British patent (No. 1947).

The process was at once put in hand in Hatton Garden, but there were many difficulties, and it was several years before satisfactory operation was achieved. Oxygen was not then available commercially, and had to be prepared on the spot from manganese dioxide, while the pressure of the London coal gas was extremely low. W. J. Cock had returned to Hatton Garden, no longer as a partner, but in his own words "to help out", and he proved an invaluable assistant to his younger colleague in getting the process going, although his health began to fail once again and he finally left in September, 1861. A few entries in Cock's diary earlier in that year indicate some of the difficulties.

"Wednesday, 6 Feb. Read Deville's last letter to Matthey. Prepared and set iron retort in furnace room and charged it with 98 lbs ox manganese.

Friday, 8 Feb. A little rain. Set retort again and lit fire. When some oxygen had come over the retort melted. Had old wrought iron retort re-fitted.

Monday, 11 Feb. A little snow. Charged wrought iron retort and set it ready in furnace for lighting tomorrow morning.

Tuesday, 12 Feb. Fine. Oxygen fire kept going all day, but ox came over very slowly.

Wednesday, 13 Feb. A little rain. Tried fusion of platinum with G.M. . . . Dr. Faraday was present."

One week later Faraday delivered his famous "Lecture on Platinum" – one of his last appearances at the Royal Institution – in which he described the new melting process and referred to "Messrs Johnson and Matthey, to whose great kindness I am indebted for these ingots and for the valuable assistance I have received in the illustrations".

In March George Matthey had to appeal to Deville. Part of his letter reads:

"For nearly two years I have been working in this manner and am afraid you will be disappointed with me. I can only assure you that my numerous disappointments have rather increased than diminished my efforts to succeed, and if you will yet suspend your judgement of me and grant me a little assistance I do hope to realise some great and profitable results".

Deville's response was to invite Matthey to Paris, but he found himself unable to leave Hatton Garden and Cock went instead. In addition, Deville sent over his "garçon de laboratoire", one Jules, who established that the major trouble lay in the impurities in the oxygen.

292

John Scudamore Sellon
1836–1918

A nephew of Johnson's wife, Sellon joined the firm as an apprentice at the age of fifteen and became a partner in 1860. His commercial abilities were of invaluable support to George Matthey during the building up of the platinum business, and in 1872, after many frustrations, he was successful in securing the very large stocks of Russian platinum for refining and fabrication

By the end of May success had been achieved. Progress continued, the size of melt increased, and in March of 1862 Deville came to London and together Matthey and he cast a huge ingot, measuring twelve inches by eight by six, and weighing 3215 ounces or 100 kilograms. It was placed on display at the Second International Exhibition of Industries in London that year together with numerous other forms of platinum, including tubes joined by fusion welding with a blow-pipe flame, and a specimen of melted iridium.

Two New Partners

During these tedious struggles with the new melting process the structure of Johnson and Matthey underwent a major change. Johnson finally retired and two further partners joined George Matthey. His young brother Edward had been apprenticed in 1850 at the age of fourteen, and in 1855 had been persuaded by George, conscious of the need for a more professional chemical and metallurgical knowledge in the business, and with a great deal more foresight than most of his contemporary industrialists, to embark on a course of study at the Royal School of Mines (then known as the Government School of Mines, in Jermyn Street). Here Edward had the benefit of instruction by Hofmann in chemistry and by the great John Percy in metallurgy, and in 1860 he was made a

293

Edward Matthey
1836–1918

A younger brother of George, whom he followed as an apprentice in 1850. Edward studied chemistry and metallurgy at the Royal School of Mines and was taken into partnership in 1860. The three partners then remained in control of the business for over fifty years. He is shown here in the uniform of the London Rifle Brigade from which he retired as Colonel in 1901

junior partner together with a nephew of Mrs. Johnson, John Scudamore Sellon, the firm now becoming Johnson Matthey and Co.

Sellon had his strongest gifts on the commercial side, where his personal character and his keen insight amounting frequently to vision were of the utmost value to the firm. He had had no scientific training, but yet had a flair for scientific matters that ensured great intimacy of co-operation with George Matthey in the utmost speed of development of new technical enterprises. And it was he who in 1872 arranged a most favourable deal with the Russian State Bank whereby Johnson and Matthey took over the entire remaining stock of platinum amounting to well over 300,000 ounces. Sellon, whose father Captain William Smith, mentioned in connection with his brother-in-law Percival Norton Johnson in Chapter 9, had adopted his mother's maiden name in 1847, also brought with him the goodwill of some influential relations including the famous surgeon Sir Benjamin Brodie and his son, also named Benjamin who, after studying under Liebig, became Professor of Chemistry at Oxford in 1855 and President of the Chemical Society from 1859 to 1861.

Edward Matthey was of entirely different character; he brought to the partnership an element of solidity and strength that was a necessary complement to the activities of his two brilliant colleagues. He provided a steadying

294

influence, a willingness to serve his colleagues at home, and a mordant wit which maintained in them both a very necessary sense of proportion. This left George Matthey with more time and energy to devote to the refining and fabrication of his beloved platinum and to seeking applications for it in the growing industries of the time. The strength of the combination is shown by the fact that it remained in control of the firm for full fifty years and in that time built up a great business that has played a leading part in the world of platinum ever since.

The Search for Applications

In his correspondence for the years 1865 and 1866, copies of which are still preserved, there are to be found letters advocating the electroplating of platinum on brass and gun metal, the supplying of platinum sheet for Grove cells, seeking business for platinum foil for the contacts on the new electric telegraph systems, as well as stimulating agents he had appointed in European countries and in the United States to greater efforts in the sale of platinum laboratory apparatus and chemical equipment.

In 1867 an International Exhibition was to be held in Paris, and George Matthey and his colleagues determined to show what a British firm could achieve. The exhibit comprised over 15,000 ounces of platinum, including two huge boilers. This exhibit created something of a sensation in showing platinum manufactures on a scale never so far imagined. It was awarded a gold medal "for perfection and improvement in the working of platinum" and George Matthey was created a Chevalier in the Légion d'Honneur. The prestige of Johnson and Matthey in their field was established beyond doubt.

The Standard Metres and Kilograms

George Matthey was entitled to take great pride and pleasure in this achievement, just fifty years after the foundation of the firm by Johnson, but it was during the course of this exhibition that another significant matter was first set in train, as mentioned in the last chapter, the production of new standard metres and kilograms in iridium-platinum. The proposal made by Deville and Debray that this alloy should be used had been accepted, but now came the problem of melting, casting and fabricating the very large amounts of alloy required and of obtaining sufficient quantities of the two metals in a high enough state of purity. It was also decided that in order to ensure uniformity of all the standards likely to be required they should be made from one ingot weighing no less than 250 kilograms. The largest casting that had ever been made so far was the 100 kilogram ingot cast by George Matthey in Hatton Garden in 1862 in the presence of Sainte-Claire Deville, and it was only natural that the latter should now turn to Matthey for advice and co-operation. He first provided the two metals, both originating from Russian mineral, and early in 1874 they were alloyed and cast into three ingots weighing 80, 85 and 90 kilograms. These were cut

George Matthey played an important part in the production of standard metres and kilograms in high purity iridium-platinum for the International Metric Commission. In 1874 three large castings, weighing 80, 85 and 95 kilograms, were cut into pieces and remelted to ensure homogeneity into a single casting measuring 142 by 18 by 8 centimetres and weighing 236 kilograms. This operation, carried out at the Conservatoire des Arts et Métiers in Paris, was conducted by George Matthey and Henri Tresca, Professor of Mechanics at the Conservatoire and Secretary of the French section of the International Metric Commission. Both Deville and Debray were present as technical advisers. This engraving, from the French magazine L'Illustration of May 16th, 1874, shows the melting operation in progress

into small pieces and on May 13th, at the Conservatoire des Arts et Métiers in Paris, they were remelted and cast into a single ingot under the direction of the secretary of the French section of the International Commission Henri Tresca, his son Gustav and George Matthey, with Deville and Debray in technical control. After cleaning the ingot weighed 236 kilograms and measured 142 × 8 × 8 centimetres. It was exhibited to the Académie des Sciences by the Director of the Conservatoire who was also Vice-President of the French section of the International Commission, General Jules Morin (1) and was then forged into a square bar 4.5 metres in length, cut into small pieces and again forged and cold drawn to yield the X-cross-section designed by Tresca with the idea of giving it maximum rigidity.

It had originally been intended to make 65 metre standards, but there had been some cracking of the forged bars and the operation of converting them to

The open X-section originally designed by Henri Tresca for the standard metres in iridium-platinum on the grounds of its maximum rigidity. This was later abandoned on the recommendation of George Matthey who advised that it was impossible to draw the section without contamination from the dies

the open X-shape by cold drawing had given rise to a considerable amount of scrap. As a result it was possible to obtain only 27 sound metres, weighing about 90 kilograms, from the working of the great block. The Committee supervising the operations met in October 1874 and expressed itself as satisfied with this result, but unfortunately at this moment Deville found that the density of the metal was barely 21.1, whereas it should have been at least 21.385. Analyses showed that iron and ruthenium were present in appreciable quantities, in spite of the fact that Deville had himself purified the metal which George Matthey had provided. By way of explanation he suggested that the iron had probably entered the metal during the forging and drawing, and also that there had been a leak in the melting crucible. As for the ruthenium, delays in getting the mineral from Russia had not left him enough time to purify it as much as he would have liked.

"Lively discussion and much research followed and Deville suggested that the whole of the metal should be remelted and subjected to a longer heating than before in order to get rid of the iron and ruthenium; this he had been led by his experiments to believe to be a possibility. Tresca and the other members of the French section were alarmed by the risks of another large melting and professed to be satisfied with what had been achieved, subject only to a remelting of the scrap. They took the view that the amount of impurity indicated did not interfere at all with the properties required in the metres and they decided that the work should proceed". (2)

In April 1875 there came into existence the Comité International des Poids et Mesures, a body of much the same complexion as the former supervising committee with two extra members. This body at once took an interest in the

297

question and did not see eye to eye with the French section. Argument continued for some time and eventually a Commission consisting of Broch, Deville and Stas, which had been appointed to analyse and determine the density of the alloy, reported that it did not satisfy the requirements laid down in 1872 by the Commission Internationale, in that it contained:

Platinum	..	87.7 per cent	Iridium	..	9.4 per cent
Rhodium	..	0.4 ,,	Palladium	..	0.1 ,,
Ruthenium	..	1.4 ,,	Copper	..	0.2 ,,
Iron	..	0.8 ,,			

In consequence of this, on September 19th, 1877, the new committee decided to reject the metres that had been made from the alloy of the French section and to call for new ones satisfying the specification of 1872. These decisions were no doubt supported by the discovery of the cracks and fissures in the metres. The committee immediately addressed itself to the question of obtaining the new metal, acting through the same committee of Broch, Deville and Stas. They at once sought and obtained the collaboration of George Matthey and also of M. Brunner, instrument-maker of Paris. On the advice of the former it was decided provisionally to abandon the X-form and to make at first only bars of rectangular section. He believed that it was impossible to achieve the X-form by drawing without introducing iron into the metal and he thought it at least doubtful whether planing would not incur the same danger. The committee therefore decided on September 15th, 1877, to have two metres made of rectangular cross section and the order for them was sent by Stas to Matthey in January 1878.

The latter immediately made a casting of 17.5 kilograms and the two metres were made out of that, together with some weights ordered by the Norwegian Government. The metres were planed, drawing being applied only to obtain the final shape. They contained only 0.23 per cent of foreign metals as compared with 2.9 per cent in those of 1874, while the density was 21.52 against 21.08, so that it was demonstrated that metres could be successfully made from pure platinum and iridium in small separate melts. This was sufficient to encourage the committee to attempt to make a metre with the X-shape from pure metals in the same way, and this was duly produced by Matthey in October 1879. In the same year he also made three iridium-platinum cylinders for kilogram standards.

In 1878 George Matthey achieved another striking success in making a standard 4 metres long for the Association Géodésique Internationale. This also was prepared in the rectangular form by rolling, with drawing applied only in the last stages. The metal had a density of 21.516 and contained only 0.33 per cent of metals other than platinum and iridium. This he described in some detail in a letter to the Association published by the Académie des Sciences, his communication being followed by comments on its density by Deville, and by further observations from Henri Tresca and finally by commendation and

298

thanks from Professor J. B. Dumas, the Secretary of the Académie (3). In giving his account of this work, later on George Matthey mentions that in 1876 he had been asked to produce a standard metre in tubular form and had done so by using his autogenous welding process. After that he had made others, of both round and square section and exhibited samples at the Paris Exhibition of 1878. (4)

With George Matthey's production of these various metres the era of experiments was past, since in the course of it he had proved that metres made of pure iridium-platinum could be produced industrially and with the X-form. The Governments of the interested countries were asked what they wanted and whether they would take it in the old or new alloy, and practically all asked for the latter. The French section, which was still officially charged with the provision of them, were not anxious to repeat their experiences and turned to Johnson, Matthey & Co for the work. An arrangement made between the company and the French Government on August 23rd, 1882 provided for the production of 30 Standards at 1.20 metres each in the X-form as well as 40 kilograms. Matthey had proved that the single casting was not necessary for the achievement of identity, but he nevertheless offered to cast 250 to 300 kilograms if it were demanded. He was left free to employ whatever methods he thought best. All the metals had to be analysed and passed by Debray as representative of the French section and by Stas as representative of the Comité International. Of the thirty metres No. 6 eventually became the standard metre, because its length at 0°C was found to be precisely equal to that of the provisional standard less 0.006 mm, and therefore it was of exactly the same length as the Mètre des Archives. At the same time one of Matthey's kilograms was designated the International Prototype Kilogram and still remains as the ultimate world standard.

The production of standard metres and kilograms went on for over twenty years, the last orders from the French Government being delivered in 1887. One result of this substantial usage of iridium–platinum was undoubtedly a greater understanding of refining and melting problems and a distinct improvement in purity, both of great value in the later applications of the platinum metals, while the long association with the French scientists, particularly Deville and Debray, brought Matthey their close friendship and from time to time their invaluable advice.

Improvements in Platinum Refining

George Matthey was disturbed by the presence of impurities in the platinum he had sent to Paris in 1874 for the preparation of the standard metres and he promptly gave his mind to improving his method of refining, turning to alloying platinum with lead which not only increases the solubility of rhodium in acidic solvents but renders the iridium almost completely insoluble in them.

In 1879 he gave a full account of his revised process in the course of a paper to the Royal Society to which his election had already been proposed by a distinguished list of supporters including Dumas, Wurtz, Sorby and Roberts-Austen. His account, "The Preparation in a State of Purity of the Group of Metals known as the Platinum Series, and Notes upon the Manufacture of Iridio-Platinum" (5) described the process of producing high purity platinum as "an operation of extreme delicacy". After melting with six times its weight of lead of known purity and granulating the alloy it was attacked slowly in dilute nitric acid. This dissolved most of the lead, taking with it a portion of the copper, iron, palladium and rhodium present. After separation of the insoluble part the bulk of the lead was removed from this solution by crystallisation as lead nitrate, the other metals being recovered "by well-known methods". The black metallic residue obtained from the nitric acid attack consisted of iridium crystals and lead compounds of platinum and rhodium with only small proportions of the other metals originally present. It was next digested in weak aqua regia, which completely dissolved the platinum together with some rhodium and all the lead. The solution after filtering and evaporation was treated with enough sulphuric acid to remove the lead as sulphate and then the platinum was precipitated as usual with an excess of ammonium chloride. The whole was then heated to about 80°C and allowed to stand for some days. This consolidated the precipitate so that most of the remaining rhodium stayed in the surface liquor to which it gave a rose tint. The precipitate was then filtered and repeatedly washed, first with a saturated solution of ammonium chloride and then with dilute hydrochloric acid. In spite of this, rhodium was still liable to be retained, and to remove it the precipitate was dried, mixed with bisulphate of potash with a little ammonium bisulphate and subjected to a gradual heat slowly brought up to a dull red. This treatment converted the rhodium into a soluble state and it could be dissolved completely by boiling with water, leaving the platinum in "a state of absolute purity of the density 21.46". There is a statement about this purity in a letter written on June 12th, 1879, by George Matthey to Deville, in which he says that he has $31\frac{1}{2}$ kilograms of platinum sponge in stock containing as sole impurity six parts per 10,000 of rhodium and some insignificant traces of iron. He also has $43\frac{3}{4}$ kilograms of sponge containing seven or eight ten-thousandths of rhodium and 31 kilograms containing fifteen ten-thousandths of it "which we propose to refine further". (6)

The iridium removed from the platinum by means of aqua regia still contained rhodium, ruthenium, iron and possibly some osmium. He alloyed it afresh with ten times its weight of lead, attacked the alloy in nitric acid and aqua regia in succession as before and then fused the product with potassium bisulphate, which almost entirely removed the rhodium. He then melted the insoluble residue with ten times its weight of dry caustic potash and three times its weight of nitre in a gold crucible, the process being prolonged for a considerable time to convert the ruthenium completely into potassium ruthenate. After solidification the cake was leached with water and the ruthenium distilled from the mixture as tetroxide, to be collected in hydrochloric acid. The fusion and

300

distillation treatments were repeated until no further ruthenium was extracted and the residual iridium oxide was ignited to metal (after further cleaning in turn with bisulphate, chlorine water and hydrofluoric acid) and if perfectly pure should possess a density of 22.39. George Matthey remarks at the end of his paper "The highest density I have yet attained is 22.38".

This paper to the Royal Society was promptly reproduced in the German and Russian scientific periodicals, and in June of the same year he was duly elected a Fellow of the Royal Society, being described as:

> "Distinguished as a Metallurgist, having special knowledge of the metals of the Platinum Group. The development of the Platinum Industry was mainly due to his efforts".

The Design of Sulphuric Acid Boilers

One of the most important products – at least as far as the consumption of platinum was concerned – was the boiler for the concentration of sulphuric acid. Reference has been made to the boiler exhibited by the French firm of Quennessen at the 1851 Exhibition and to George Matthey's determination to compete. The means he adopted to achieve success in the field can now be detailed.

In Chapter 9 the activities of Wollaston in producing the first vessels of this kind were described, and in Chapter 10 the work of Jean Bréant that followed. The object was to concentrate the acid produced in the chamber process at 77 and 78 per cent strength up to oil of vitriol at 95 to 98 per cent, or even higher. Originally this operation was conducted in comparatively small glass vessels but their fragile nature and the continual breakages caused frequent losses and accidents. The early platinum boilers were operated on a batch principle, the acid being boiled down to the correct strength and then siphoned off before another charge could be treated. This intermittent process subjected the boilers to great wear and tear and led to the opening of the many joints in vessels fabricated from the gold-soldered small sheets at that time available.

By the time that George Matthey began to produce these pieces of equipment the size of his platinum ingots had reached 500 ounces and the rolling of these into sheets had to be contracted out to a firm in Birmingham. Matthey realised that if progress was to be made, and if platinum was to replace glass, he had to offer continuous operation and an apparatus economical in metal, in time and in fuel consumption. To this end he retained the services of one of the very earliest chemical engineers, William Petrie of Charlton in Kent, the father of the famous archaeologist Sir Flinders Petrie. After taking a course in chemistry under Professor Daniell at King's College in London, Petrie had spent some time in Frankfurt studying electricity and on his return to London he took up this latter subject professionally, becoming associated with William Edwards Staite (1809–1854), one of the pioneers of electric lighting. They were unable to secure adequate financial support, however, and Petrie turned to electro-

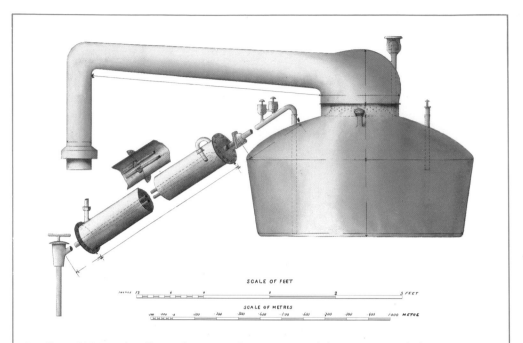

SCALE OF FEET

SCALE OF METRES

From 1855 until well into the twentieth century one of the major uses of platinum was for the construction of boilers to concentrate the weak sulphuric acid produced in the lead chamber process. George Matthey was primarily responsible for the growth of this activity and several hundred boilers were supplied to acid manufacturers throughout the world. Each quotation was accompanied by a large coloured drawing such as this, much reduced in reproduction here, showing a design of 1867 capable of treating five tons of acid per day

chemistry and chemical engineering. He became associated with the sulphuric acid manufacturing firm Thomas Farmer of Kennington which it will be remembered was one of Wollaston's first customers for a platinum boiler in 1809. Here he introduced improvements in the operation of the plant and in 1854 began to collaborate with George Matthey.

The first result of their joint endeavours was shown at the Paris Exhibition of 1855, the boiler having a much larger bottom surface for exposure to the heat with the result that it operated with a much less depth of acid. Heat transfer was also improved by heating "by means of direct radiation from the surface of a clear fire, as distinguished from heating by draught through flues in part out of sight of the fire". Furthermore, the boiler was arranged for continuous operation by direct feed and withdrawal by siphon. Its capacity was 40 gallons and its output 34 cwt. of strong acid per day.

After 1860 an entirely new factor was introduced into the construction of the boilers by George Matthey's invention of the fusion welding of platinum by an

William Petrie
1821–1908

Educated under Daniell at King's College and then in Frankfurt, Petrie first engaged himself heavily but unsuccessfully in the early years of electric lighting and then turned to industrial chemistry and the technology of sulphuric acid manufacture. For some twenty years, beginning in 1854, he collaborated with George Matthey in the design of platinum boilers, applying sound principles of chemical engineering and helping to establish a leadership in this field that continued until the early years of the twentieth century

oxy-hydrogen blow-pipe which made possible fully autogenous and therefore much stronger joints as well as reducing the cost of the boilers.

By now energetic developments were taking place in chemical engineering. In a letter to one of his American agents in 1859 George Matthey wrote that he had come to the conclusion that in designing these vessels the most attention should be paid to the area of the bottom and the sides rather than to the volume:

"the gallons of liquid per hour which the vessel will rectify or distil depends on the number of square feet of heat-absorbing surface of the vessel, while the bulk of liquid contained at one time is quite unimportant if the system of continuous running in and out at the same time be adopted". (6)

Further improvements were made in boiler design by William Petrie, covered in a patent filed in 1862 (7). This was for a vessel constructed in iridium-platinum in which the concentrated acid left the boiler by a tube passing through the side from a separate compartment in the centre, it then being cooled by arranging for it to pre-heat the cold feed of incoming acid. This vessel, shown at the London Exhibition of 1862, would yield two tons of concentrated acid per day. Another refinement devised by Petrie was the introduction of a gauge or indicator to show the degree of concentration of the acid. This comprised a bi-

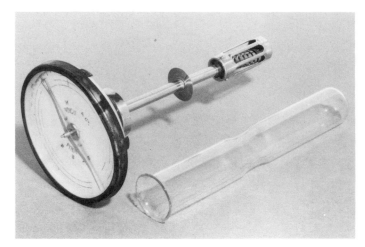

The indicating pyrometer designed by William Petrie to determine the boiling point and thence the concentration of the sulphuric acid. The element consisted of a bi-metallic helix of iridium-platinum and platinum-gold. The instrument has now been placed in the keeping of the Science Museum

metallic helix of a platinum-gold alloy laminated to an iridium-platinum alloy of different co-efficient of expansion attached to a pointer that would indicate the boiling point of the acid and thus its degree of concentration.

The capacity of the boilers was also increased, and in 1865 a quotation was made for one able to treat 5 tons per day, while a further improvement was the upward instead of the downward slope of the swan-neck to the condenser so that any splashes were returned to the vessel. Then in 1867 a boiler was shown at the Paris Exhibition able to handle 8 tons per day and was sold to a sulphuric acid manufacturer in Rouen under the noses of the French competition.

The next development was due to Manning Prentice (1846–1898) who had studied chemistry at University College, London, before joining his family firm of Prentice Brothers in Stowmarket in Suffolk, manufacturers of gun cotton and of sulphuric and nitric acids. In 1875 Prentice conceived the idea that by making the bottoms of the retorts corrugated he would both stiffen the vessel and increase the heating surface. His patent (8) was acquired by George Matthey, and the principle was also applied to the open platinum used pans ahead of the boilers for the preliminary concentration of the weak chamber acid.

Almost immediately afterwards there came the third major development in design. This was brought about by Gustav Delplace (1845–1913), born in Namur in Belgium the son of a plumber who, after following his father's trade, progressed into the construction of lead chambers for sulphuric acid and then established a chemical plant manufacturing business in Frankfurt (9). Delplace proposed the use of a very flat vessel and a thin layer of acid, resulting in considerable economy in the weight of platinum used and still greater evaporating power. Again Matthey secured the right to use this idea, while he also incorporated Prentice's corrugated bottom in a new series of boilers. An installation of this type, illustrated here, was shown at the Paris Universal Exhibition in

304

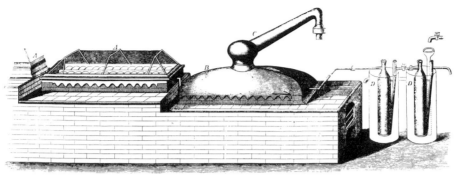

A Johnson Matthey sulphuric acid boiler of 1876 incorporating both the corrugated bottom and the open pans of Manning Prentice and the square shallow form proposed by Gustav Delplace, this design providing increased strength and economy in operation with high purity of acid. Until the contact process for the manufacture of sulphuric acid came into full use some hundreds of boilers of this type were supplied to acid makers throughout the industrialised countries of the world

1878. Among the visitors to this exhibition was Professor Benjamin Silliman from the United States who recorded in his review of the exhibits:

"Now, by very simple modification in the form and mode of using the platinum boilers, their cost is greatly reduced and the daily product of concentrated acid at the same time very much increased. The new boilers, first introduced by Messrs Johnson Matthey and Co., are rectangular in shape, with corrugated bottoms which offer extended surface with additional strength and evaporating power so that the economy of the new form of boiler is shown to be about fifty per cent each in the first weight of platinum and in the consumption of fuel, with an important saving in the cost of attendance and labour". (10)

For an output of four to six tons of acid per day the still measured three feet by one foot six inches and usually two were worked in series "to provide a degree of concentration higher than any before obtained commercially in platinum apparatus worked continuously".

From this time onwards there was little change in the general design of boilers other than increases in size; in 1893 one was shown at the Chicago Exhibition capable of concentrating ten tons per day. They remained a prominent feature of Johnson Matthey's work for the rest of the century and into the early years of the next until the contact process using a platinum catalyst came into full use. The last order received for platinum boilers was from the well known makers of sulphuric acid Spencer Chapman and Messel of Silvertown in East London in August 1914. This was for two boilers for the concentration of their chamber acid, each with a capacity of ten tons a day of 99.4 per cent acid, weighing 2396 ounces together. The last recorded case of a boiler being taken out of service was as late as 1926 when the last two vessels from an original set of twenty-six built in 1895 were returned as scrap by the African Explosives plant at Modderfontein. (11)

The success of these boilers, and their adoption in great numbers throughout the industrial world, were undoubtedly due to the application of sound chemical engineering principles by George Matthey and his three collaborators in a period when such concepts were only just beginning to be appreciated.

The Growing Uses of Platinum

The sulphuric acid boiler was not of course the only product to come from Johnson Matthey. Apart from an increasing output of platinum crucibles, dishes, spoons, foil and wire to meet the demands of the rapidly growing number of laboratories, there now began to emerge several industrial developments that called for platinum in one form or another – the electric telegraph, the incandescent electric lamp, the first motor cars, and even photography.

The invention of the electric telegraph – the forerunner of course of the telephone, the radio and all modern telecommunication systems – has been described many times, together with the disputes and litigation between the early workers Wheatstone, Cooke, Henry, Morse and others. Its success depended upon the concept of a relay by which an electric current could be actuated at a great distance. The first such device, proposed by Professor Joseph Henry at Princeton in 1833, utilised a U-shaped piece of wire that could be caused to dip into a cup of mercury by exciting an electromagnet, but by 1837 a more reliable relay – or "electrical renewer" as he called it – has been developed by Edward Davy, a chemist of Fleet Street in London and distantly related to Sir Humphry and his family. Concerned at the advances made by Cooke and Wheatstone, Davy filed a patent for his invention (12), the first such device to employ metallic make-and-break contacts, and his sketch of the mechanism is reproduced here. He had demonstrated his telegraph system in Regent's Park and held an exhibition in a hall in London, but just as his success seemed to be assured the breakdown of his marriage prompted him to leave England for

One of the major uses of platinum and its alloys, for make-and-break contacts in sensitive electrical devices, began with the design of the first electromagnetic relay for the electric telegraph by Edward Davy. This is his diagram taken from the patent he filed in 1838

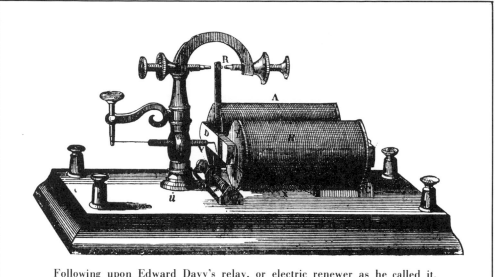

Following upon Edward Davy's relay, or electric renewer as he called it, Samuel Morse employed a similar device in his electric telegraph in 1840. This shows one of his early designs, with the platinum contact points indicated at R

Australia where he spent the remainder of his life, leaving the field entirely open to his rivals.

His idea was shortly taken up by Samuel Morse in America in 1838 and became an essential feature of telegraphy (13). One of Morse's early relays with platinum contact points is illustrated here. It was the prototype of the many millions of more sophisticated relays that have incorporated platinum or platinum-alloy contacts to ensure their reliability.

The Electric Light

As early as 1845 the use of a coil of platinum wire sealed into a glass container to form an incandescent electric lamp had been proposed by W. R. Grove (14), while W. E. Staite had devised a lamp containing an arch of iridium-platinum wire in 1848 (15), but many years were to pass before a reliable means of illumination could be produced. Joseph Wilson Swan (1828–1914) who had attended a lecture in his native Sunderland by Staite that first set him upon the road to success had, however, to await the invention of the vacuum pump by Hermann Sprengel in 1865 before this could be achieved with a lamp having a carbon filament supported on platinum wires and with thinner platinum lead-in wires fused through the wall of the glass. This he was able to exhibit to the Newcastle-upon-Tyne Chemical Society in 1878, and the Swan Electric Lamp Company was formed in 1881.

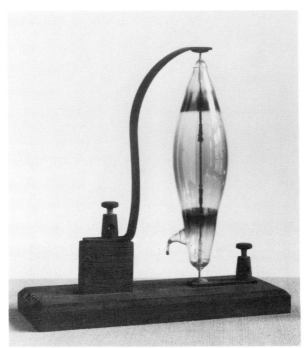

The first practical incandescent lamp made by Joseph Wilson Swan in 1878. His carbon filament was held between two platinum supports that served also as lead-in wires. Thomas Edison, after experimenting with platinum and iridium filaments, also turned to carbon held in platinum supports and for many years all electric lamps employed platinum lead-in wires sealed through the glass envelope

In the meantime Thomas Edison (1847–1931) in America had been carrying out similar experiments but using platinum, iridium-platinum alloys and even iridium filaments; a letter from him dated January 9th, 1879, still exists in the Johnson Matthey archives in which he enquired the price of "sticks" of iridium 1/64 inch in diameter and 1 inch long and also "the price of ingots". Hearing of Swan's success, however, Edison abandoned these and turned to carbon, their two English enterprises then being merged into the Edison and Swan United Electric Light Company Limited. For many years thereafter the supply of platinum leading-in wires, based upon the fact that the co-efficient of expansion of platinum is very close to that of glass, continued to be a feature of Johnson Matthey's business until cheaper substitutes were eventually found.

In 1897 Carl Auer von Welsbach (1858–1929) of Vienna, after inventing his gas mantle, developed a process for preparing osmium in the form of filaments for an incandescent lamp but the lamps were expensive and the filaments very fragile and production ceased after a few years.

The Thermionic Valve

The development of the incandescent lamp, important enough in itself, led to another most significant discovery, the emission of electrons from a hot metal electrode in a vacuum. This was first observed by Johann Wilhelm Hittorf

(1824–1914), Professor of Chemistry and Physics at the University of Münster, who found in 1884 that when an electrode of platinum or iridium was heated inside an evcacuated glass envelope the conductivity of the gaseous space rose rapidly with increasing temperature (16).

Independently two other German scientists, Julius Elster (1854–1920) and Hans Geitel (1855–1923), both of the Wölfenbuttel Gymnasium, showed that "electrified particles" were given off by a platinum wire in an exhausted glass bulb (17). During his work on the incandescent lamp Edison had already noticed in 1883 that a platinum spiral heated in a vacuum gave rise to a deposit of the metal on the inner surface of his glass bulb, and also that a carbon filament produced the same result except where the glass was screened by the positive limb of the filament, a phenomenon that became known as "the Edison Effect" (18). This was further investigated by J. A. (later Sir Ambrose) Fleming, the first Professor of Electrical Engineering at University College, London. Since 1882 he had also been scientific adviser to the Edison Electric Light Company (which merged with Swan's company a year later) and was thus fully acquainted with the problems in the early days of electric lighting, and he demonstrated that a plate or a wire zigzag of platinum, placed before the negative leg of the carbon filament would almost stop the effect occurring (19).

Then in 1899 Fleming took up another consulting post, this time to Marconi's Wireless Telegraph Company, and in seeking a means of converting the feeble oscillations produced in the receiving aerial into a unidirectional current his thoughts turned back to these earlier experiments. By 1904 he had invented the first thermionic valve, using either a platinum or an aluminium cylinder surrounding the filament (20), a device that was immediately adopted by Marconi and that led to all the revolutionary developments in electronics.

The carbon filament was replaced by tungsten in 1908, but another type that came into use was the oxide-coated filament due to Arthur Rudolph Wehnelt (1871–1944), Professor of Physics at Erlangen. This consisted of an iridium-platinum alloy wire having a coating of barium or strontium oxide, this giving greater emission on heating (21). Some years later Henry Joseph Round of the Marconi Company patented a platinum cathode in tubular form, coated with calcium oxide (22).

Early Automobile Ignition

The birth of the internal combustion engine operating on light oil or petrol, chiefly associated with Gottlieb Daimler (1834–1900), brought with it yet another application for platinum. Earlier gas engines had employed a small tube of iron or nickel closed at its outer end, screwed into the top of the cylinder and externally heated by a flame. In 1878 Sir Dugald Clerk produced the first two-stroke engine and filed a patent for the use of "a small quantity or bundle of platinum which is sufficiently heated to ignite the mixture of air and gas" (23). This was later modified to consist of a small cage or box of platinum with a

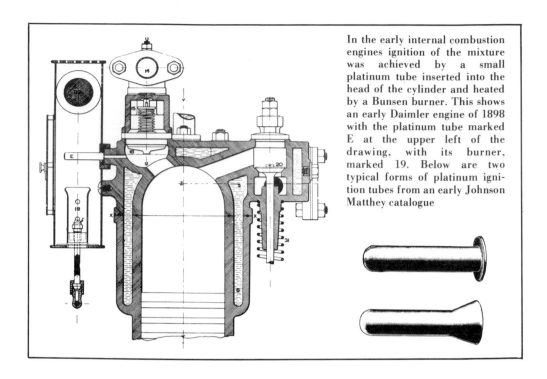

In the early internal combustion engines ignition of the mixture was achieved by a small platinum tube inserted into the head of the cylinder and heated by a Bunsen burner. This shows an early Daimler engine of 1898 with the platinum tube marked E at the upper left of the drawing, with its burner, marked 19. Below are two typical forms of platinum ignition tubes from an early Johnson Matthey catalogue

number of ribs, also of platinum, running across it, housed in a cavity at the side of the cylinder (24). When in 1884 Daimler produced his small high speed engine (25) he incorporated a small platinum tube at the top of each cylinder heated externally by Bunsen burners! Incredible as this may now seem, the device was used in automobiles from 1885 until replaced many years later by the sparking plug and the magneto, and Johnson Matthey and their competitors in Europe produced great numbers of these platinum tubes.

The progression of the original low voltage magnetos of about 1895 to the high voltage design of around 1902 is primarily associated with the names of the well-known British engineer Frederick Richard Simms (1865–1944) and of Robert Bosch (1861–1942) of Stuttgart. While the former type relied upon copper and silver, the high tension magneto first employed platinum contacts but when further improvements in design raised the voltage to 45,000 this was found to be too soft to withstand the conditions. As a result Simms appealed to Johnson Matthey to supply him with a much harder material, a 25 per cent iridium-platinum alloy, for his contacts, this continuing in use for a number of years until its replacement by tungsten. (26)

Platinum in Photography

During the early years of the evolution of photography there was considerable concern about the fading of prints and a great deal of effort went into the

310

search for a means of securing permanence. The problem was tackled by William Willis among others who set out to find a metal that would withstand any conditions and produce prints that would out-last those based upon silver. After some years of experimentation in his laboratory at Bromley in Kent he invented and patented a process in which the paper was sensitised with a mixture of ferric oxalate and potassium chloroplatinate. On exposure to light the oxalate was reduced to the ferrous state and this in turn reduced the platinum salt to metal (27). By this means permanent prints of excellent quality were produced with the most delicate shades of half-tones, while the procedure was relatively simple to manipulate. By 1879 Willis had formed the Platinotype Company which licensed photographers for only five shillings and supplied the sensitised paper. For many years until the outbreak of the first World War, when platinum supplies were commandeered by the government, this process was used for the highest class of portraiture and for documentary photography and the supply of potassium chloroplatinate by Johnson Matthey to Willis reached ten thousand ounces a year.

Platinotype paper is no longer available commercially, but in recent years the process has been revived by creative enthusiasts who appreciate the quite remarkable results that it can offer.

William Willis
1841–1923

One of the pioneers of photography, Willis first worked in engineering and banking before leaving to work with his father, the well-known landscape engraver who originated the aniline process for copying engineering drawings, on a new photographic silver process. He was dissatisfied by the lack of permanence of the silver process, and worked for many years to develop a more satisfactory photographic printing process using platinum. He obtained several patents for his Platinotype process and received awards from the Royal Photographic Society and the International Inventions Exhibition. His many interests included the spectographic analysis of minerals and metals, undertaking much original work on them, but which he never published

311

The First Artificial Fibres

Another application of platinum that originated in this period – and one that continues and increases – was in the production of the first man-made fibre, artificial silk, or as it became known later, viscose rayon. The initial invention was due to Charles Frederick Cross, Edmund John Bevan and Clayton Beadle, consulting chemists in London, who first produced cellulose xanthate from wood pulp in 1892 (28). Their patent was acquired by the old-established silk weaving firm of Courtaulds who, after some years of further development, began the manufacture of yarn in Coventry in 1904. The heart of the process was a spinning jet, a top-hat shaped device having a great number of very small and accurately drilled holes through which the highly alkaline viscose solution is extruded into fibres in the sulphuric acid coagulating bath. To withstand these aggressive conditions, and to avoid attrition of the holes that govern the size and quality of the fibre, platinum became the obvious choice for the spinneret material and although several alloys of platinum with either rhodium or gold have since replaced the pure metal the application remains the same and many thousands of such spinnerets are in use at any one time in a modern viscose rayon plant. For some time Johnson Matthey provided only the platinum in sheet form but later took up the manufacture of the spinnerets. The development of the presses has been described in more detail by J. W. S. Hearle and A. Johnson (29).

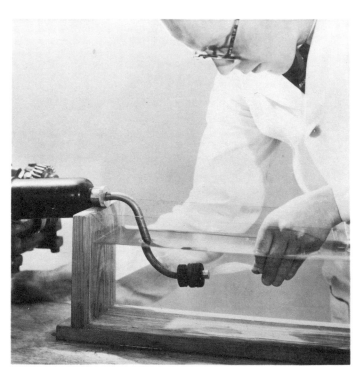

This laboratory demonstration of the formation of a multifilament viscose rayon yarn shows the pumping of the alkaline cellulose xanthate solution through the holes in a spinneret immersed in the acidic coagulating solution. Only platinum or its alloys can withstand these highly corrosive conditions. Beginning in 1904, the manufacture of rayon now requires many thousands of such spinnerets in any one viscose rayon plant

By courtesy of Courtaulds Limited

312

The Melting and Working of Iridium

The melting of platinum and its alloys in the lime-block furnace had been successfully established, but iridium melting at 2443°C, still presented difficulties. In 1882 work on this problem was undertaken by Henry Andrew Kent who had come over from John Harrison of Clerkenwell, the firm employed for platinum fabrication work until the death of the owner in 1874. Kent, encouraged by Sir William Crookes (1832–1919) the prominent Victorian scientist and the founder of *The Chemical News*, eventually succeeded in melting pure iridium in some quantity and in forging and rolling it to sheet, and at Crookes' request made a number of crucibles. Later Crookes read a paper to the Royal Society (of which he became President in 1913), "On the Use of Iridium Crucibles in Chemical Operations" beginning:

> "I should like to draw the attention of chemists to the great advantages of using crucibles of pure iridium instead of platinum in laboratory work. Through the kindness of Messrs. Johnson and Matthey I have had an opportunity of experimenting with crucibles of wrought iridium, and have used one for several months in the usual operations of quantitative analysis in my laboratory" (30).

He reported that his iridium crucible resisted the fusion of many fluxes, including caustic soda, and were unattacked by molten lead, zinc, nickel, iron and gold. Crookes was well in advance of his time in advocating the use of iridium for this purpose; in the last twenty years they have been used extensively for the growing of single crystals from oxide melts for use in electronics.

One of the iridium crucibles made by Johnson Matthey for Sir William Crookes still survives in the company's possession. They were made by Henry Andrew Kent, an assistant to John Sellon, who played a prominent part in the early period of platinum fabrication from 1875 to 1893

Conclusion

George Matthey finally retired in 1909 after a career spanning no less than seventy years. From a laboratory scale operation he had transformed the refining and fabrication of platinum into an important branch of industry. The earliest recorded figures show the sales of platinum by Johnson Matthey as some 15,000 ounces in 1860, increasing to 75,000 ounces by 1880.

He died in 1913 at the age of 87, and a leading article in *The Times* included the observation:

> "The death of Mr. George Matthey, F.R.S., serves to remind us of the part which the rare and precious metals have played in the general advance of practical science within the last half -century".

and continued with the sentence quoted at the head of this chapter.

He had been succeeded as chairman of the company by John Sellon, a man who had shared his resolve to make Johnson Matthey pre-eminent in the platinum industry, until his own death in 1918, the year in which Edward Matthey also died.

One tribute to the work of these three partners for so many years was made by Sir William Roberts-Austen in the course of a lecture on the rarer metals given to the Royal Institution in 1895. Referring to a range of exhibits of the platinum metals on view to the audience, he said:

> "We are indebted for this magnificent display to my friends Messrs George and Edward Matthey and to Mr. Sellon, all members of a great firm of metallurgists. You should specially look at the splendid mass of palladium, extracted from native gold, at the melted and rolled iridium, and at the masses of osmium and rhodium. No other nation in the world could show such specimens as these and we are justly proud of them" (31).

The succession was continued by George's son Percy St. Claire Matthey (1862–1928), named after his father's friend and collaborator Sainte-Claire Deville and educated for a time at the École Normale, and then on his death by Edward's son Hay Whitworth Pierre Matthey (1876–1957) who had worked closely with his cousin Percy on platinum and who served as chairman until the close of his life.

On the foundations that George Matthey had laid so painstakingly his successors were able to build and expand to an even greater extent. They were not, however, without active competitors in continental Europe as the following chapter will show.

314

References for Chapter 16

1   A. J. Morin, *Comptes rendus*, 1874, **78**, 1502–1506

2   A. Perard and C. Volet, Less Mètres Prototypes du Bureau International, Paris, 1945, 7–25

3   G. Matthey, *Comptes rendus*, 1876, **83**, 1090–1091; H. Sainte-Claire Deville, *ibid.*, 1091–1093; H. Tresca, *ibid.*, 1093–1096; J. B. Dumas, *ibid.*, 1096–1097

4   G. Matthey, *Chem. News*, 1879, **39**, 175–177

5   G. Matthey, *Proc. Roy. Soc.*, 1879, **28**, 463–471

6   Letters of George Matthey in the possession of Johnson Matthey and Co. Limited

7   W. Petrie, British Patent 1528 of 1862

8   M. Prentice, British Patent 4391 of 1875

9   A. Nemes, *Chem. Zeitung*, 1913, **37**, 237

10   B. Silliman, *Eng. and Min. J.*, 1878, August 31, 147–148

11   M. S. Salomon, A.E. and C.I. Reporter, 1926, February

12   E. Davy, British Patent 7719 of 1838

13   S. F. B. Morse, U.S. Patent 1647 of 1840

14   W. R. Grove, *Phil. Mag.*, 1845, **27**, 442–445

15   W. E. Staite, British Patent 12,212 of 1848

16   A. W. Hittorf, *Ann. Phys. (Wiedemann)*, 1883, **20**, 705–755; **21**, 90–139

17   J. Elster and H. Geitel, *Ann. Phys. (Wiedemann)*, 1887, **31**, 109–126

18   T. A. Edison, *Engineering*, 1884, December 12, 553

19   J. A. Fleming, *Phil. Mag.*, 1885, **20**, 141–144

20   J. A. Fleming, *Proc. Roy. Soc.*, 1905, **74**, 466–487

21   A. R. Wehnelt, *Ber. Phys. Med. Soz. Erlangen*, 1903, 150–158

22   H. J. Round, British Patent 6476 of 1915

23   D. Clerk, British Patent 3045 of 1878

24   D. Clerk and G. A. Burls, The Gas, Petrol and Oil Engine, London, 1913, Vol II, 271–272

25   G. Daimler, British Patent 4315 of 1885

26   F. R. Simms, personal communication, July 1942

27   W. Willis, British Patents 2011 of 1873, 2800 of 1878, 1117 of 1880 and 1681 of 1887

28   C. F. Cross, E. J. Bevan and C. Beadle, British Patent 8700 of 1892

29   J. W. S. Hearle and A. Johnson, *Platinum Metals Rev.*, 1961, **5**, 2–8

30   W. Crookes, *Proc. Roy. Soc.*, 1908, **80A**, 535–536

31   W. C. Roberts-Austen, *Proc. Roy. Inst.*, 1895, **14**, 497–520

Wilhelm Carl Heraeus
1827–1904

The descendant of a long line of apothecaries, Heraeus studied chemistry under Professor Wöhler at Göttingen and in 1851 established his refinery, Platinschmelze W.C. Heraeus, in Hanau, remaining in control until his retirement in 1889

17

The Development of the Platinum Industry in Continental Europe

"I have been refining and working platinum now for sixteen years, always by the method that the renowned Dr. Wollaston brought to such perfection."

PIERRE AUGUSTINE CUOQ, 1833

The establishment of an effective platinum industry in France originated with the acquisition, shortly after the fall of Napoleon and the restoration of the Spanish monarchy, of the accumulated stock of native platinum for which, without Chabaneau and Proust, the Spaniards had no outlet, by one Pierre Augustine Cuoq (1778–1851). Born in the village of Tence south of St. Etienne, Cuoq served in the French Revolutionary Army on the Rhine and then in Napoleon's Italian campaign. After the peace of Campo Formio he returned to civilian life in local administration and then in 1801 as a lawyer in Lyons where he met his future partner Couturier with whom he joined forces in a merchanting business in 1807. Cuoq travelled widely in the course of his commercial activities, visiting Germany, the Near East, Italy and above all Spain where he spent several periods of time.

It was on one of these last visits that he negotiated the purchase of about 1000 kilograms of the native platinum that had earlier been imported from New Granada by the Spanish authorities and for which they now had neither a method of refining or any forseeable use.

There is some evidence that Cuoq was acquainted with Vauquelin, at the time associated with the Mint in Paris, and that he provided him with a number of samples of this native metal. It is also clear that the firm of Cuoq and Couturier moved to Paris; both men are listed as members of the Société d'Encouragement in 1817, described as merchants with the same address in the Rue de Menard. Their arrangement for the refining of this immense quantity of platinum with Jean Robert Bréant (1775–1850), also an assayer at the Mint and well known to and advised by Vauquelin, has been described in Chapter 10. It led to the production of large ingots and of the first sulphuric acid boilers to be made in France and in only a short period of time to the foundation of the leading manufacturers of platinum anywhere in the world, and in France to a position of dominance that they and their successors held for almost a century.

Bréant was not only successful in his refining treatment, he also evolved a technique of forge welding to avoid his earlier procedure of soldering with gold, and by the February of 1817 he and Cuoq and Couturier were able to show the Société d'Encouragement a sulphuric acid boiler holding 162 litres, both parties receiving the commendation of the society for their achievements as well as for having reduced the price of platinum (1).

This association was not to last long, however. In 1819 Bréant established his own small works and rolling mill in the Rue de Montmartre, continuing to manufacture boilers, with a larger refinery in the Place du Commerce in the suburb of Grenelle in premises next to those of a M. H. Desmoutis who shortly entered into partnership with Bréant. Cuoq and Couturier continued to operate in the Rue du Richelieu, later moving to the Rue Lulli. In 1820 Wollaston made an entry in one of his notebooks giving the price of platinum from "Cuveq, Couturier and Co" and mentioning a vessel they had made holding 300 litres and weighing 776 ounces (2).

In 1833 a letter from Cuoq and Couturier in the *Journal für technische chemie* (3) rebutted an allegation that their platinum contained arsenic and included the phrase quoted at the head of this chapter. This statement cannot have been altogether true, as Wollaston's process was not of course made public until 1829, and no doubt in their earlier years they were basing their fabrication methods on those of Thomas Cock and Richard Knight.

In 1837 Bréant no longer appears, being no doubt more fully occupied at the Mint, where he eventually became Director in 1846, and he was succeeded by one Montrelay while Cuoq had set up his own merchanting business in Marseilles in the previous year and had also become a Deputy for his native Department de la Haute Loire. In 1845 the firm became known as Desmoutis, Morin and Chapuis, successors to Bréant, Montrelay, Cuoq and Couturier, continuing in this style until 1854.

Their subsequent history is somewhat complicated and difficult to piece together, the firm undergoing several changes of name and partners. Nothing is known of Morin; there were two brothers Prosper and André Chapuis, the second one also a fabricator of platinum who continued in business until the 1860s, exhibiting a number of platinum items at the London Exhibition of 1862.

The Two Quennessens

In about 1856 the Desmoutis company was joined by François Adrien Quennessen (1813–1889) who was to remain a most effective partner and later, followed by his son, to control the business for very many years. Quennessen had earlier been fabricating platinum on his own account in the Rue de Bouloi and it will be remembered from Chapter 16 that it was he who greatly surprised George Matthey with his display of chemical plant at the 1851 Exhibition. The report of the Jury on that occasion reads: (4)

"Quennessen (France. No. 1683). This exhibitor stands first in the exhibition of chemical apparatus having exhibited a platina alembic for sulphuric acid containing 250 pints, *made in one piece without seam or solder*; also long platina tubes made without seam, besides crucibles capsules etc. all of which are executed with the greatest care and appear to be of the most finished and exquisite workmanship. Among the articles exhibited by M. Quennessen is an apparatus for the distillation of hydrofluoric acid of a very complete and perfect kind. A Council Medal is awarded to M. Quennessen".

In the same year Quennessen had obtained a French patent for the fabrication of a platinum siphon by autogenous welding, (5) much less expensive than those made by soldering, for the decantation of sulphuric acid or other acids. His medal from the London Exhibition was followed on his return home by the award of the Cross of the Legion of Honour and he was clearly established as a leading fabricator of platinum. In 1855 he exhibited a platinum boiler at the Paris Exhibition under the name of Adr. Quennessen et Cie (6), but about this time he joined the Desmoutis company as an employee, the firm later becoming Desmoutis, Chapuis and Quennessen, still in the Rue Montmartre.

In Deville and Debray's earlier papers on the melting of platinum in 1857 they expressed their gratitude to Desmoutis and Chapuis and separately to Quennessen for their generous provision of platinum, while in 1859 they record the licencing of their patents to Desmoutis, Chapuis and Quennessen as one entity. In 1862 under this last name they exhibited at the London International Exhibition "a magnificent case of platinum, melted, hammered and wrought and in chemical combination" (7).

In the 1870s the name was changed again to Desmoutis Quennessen and Le Brun, and under this last name they were producing sulphuric acid boilers incorporating some of the improvements in design already introduced by George Matthey and his team of chemical engineers, exhibiting them in Paris in 1878 and 1889. The French company was active in the production of platinum for all the developing applications, the ignition tubes for the early automobiles, lead-in wires for incandescent lamps and platinum compounds for photography, as well of course as meeting the normal demands for laboratory apparatus, while they were also producing palladium in various forms and rhodium, iridium, osmium and ruthenium for use by researchers.

While the French company was in competition with Johnson Matthey to only a limited extent, particularly in the sulphuric acid boiler field, the two were in direct and fairly fierce competition in their needs for native platinum. Apart from an occasional shipment of mineral from Colombia, the only source was Russia, where it will be remembered coining had ceased in 1846 and the remaining stocks of platinum had been transferred to the State Bank. When George Matthey made his agreement in 1851 with Count Demidov there was a proviso that a similar arrangement should be made with a French refiner, and both Johnson and Matthey went out of their way to draw this to the attention of Desmoutis, Morin and Chapuis with the suggestion that they should make application. After some hesitation they decided so to do, and came to a similar

319

PLATINE

QUENNESSEN, DE BELMONT, LEGENDRE & CIE

SUCCESSEURS DE DES MOUTIS & CIE

56, Rue Montmartre, PARIS

The French company, first established by Bréant in the Rue Montmartre in 1819, underwent many changes of partnership over the years, eventually being headed by Adrien Quennessen. After the latter's death his son Louis, who had carried out research on the platinum metals at the École Normale, joined the firm which then adopted the style shown here

arrangement with Demidov which lasted satisfactorily until the outbreak of the Crimean War in 1854. When peace returned in 1856 business with Russia remained extremely difficult to conduct from England, and Count Demidov concluded a series of two-year contracts to supply the Paris company. It was not until 1872, as recorded in the last chapter, that John Sellon was able to close a most favourable deal with the Russian State Bank by which Johnson Matthey took over the whole of the remaining stocks of platinum in all forms including coin and residues. Adrien Quennessen had been seeking a similar arrangement, but failed to achieve it and his firm was now faced with a great problem of supply. Coming to London to negotiate, he was offered and accepted half the great quantity of metal that had been purchased, this leading to a lasting friendship between the two houses, who now had adequate supplies at their disposal.

Adrien Quennessen had a son Louis who studied chemistry under Professor Emile Leidié (1855–1904) at the École Normale, staying on to do post-graduate research, and in this he was most fortunate as he became part of an almost apostolic succession in the chemistry and metallurgy of the platinum metals. Quennessen senior was in and out of Deville's laboratory during the development of the melting process for platinum and its alloys and was well known to him and to Debray. One of Deville's students, Alphonse Joly (1845–1897) remained to become director of the chemical laboratory at the École Normale, responsible to Debray after Deville's death, and contributed a number of research papers to the Académie des Sciences over the next few years. The first of these, jointly with Debray, was on ruthenium and its compounds in 1888, while his own papers also dealt with the same subject and with the atomic weight of ruthenium, with compounds of iridium and the atomic weight of the metal and with osmium compounds (8). Leidié, after obtaining his doctorate with a thesis on the chemical compounds of rhodium in 1889 (9), began to collaborate with Joly in a series of researches, first on the separation of the platinum metals, on the atomic weight of palladium and then on the complex compounds of the platinum metals (10). After Joly's death at the age of only fifty-two Leidié continued his studies on the chemistry of the platinum metals and their separation (11) being joined in his researches by the young

Quennessen in 1901, three papers on extraction metallurgy and analytical methods resulting from their collaboration (12). Unfortunately Leidié also died at an early age, but Louis Quennessen persued the same line of investigation on his own, submitting several papers to the Société Chimique de Paris on the absorption of hydrogen by palladium, on the separation of iridium from platinum and on the compounds of iridium (13).

He was thus well equipped to join the family business, which had again changed its name, the elder Quennessen having died in the 1880s, to Desmoutis Lemaire, the latter representing the widowed daughter of Adrien Quennessen. In 1907 the firm was re-organised as "Quennessen, de Belmont, Legendre et Cie, Successeurs de Desmoutis et Cie". The de Belmont name was that of the son-in-law of Desmoutis, a man of wealth but no technical knowledge, while Legendre was the company secretary, and Madame Lemaire remained a substantial shareholder.

The Comptoir Lyon-Alemand

Unfortunately the outbreak of war in 1914 came as a very severe blow to the company. The refining and working of platinum was their only activity and Russia their only supplier of raw material. Small quantities were secured for a time, but the Russian revolution put an end to this and in 1917 they went into voluntary liquidation. Louis Quennessen's last contribution to the literature was a long paper summarising the history of the platinum metals, their occurrence, mineralogy and methods of analysis, published in the same year (14).

In 1919 the shares of Quennessen, de Belmont and Legendre, but not those of Madame Lemaire, were taken over by the Comptoir Lyon-Alemand. This had been established in Paris in 1800 by Joseph Alemand who was joined by his daughter who, married to one Lyon, succeeded her father in 1813, continuing in the business together with her three sons until 1826, the firm then becoming Comptoir Veuve Lyon-Alemand until 1880 when it was incorporated under its present name. In 1925, on the death of Madame Lemaire, her holding in the Quennessen company was also acquired, and in the same year the Comptoir Lyon-Alemand also acquired another bullion and platinum company, Marret Bonnin Lebel et Guieu, founded in 1810, becoming the sole French refiner of the platinum metals. Located in the Rue de Montmorency in Paris, they continue in operation as one of the leading platinum companies of the world.

The House of Heraeus

The year 1851 was a remarkable one in the history of platinum. Not only did George Matthey enter into his partnership with Johnson, only to be stirred into greater activity by the exhibition mounted by his French competitor Quennessen at the Great Exhibition in London, but a new entrant appeared in the industry, this time in Germany, in the person of Wilhelm Carl Heraeus.

321

Since 1660 his ancestor Isaac Heraeus had been established as a pharmacist in Hanau, trading under the sign of the White Unicorn in the Neustädter Markt and serving the Court of the Count of Hanau. The young Wilhelm Carl was first trained as a pharmacist and was then sent by his father to study chemistry under Professor Friedrich Wöhler at Göttingen. On his return to take charge of the family business he found that his mind was not sufficiently occupied by the pharmaceutical practice and he turned first to chemical preparation, building himself a laboratory, and then to the refining of the precious metal scraps and residues from the numerous old established goldsmiths and silversmiths of the town, returning to them the pure metals including small amounts of platinum.

His interest in this metal prompted him to establish the W. C. Heraeus Platinum Refinery in 1851 when only twenty-four years of age, and six years later he became aware of the new melting techniques proposed by Deville through the latters' friend and his old teacher Wöhler. In that year there appeared a modest advertisement on the last page of Poggendorff's *Annalen*;

W. C. Heraeus in Hanau

Owners of a Refinery for Platinum, Palladium, Gold and Silver; delivers worked platinum 20 per cent cheaper than has hitherto been the practice in Germany, namely at 466⅔ Gulden per Kilogram, with a discount of 3 per cent on the order of a whole kilogram and 5 per cent on 3 kilograms. Fabrication is charged for at cheap rates (15).

For his refining process Heraeus dissolved native platinum in aqua regia in a closed vessel under pressure, this method effecting a more rapid and more nearly complete dissolution. The solution was then evaporated to dryness and the residue heated to 125°C, reducing the chlorides of palladium and iridium to a lower valency and ensuring that they were not precipitated when treating the re-dissolved material with ammonium chloride. The precipitate was pressed, broken into pieces and then melted in the lime furnance (16).

As early as 1862 Deville and Debray reported:

"A new process for casting platinum is now used, invented by M. Heraeus, a manufacturer at Hanau who by the advice of his illustrious master M. Wöhler, has adopted for several years the processes we published on the treatment of platinum which in the hands of a clever manufacturer and an enlightened chemist have undergone already, as we anticipated, simplification and improvement. M. Heraeus runs the platinum into moulds of iron, which we have given up, but he obviates all the inconvenience arising from the fusibility of iron by placing at the bottom of the mould a sheet of platinum 1 millimetre thick which withstands the first contact of the molten metal" (17).

The platinum ingots, weighing about 2 kilograms each, were forged by a local smith on his anvil, the rolling of sheet then being carried out in his laboratory by Heraeus and his workpeople, "a work of great hardship", while the making of crucibles and other pieces of apparatus was turned over to coppersmiths and goldsmiths in the town (18). As early as 1857 Heraeus made his first export sale of 30 kilograms of rods, sheet and wire to a firm in New York,

322

The old apothecary, "At the Sign of the White Unicorn", in which the Heraeus family had conducted their business since the seventeenth century, became the first home of the platinum refinery in 1851. Later more modern premises were erected in Hanau

while his total sales for 1859 amounted to 59 kilograms, growing to 400 kilograms by 1879 and to over 1000 ten years later.

Heraeus was not satisfied with the purity of his platinum and he investigated methods of making a much purer product and of preparing the other metals of the platinum group separately and in quantity. In this work he appears to have run along similar lines to those adopted by Deville and Debray and by George Matthey, described in the last two chapters, but his progress was rather slow and it was not until 1891 that he was able to report success, but unfortunately he did not provide any analyses to compare with those of Deville and Matthey. This work was carried out with the co-operation of the Physikalisch-Technischen Reichsanstalt recently founded at Charlottenburg by Werner Siemens and directed by Hermann Helmholtz (19). In the course of this investigation he also studied the alloys of platinum with iridium and rhodium and he succeeded in drawing into wire metal containing as much as 40 and even 50 per cent of the alloying elements.

323

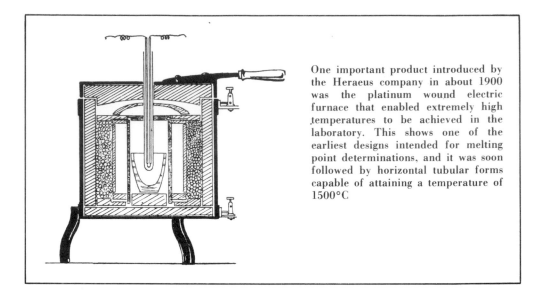

One important product introduced by the Heraeus company in about 1900 was the platinum wound electric furnace that enabled extremely high temperatures to be achieved in the laboratory. This shows one of the earliest designs intended for melting point determinations, and it was soon followed by horizontal tubular forms capable of attaining a temperature of 1500°C

Competition from Heraeus became important soon after he began his operations and joined the British and French companies in the manufacture of platinum boilers for the concentration of sulphuric acid, believing however, that better results could be obtained by lining the platinum with gold. This he achieved by pouring molten gold on to the surface of platinum blocks heated to a temperature above the melting point of gold and rolling into sheet (20).

Wilhelm Carl Heraeus had two sons, Wilhelm (1860–1948) and Heinrich (1861–1910) who took over the business on their father's retirement from active work in 1889 and continued to build on the foundations he had laid. Both had qualified as pharmacists; Wilhelm had also studied chemistry and Heinrich metallurgy, and they had both already absorbed the techniques of refining and working platinum in their father's business. In 1890 they were joined by Dr. Richard Küch (1860–1915), a physicist who was to make important scientific contributions to the company, including the production of high purity platinum.

For a very long period the platinum work had been carried out in the old apothecary building, illustrated on page 323, in which Isaac Heraeus had established himself in 1660, but naturally this had rapidly become much too small and in the years 1891 to 1896 a new plant was built on the outskirts of Hanau. Here they were able to expand the mechanical production of platinum and its alloys and to meet the new demands arising from the early telegraph, the electric light and the ignition of Daimler's early motor car engines. A little later, by 1901, they had pioneered the design and construction of platinum resistance elements in electric furnaces for laboratory use at temperatures up to 1500°C (21). A detailed account of this development has recently been given by Dr. R. C. Mackenzie in *Platinum Metals Review* (22).

324

In 1919 the business was incorporated as a limited company with the two brothers, together with Richard Küch and Wilhelm's brother-in-law Charles Engelhard, their American agent, as directors.

In 1951 the Heraeus company celebrated its centenary with the publication of a small book (18) and a larger technical work containing contributions by a number of their research scientists and opening with a commemorative paper by Professor Walther Gerlach, Rector of the University of Munich. The company continues to be among the leading refiners and fabricators of platinum with members of the original family still represented among its directors.

The Roessler Family Business

Another great German concern, established some years later than Heraeus but also with antecedents going back to much earlier times, originated with Friedrich Ernst Roessler, warden of the mint then being re-opened by the city of Frankfurt-am-Main in conjunction with the setting up of a state-operated refinery. He had been appointed to this post in 1841 after consultations with his father, Johann Hector Roessler (1779–1863) who directed the Grand Ducal Mint in Darmstadt to which the free Imperial City of Frankfurt had been constrained to place their orders for coinage as their own mint had become obsolete.

The younger Roessler had studied the natural sciences and engineering in Munich and had benefited from the experiences of his brother-in-law Franz Xavier Handl, the chief refiner of the Royal Bavarian Mint. He had also worked in the Darmstadt mint and had spent some time at those in Vienna and Paris. He was therefore well qualified for his new appointment, and only two years later the City Council decided to lease to him the plant and buildings of their gold and silver refinery, these to be operated for his personal account while he continued to be responsible to them for the minting of coin. The minutes recording this unusual division of responsibility describe Roessler as "a man whose reputation and credit leave nothing to be desired", going on to describe the refining business as

"such activities are always operated with more enthusiasm for the account of private persons rather than by nationalised undertakings".

The refining business soon increased to such an extent that Roessler had to find a new site in order to expand and he erected a chemical plant on a plot of land he had bought for this purpose – a plot still identifiable in the heart of the company's present headquarters in Frankfurt. The new refinery handled increasing quantities of demonetised coin as well as crude silver containing gold from the lead mines on the Rhine. Recovery of the small amount of platinum contained in the gold and silver coins began during the 1850s and some years later several kilograms were recovered from the slag produced in smelting the gold with nitre although some of the platinum metals were lost when the gold

Friedrich Ernst Roessler
1813–1883

Master of the mint then operated by the Free City of Frankfurt, Roessler was invited by the City Council in 1843 to take over for his own account the associated refinery while continuing to remain responsible for the coinage. The refinery prospered and later became the nucleus from which developed the Deutsche Gold-und Silber Scheideanstalt, founded in 1873 and now known as DEGUSSA

was dissolved in aqua regia and reprecipitated with ferrous sulphate. In addition, platinum was being recovered from discarded laboratory apparatus.

The successful campaign of 1866 designed by Bismarck to bring about the unification of Germany under the Hohenzollern dynasty brought about a change in Roessler's affairs. Frankfurt lost its status as a free city, the mint passed under the control of the Prussian authorities, and for a time it seemed that the refinery would cease to operate, but in view of its importance to the economic well-being of Frankfurt it was finally reprieved but its control had to be separated from that of the mint. Roessler accordingly purchased the refinery and immediately transferred it to his two elder sons, retaining his post in charge of the mint until his retirement at the age of sixty in 1873. Six years later the mint was finally closed down.

The refinery was now re-established under the name of Friedrich Roessler Söhne, the two sons being Johann Hector Roessler (1842–1915) and Heinrich (1845–1924). Hector, a graduate of the Mining Academy of Freiberg, had been managing a small chemical works since 1863, while Heinrich had also spent some time at Freiberg and had then studied chemistry under Wöhler at Göttingen, obtaining his doctorate in 1866 with a dissertation on the double cyanides of platinum and palladium. A number of technical improvements were made, the volume of business increased, and by 1873 it was decided to form the business into a limited company, to be known as the Deutsche Gold-und Silber-Scheideanstalt vormals Roessler.

326

An advance in the recovery of the platinum metals stemmed from a suggestion put forward by Professor Max Pettenkofer (1818–1901) of the University of Munich who had earlier served in the mint there and who had become a friend of Friedrich Roessler. By substituting ferrous chloride for ferrous sulphate the platinum metals were separated more readily and in greater amount from the gold refining operations. An account of the whole complex process was given by Ludwig Opificius (1849–1910) who had started as a boy under Friedrich Roessler in 1865, becoming refinery manager ten years later (23). By the end of the 1880s some twenty kilograms of platinum and approximately one kilogram of palladium were being obtained each year.

The adoption of the Wohlwill electrolytic process for gold refining in 1896 marked another step forward in the recovery of the platinum metals. These collected in the anode slimes, which were then enriched and the individual metals separated and refined to a state of high purity. Meanwhile attention had been given to the fabrication of platinum and its growing uses in industry and in research. One outlet in which Heinrich Roessler took an interest was the preparation of solutions containing the platinum metals for use in the decoration of pottery and glass alongside the older gold preparations, and in 1885 he described his formulations in a paper to the ceramics journal *Sprechsaal* (24).

In 1901 the two brothers Heinrich and Hector withdrew from the board of management, although the former remained on the supervisory board until his death in 1924, but the Roessler family continued to be represented for many years by a son and a nephew.

Heinrich Roessler
1845–1924

The second son of Friedrich, Heinrich Roessler first studied metallurgy at the Freiberg Mining Academy and then chemistry under Wöhler at Göttingen. With his elder brother Hector he took over the refinery in 1867, both remaining active as directors of the limited company formed in 1873 until 1901

In 1928 it seemed to the then directors that the rather long and complicated name under which they operated could well be simplified, and it was then that the present name of DEGUSSA was coined. The further progress of this international enterprise formed the subject of several publications on the occasion of their centenary in 1973 (25).

The Siebert Platinum Refinery

Another establishment that started as a service to the Hanau jewellery industry began to undertake platinum refining in 1881. This development was initiated by Wilhelm Siebert (1862–1927), the son of a cigar-box manufacturer, Georg Siebert (1835–1909) who had begun to treat the local goldsmith's scrap in 1864. Wilhelm had become familiar with the methods of assaying during a period spent with a company in Pforzheim, Dr. Richter and Co., and on his return home he began to experiment in the processing of platinum, first from the goldsmiths' residues and then from a share of the Russian coinage and other material that had been acquired by Johnson Matthey.

The new refinery was successful, and prospered further after 1884 when Wilhelm Siebert visited Russia and secured the promise of deliveries of platinum concentrates from the Urals on a long term basis.

He recorded later that:

"We set about the refining of native platinum and the production of platinum in the highest possible purity and in maximum quantity. This gave us a great deal of work and anxiety with many set backs to be overcome" (26).

In 1889 Wilhelm's younger brother Jean Siebert (1870–1925) joined the business and succeeded in expanding the market for platinum products, including of course lead-in wires for electric lamps, contacts for the telegraph and telephone industry and ignition tubes both at home and abroad. The fabrication of sulphuric acid boilers was also undertaken and a paper describing an improved design of a cascade apparatus was published in 1893 (27).

Contacts between the Siebert company and the Deutsche Gold-und Silber-Scheideanstalt had been close since the beginning of the former's operations, and in 1906 the latter became a shareholder. This enabled the Sieberts to expand, while it provided the Roesslers with greater access to the market for fabricated platinum. In 1921 this share was increased to 50 per cent and the firm was incorporated as a limited company, G. Siebert G.m.b.H., Hanau, which was wholly acquired by DEGUSSA in 1930. A year later, to commemorate their fiftieth anniversary, the company published a valuable "Festschrift" containing many scientific contributions both from their own staff and from distinguished academics such as Ostwald and Tamman (26).

The Hanau plant became the headquarters of Degussa's metallurgical work, with an extensive range of products fabricated from platinum and its allied metals. Then in 1975 an entirely new refinery was built at Wolfgang near Hanau and the pyrometallurgical and chemical work was transferred there, together

Wilhelm Siebert
1862–1927

Practising first as an assayer in Pforzheim, Siebert began to work in platinum in 1881 and after a visit to Russia in 1884 to secure supplies of native metal from the Urals he developed the refining and fabricating activities most successfully. The firm of G. Siebert was later absorbed by DEGUSSA

with a modern research department, providing the largest facility of its kind in continental Europe.

Developments in the Low Countries

For many years the other European countries were supplied with their needs of platinum by the established sources already described, but later on two other enterprises entered the industry in Holland and Belgium respectively, both stemming originally from gold and silver refining.

In 1827 Hans Halbes Drijfhout set up a small gold refinery in the village of Balk in Friesland, to be followed by his son Willem who moved to Amsterdam in 1886, the business then becoming H. Drijfhout and Zoon and including rolling and wire drawing facilities. Later he began to undertake the refining and working of platinum on a relatively small scale, but in 1927 the company was absorbed by the Comptoir Lyon-Alemand of Paris and continues as their subsidiary in Amsterdam.

329

Brussels was also a centre of the jewellery industry in the nineteenth century, and in 1864 Antoine Pauwels foresaw the need for a service of recovery of their scrap gold, silver and platinum and established such a business in 1864, supplying the jewellers with refined metals. Antoine died in 1906 and his sons Ferdinand and François succeeded him, developing and expanding the business, now known as Pauwels Freres, and including the manufacture of platinum and its alloys and compounds in all the forms required by industry. In 1961 the company became part of the Johnson Matthey organisation.

An International Research Institute

In 1921 a joint initiative by three interested groups, the city of Schwäbisch Gmünd, the state of Württemberg and the industrial companies engaged in the refining and fabrication of the noble metals in Germany, led to the foundation of a research organisation uniquely associated with these metals and the only one of its kind in the world, the Forschungsinstitut für Edelmetalle.

The city of Schwäbisch Gmünd, some 50 kilometres east of Stuttgart, was chosen as the home of the Institute primarily because it was one of the three German locations of the noble metals industry, the others being Hanau and Pforzheim, and secondly because there already existed in the town a training school for craftsmen, the Staatliche Höhere Fachschule für die Edelmetallindustrie, which could house the newly formed institute and in which members of its staff could take part in teaching.

The first three directors of the Institute in its early years were Dr. Rudolf Vogel from the University of Göttingen, then Dr. Hans Moser, the director of the State Mint at Stuttgart, and later Dr. J. A. A. Leroux.

In 1928 a young metallurgist who had just obtained his Ph.D at the University of Münster joined the Institute as a research associate; six years later he was appointed its Director and his association with the Institute continues to this day. This was of course Dr. Ernst Raub, who has an exceptionally wide international reputation in his field.

Under Dr. Raub's direction the great contributions to the physical metallurgy of platinum and the other noble metals began, and the first of a long series of investigations of the equilibrium diagrams of platinum alloys was published in 1935. Since that time a great many papers have emerged from the Institute both on alloy systems and on the electrodeposition of the platinum metals. The Institute continues to carry out both basic research supported by government departments and specialised contract research funded by national and international companies, such as investigations on the development of new electrical contact materials and their performance in service. Collaboration with industry has been most successful, and there are many products now commercially successful in which the Institute has played a decisive role in their development. Its unique character has also led to the establishment of valuable connections with both industry and the universities throughout the world.

References for Chapter 17

1   J. F. L. Mérimée, *Bull. Soc. Enc. Ind. Nat.*, 1817, **16**, 33–36

2   W. H. Wollaston, MSS 7736, Cambridge University Library

3   Cuoq Couturier et Cie., *J. Tech. Chem.*, 1833, **16**, 376–377

4   Reports of the Juries, Exhibition of 1851, London, 1852, 296

5   A. Quennessen, French patent 5530 of 1851

6   Exposition des Produits de l'Industrie de toute les Nations 1855, Catalogue Officiel, French Empire, Class 16, **III**, item 4934

7   W. W. Smyth, Reports of the Juries, International Exhibition, London 1862, Class II, Section A, 34

8   A. Joly, *Comptes rendus*, 1888, **106**, 328–333; 1888, **107**, 994–997; 1890, **111**, 969–972

9   E. Leidié, *Ann. Chim.*, 1889, **17**, 257–313

10 A. Joly and E. Leidié, *Comptes rendus*, 1891, **112**, 793–796; 1259–1261; 1893, **116**, 146–148; 1894, **118**, 468–471; 1885, **120**, 1341–1343; 1898, **127**, 103–106

11 E. Leidié, *Comptes rendus*, 1899, **129**, 214–215; 1899, **129**, 1249–1251; *Bull. Soc. Chim.*, 1900, **23**, 898–899

12 E. Leidié and L. Quennessen, *Bull. Soc. Chim.*, 1901, **25**, 840–842; 1902, **27**, 179–183; 1903, **29**, 801–807

13 L. Quennessen, *Bull. Soc. Chim.*, 1905, **33**, 191–193; 875–879; 1308–1310; 1906, **35**, 619–621

14 L. Quennessen, *L'Industrie Chimique*, 1917, **4**, 752–754; 774–775; **5**, 6–7

15 *Ann. Phys. (Poggendorff)*, *1857*, **101**, 644

16 J. Philipp, *Poly J. (Dingler)*, 1876, **220**, 95–96

17 H. Sainte-Claire Deville and H. Debray, *Comptes rendus*, 1862, **54**, 1139–1141

18 O. Heraeus and F. Küch, Hundert Jahre Heraeus, Hanau, 1951

19 W. C. Heraeus, *Z. Instrument Kunde*, 1891, **11**, 262–264

20 W. C. Heraeus, *Z. angewand. Chem.*, 1892, 300–301

21 H. Danneel, *Z. Elektrochem*, 1902, **8**, 822–824; E. Haagn, *ibid*, 509–512; R. C. Mackenzie, *Anal. Proc. Chem. Soc.*, 1980, **17**, 79–81

22 R. C. Mackenzie, *Platinum Metals Rev.*, 1982, **26**, in the press

23 L. Opificius, *Poly. J. (Dingler)*, 1877, **224**, 414–417

24 H. Roessler, *Sprechsaal*, 1885, 385; *Poly. J. (Dingler)*, 1885, **258**, 275

25 Aller Anfang ist Schwer, Bilder zur Hundertjahringen Geschichte der Degussa, Frankfurt, 1973; Edelmetall und Chemie, Frankfurt, 1973

26 H. Houben, Festschrift zum Fünfzigjahrigen Bestehen der Platinschmelze G. Siebert, Hanau, 1931

27 G. Siebert, *Z. angewand. Chem.*, 1893, 346–347

Dmitri Ivanovich Mendeleev
1834–1907

Born in Tobolsk in Siberia, Mendeleev came to St. Petersburg in 1850 to study at the teacher training institute of the University. In 1859 he travelled to Paris to work with Regnault and then to Heidelberg under Kirchhoff and he attended the Karlsruhe Conference in 1860. In 1866 he was appointed Professor of Chemistry at St. Petersburg and then in 1869 described his first Periodic Table to the newly formed Russian Chemical Society

From a portrait by I. A. Repin

18

The Platinum Metals in the Periodic System

"The six known platiniferous metals, from a certain point of view, may be rightly considered as forming a separate and well-defined group."

KARL KARLOVICH KLAUS, 1860

The gradual increase in the number of elements being discovered and isolated during the early part of the nineteenth century led to a number of attempts at their classification. As early as 1816 the great physicist André Marie Ampère (1775–1836), Professor of Mathematics and Mechanics at the École Polytechnique but at this stage of his career very interested in chemistry and in the whole concept of classification, put forward a scheme of ordering the elements that would bring out "the most numerous and essential analogies and be to chemistry what the natural methods are to botany and zoology" (1). All the elements then known were classified into five groups, one of these being called the "Chrysides", derived from the Greek word for gold, and including palladium, platinum, gold, iridium and rhodium. Osmium, however, he grouped with titanium. Some of the similarities between the platinum metals were thus recognised at this early date, but Ampère's method contained no numerical concept.

Döbereiner's Triads

That such a quantitative component was necessary was first recognised by J. W. Döbereiner who noticed in 1817 that the molecular weights for calcium oxide, strontium oxide and barium oxide formed a regular series or triad with that of strontium being the arithmetic mean of the other two. Twelve years later he published his paper on the Classification of the Elements in Poggendorff's *Annalen der Physik und Chemie*, curiously immediately following an abridged translation of Wollaston's paper on the production of malleable platinum given to the Royal Society in 1828 (2). Expressing first his great interest in the atomic weights of Berzelius, Döbereiner again showed that when the elements were arranged in groups of three resembling each other chemically the atomic weight of the middle one was the mean of the other two. After discussing the halogens, the alkaline earths and the group of sulphur, selenium and tellurium among others, he turned to the similarities between iron, nickel and cobalt and then to the platinum metals:

"The interesting series of analogous metals that occur in native Platina, namely Platinum, Palladium, Rhodium, Iridium, Osmium and Pluran, fall according to their specific and atomic weights into two groups. To the first belong Platinum, Iridium and Osmium, to the other Palladium, Rhodium and Pluran, which last corresponds with osmium, as rhodium does with iridium and palladium with platinum".

His Pluran, to which he referred in a footnote ("The existence of Pluran is however somewhat doubtful") was one of the supposed elements discovered by Osann in 1827 in native platinum from the Urals and given that name from the two initial letters of Platinum and Urals. Only in 1844 was the true sixth member of the group, ruthenium, discovered by Klaus, as recorded earlier in Chapter 12.

Very little was heard of Döbereiner's triads. Not until 1853 in fact did any serious notice appear to have been taken of them, but in that year John Hall Gladstone (1827–1902), a former student of Thomas Graham and Liebig, then a lecturer in chemistry at St. Thomas' Hospital in London and later Fullerian Professor of Chemistry at the Royal Institution, published a paper in *The Philosophical Magazine*, On the Relations between the Atomic Weights of Analogous Elements. In the course of this he commented:

"Who has failed to remark that the platinum group has double the atomic weights of the palladium group" (3).

Four years later Ernst Lenssen, one of the young assistants in Professor Fresenius' analytical laboratory in Wiesbaden, also speculated on the triads, grouping the elements by their chemical and physical characteristics and even by the colour of their oxides (4) included one consisting of palladium, ruthenium and rhodium (in that order) and another comprising osmium, platinum and iridium, again incorrectly arranged by their then atomic weights, or rather the equivalents, that he employed.

The Schemes of Odling and Newlands

A more comprehensive scheme for the classification of the elements was also published in 1857 by William Odling, at that time Professor of Chemistry at Guy's Hospital in London. In this he arranged forty-nine elements into thirteen groups of which the last contained the platinum metals and gold. He wrote:

"The propriety of associating gold with the platinum group is very questionable. Palladium appears to present a relation of parity with rhodium and ruthenium, platinum with iridium and possibly with osmium, though indeed many osmic reactions are altogether special" (5).

During their work on the platinum metals described in Chapter 15, Deville

William Odling
1829–1921

The son of a London doctor, Odling entered Guy's Hospital to study medicine and chemistry, becoming a demonstrator in the latter subject in 1850. After a period of study under Gerhardt in Paris he was appointed a lecturer and then Professor of Practical Chemistry in 1856. The first of his several papers on the classification of the elements appeared a year later. In 1859 he was elected a Fellow of the Royal Society and in the following year he attended the Karlsruhe Congress. In 1868 he succeeded Faraday as Fullerian Professor of Chemistry at the Royal Institution, moving to Oxford as Waynflete Professor of Chemistry in 1872. In that year he married the daughter of Alfred Smee, the surgeon to the Bank of England, whose work on the electroplating of the platinum metals has been described in Chapter 11

and Debray also emphasised the resemblances between these elements. In 1859 they wrote:

"The family of the platinum metals has a particular character, completely apart from the more or less natural families formed by the other metals. It is true that they are not entirely analogous on every point, but they have their own character, a common appearance that separates them, while from the point of view of a rational classification one should separate them from the diverse families of elements" (6).

Odling returned to this subject later, revising and extending his classification in 1861 and again in 1864, but in the interval Karl Klaus presented a paper on the platinum metals to the Academy of Science in St. Petersburg in which he also recognised them as a distinct group of elements (7):

"These metals may be arranged in two superimposed series, the superior horizontal which I designate the principal series because the metals which constitute it predominate in the various platinum ores. This series is characterised equally by an elevated atomic weight and by almost the same specific gravity ... The second horizontal series contains the remainder of the platiniferous metals, which also possess almost identical atomic and specific weights, but have in this respect but half the quantities of the principal series".

335

Klaus went on to show that the metals vertically above one another in his table resembled each other, the pairs ruthenium and osmium, rhodium and iridium, and palladium and platinum having identical reactions in the formation of their compounds.

Klaus's Horizontal Series

Principal Series Secondary Series	Osmium Ruthenium	Iridium Rhodium	Platinum Palladium

It will be seen that Klaus had his metals in the correct order as established much later on.

In his famous Lecture on Platinum, given to the Royal Institution in February 1861, Faraday clearly accepted these conclusions and quoted Klaus almost verbatim (8).

Odling's revised and enlarged scheme of 1861 included fifty-seven elements arranged in seventeen groups, the last two being very similar to those of Klaus (9),

336

and then in 1863 the first of a long series of papers by J. A. R. Newlands appeared in *Chemical News*, followed by several more in the next three years (10). In his final table of the elements, arranged numerically in the order of their atomic weights, he pointed out that

> "the numbers of analogous elements generally differ either by seven or by some multiple of seven; in other words members of the same group stand to each other in the same relation as the extremities of one or more octaves in music . . . This relationship I propose to provisionally term the Law of Octaves".

Newlands was uncertain how to deal with the platinum metals and he achieved his arithmetical symmetry only by assigning one number to each of the pairs rhodium and ruthenium in the earlier series and to platinum and iridium in the later, while he placed osmium alongside tellurium in another group. He also predicted that another element should exist between iridium and rhodium and another between palladium and platinum. Unfortunately for Newlands the Chemical Society declined to publish his paper, Odling, later the President, explaining that they "made it a rule not to publish papers of a purely theoretical nature".

Newlands continued to interest himself in arranging the elements so as to emphasise the family relationships, assigning consecutive atomic or "ordinal" numbers to them and leaving blanks for elements as yet to be discovered, and in a small book he produced in 1884 he claimed with some justification to have been the first to publish a list of the elements in the order of their atomic weights and to have described the periodic law (11).

Meanwhile in 1864 Odling, probably unaware of Newlands' later publication, contributed a paper "On the Proportional Numbers of the Elements" to the *Quarterly Journal of Science* in which he listed sixty-one elements in increasing order of atomic weight (12). In this he gave rhodium, ruthenium and palladium in that order and then platinum, iridium and osmium.

The Karlsruhe Congress

The accuracy of the atomic weights so far determined was in grave doubt and the subject of much controversy. Some values were only one half of their now established figures while some were twice as great. Friedrich Wöhler had complained that "the confusion can be tolerated no longer". From this state of chaos order was restored by the well-known paper from Stanislao Cannizzaro, (1826–1910), Professor of Chemistry at Genoa, given at the Karlsruhe Congress in 1860 (13). This famous gathering of more than 120 chemists, the first international scientific conference, was proposed by August Kekulé and some of his colleagues to secure more precise definitions of the concepts of atoms and molecules and to bring uniformity into the values of atomic weights. William Odling, one of the very few who had already read Cannizzaro's paper was among the signatories calling this meeting and he was present for the discussion.

337

Earlier he had studied chemistry in Paris under Gerhardt, had translated Laurents' Méthode de Chimie into English in 1855, and supported their unitary theory, the atomic weight to be taken as the smallest quantity of an element present in the molecular weight of any of their compounds. But it was Cannizzaro, whose paper, a reprint of an earlier contribution given in 1858, was distributed by his colleague Angelo Pavesi, Professor of Chemistry at Pavia, after the close of the meeting that settled the whole problem of atomic weights based upon the earlier proposals of Gerhardt and Laurent. Half a century of confusion was cleared up and it was now possible to ascribe the correct atomic weights to all the known elements.

Lothar Meyer and Mendeleev

Among those attending the Karlsruhe Congress was Julius Lothar Meyer, at that time Professor of Chemistry at Breslau, and he recorded later how on reading Cannizzaro's paper during his return journey "the scales fell from my eyes, doubt vanished, and was replaced by a feeling of peaceful certainty". When preparing a text-book Meyer was thus able to take account of the numerical relationships between the elements and in his Die Modernen

Julius Lothar Meyer
1830–1895

A native of the small town of Varel near Oldenburg in north Germany, Meyer was far from robust as a child and was given an out-door education under the head gardener to the Duke of Oldenburg. He began his higher education at Zürich and then transferred to Würzburg. After graduation he went to Heidelberg to study under Bunsen and Kirchhoff and in 1864 published his book on the modern theories of chemistry, this containing a table of most of the elements arranged in order of their atomic weights. He became Professor of Chemistry in the Technische Hochschule at Karlsruhe in 1869 and in 1876 accepted a similar chair in the University of Tübingen. In 1882 he and Mendeleev were jointly awarded the Davy Medal by the Royal Society for their development of the periodic system

Theorien der Chemie, written in 1862 but not published until 1864, he included a table of most of them (14). Here he placed correctly the three lighter members of the platinum group, ruthenium, rhodium and palladium, but wrongly gave the heavier three in the order platinum, iridium and osmium, an error that he was later to rectify.

In the meantime, however, the most clear and comprehensive treatment of the elements and their classification was devised, as every chemist knows, by the outstanding genius Dmitri Ivanovich Mendeleev, the great Russian scientist whose name has ever since been firmly associated with the Periodic Table. Mendeleev had studied at St. Petersburg and he had been more than fortunate in his teachers. The senior of these, Professor Nikolai Nikolaevich Zinin (1812–1880) had travelled widely in western Europe, spending a year with Liebig at Giessen and returning to the University of Kazan where in 1841 he had been a close colleague of Klaus before being called to St. Petersburg in 1847. He accepted the new concepts of Gerhardt and Laurent, the first to do so in Russia, and he attended the Karlsruhe Congress in 1860. Mendeleev's other mentor, who became closely attached to his brilliant student and gave him private lessons during a period of illness, was Professor Aleksyei Andreivich Voskressenskii (1809–1880) who had also spent some time with Liebig and who was affectionately known to his students as "the grandfather of Russian chemistry". He was also a disciple of Gerhardt and Laurent. Mendeleev, after graduating, visited Paris to study under Victor Regnault and then spent a period in Heidelberg where he opened a private laboratory. It was from there that he travelled to Karlsruhe and his appreciation of those discussions is shown in a letter he wrote to Voskressenskii that was published in the St. Petersburg Gazette.

This began:

"The chemical congress just ended in Karlsruhe produced such a remarkable effect on the history of our science that I consider it a duty, even in a few words, to describe all the sittings of the congress and the results that it reached".

After giving a brief account of these discussions he concluded:

"Cannizzaro spoke heatedly, showing that all should use the same new atomic weights. There was no vote on the question, but the great majority took the side of Cannizzaro".

Mendeleev had devoted long years to the accumulation of evidence for his developing ideas on the classification of the elements, carrying out hundreds of experiments, reading widely in the literature and corresponding with chemists throughout Europe to collect appropriate data. All this information, on their physical and chemical properties, on the nature of their combinations and on the isomorphism of their compounds, was then inscribed on to small white cards which he arranged until he was satisfied with their sequence. Early in 1869 he distributed privately a pamphlet entitled "An Experimental System of the Elements based on their Atomic Weights and Chemical Analogies", and then in the following March, at a meeting of the Russian Chemical Society which he had done so much to organise in the previous year, and with Zinin presiding, a paper from Mendeleev was read by his friend and colleague Professor Nikolai Menschutkin because the author had been taken ill. This was published in the first volume of the Society's transactions (15), and was briefly referred to in the German periodicals, giving their readers some indication of Mendeleev's ideas. It was not, however, a complete periodic table as we know it but rather a preliminary study in which he merely arranged the elements in six columns and, as he later emphasised, he was unaware of the publications of Meyer and Newlands and only of Odling's first communication of 1857 and Lenssen's even earlier work.

As with his predecessors, Mendeleev had difficulties with the platinum group of metals on account of their close similarity and the very small differences in their atomic weights as then determined. In this first system he arranged them:

Rh	104.4	Pt	197.4
Ru	104.4	Ir	198
Pd	106.6	Os	199

In a second paper read to a meeting of Russian scientists in Moscow in August 1869, "On the Atomic Volume of Simple Bodies", (16), he produced a clearer table, a prototype of his final version, in which he showed the eighth group in the same order as before but he now assumed the presence of an empty period between the ruthenium and the osmium groups.

Then in 1870 Lothar Meyer contributed a paper to Liebig's *Annalen*, "The Nature of the Chemical Elements as a Function of their Atomic Weights", (17) in which he arranged the platinum metals in their correct order but with some uncertainty:

Ru	103.5	Os	198.6?
Rh	104.1	Ir	196.7
Pd	106.2	Pt	196.7

He commented:

"To obtain this arrangement, some few of the elements whose atomic weights have been found to be nearly equal, and which have probably not been very carefully determined, must be rearranged somewhat, Os before Ir and Pt, and these before Au. Whether this reversal of the series corresponds to the properly determined atomic weights must be shown by later researches".

Partly arising from this paper of Meyer's, Mendeleev published a further account of his system a year later, and this was translated in full in the *Annalen* (18). Running to almost a hundred pages in the German version, this gave a much clearer and more comprehensive description of his periodic system – the first time he actually used the phrase – and dealt at length not only with the properties of the elements but also with their compounds. More courageous than

340

		Группа I.	Группа II.	Группа III.	Группа IV.	Группа V.	Группа VI.	Группа VII.	Группа VIII. Переходъ къ груп. I,
		H=1							
Типическіе элементы.		Li=7	Be=9,4	B=11	C=12	N=14	O=16	F=19	
Первый періодъ	Рядъ 1-й	Na=23	Mg=24	Al=27,3	Si=28	P=31	S=32	Cl=35,5	
	2-й	K=39	Ca=40	—=44	Ti=50?	V=51	Cr=52	Mn=55	Fe=56, Co=59, Ni=59, Cu=63
Второй періодъ	3-й	(Cu=63)	Zn=65	—=68	—=72	As=75	Se=78	Br=80	
	4-й	Rb=85	Sr=87	(?Yt=88?)	Zr=90	Nb=94	Mo=96	—=100	Ru=104, Ru=104, Pd=104, Ag=108
Третій періодъ	5-й	(Ag=108)	Cd=112	In=113	Sn=118	Sb=122	Te=128?	J=127	
	6-й	Cs=133	Ba=137	—=137	Ce=138?	—	—	—	—
Четвер. періодъ	7-й								
	8-й	—	—	—	—	Ta=182	W=184	—	Os=199?, Ir=198?, Pt=197, Au=197
Пятый періодъ	9-й	(Au=197)	Hg=200	Tl=204	Pb=207	Bi=208	—	—	
	10-й	—	—	—	Th=232	—	Ur=240	—	
Высшая соляная окись.		R²O	R²O² или RO	R²O³	R²O⁴ или RO²	R²O⁵	R²O⁶ или RO³	R²O⁷	R²O⁸ или RO⁴
Высшее водородное соединеніе.				(RH⁴?)	RH⁴	RH³	RH²	RH	

The revised and comprehensive Periodic Table published by Mendeleev in the Journal of the Russian Chemical Society in 1871. The platinum metals were now placed in their correct order in Group VIII, although this also included copper, silver and gold. A space was left for as yet undiscovered members of the platinum group between the lighter and the heavier triads and the atomic weights were given only in round figures because of Mendeleev's uncertainty about their accuracy

Lothar Meyer, he ascribed different atomic weights to a number of elements, based solely upon their chemical analogies, calling for new determinations to be made. In the case of the three heavier platinum metals he wrote:

"Three elements stand in the system in succession between $W = 184$ and $Hg = 200$. Their atomic weights are actually smaller than W, but the succession does not correspond to expectations, for in considering that Os, Ru, Fe are similar, but that Ru and Fe have smaller atomic weights than Pd and Ni, it is to be expected that the atomic weight of Os is smaller than that of Pt, and that Ir, standing between Pt and Os, has a middle value of atomic weight. Moreover, the inaccuracy of the atomic weight determinations of the Pt metals is readily understood, not simply because their separation from one another is difficult but also because their compounds that have been used for atomic weight determinations are not of great stability".

Mendeleev's table of 1871 is reproduced here from the original Russian version. It will be seen that in addition to iron, cobalt and nickel and to the platinum metals he had included copper, silver and gold in Group VIII but had left alternative positions for them in Group I, and that, unaware of most of the rare earth elements, he had again left gaps for an extra series between the two

341

Karl Friedrich Otto Seubert
1851–1921

The son of a Professor at the Karlsruhe Technische Hochschule, Seubert first studied pharmacy there and then turned to chemistry, serving for several years as assistant to Lothar Meyer and accompanying him to Breslau and then to Tübingen over a period of twenty years and succeeding him as Professor. His doctoral thesis was on the atomic weight of iridium and he went on to re-determine the atomic weights of all the other platinum metals, confirming the order in which Mendeleev had placed them

platinum metal triads. This had the inevitable result of prompting the re-examination of native platinum for the apparently missing elements and led to a number of "discoveries" that will be referred to a little later. Also, as is well known, he successfully predicted from a knowledge of their adjacent elements all the essential properties of as yet undiscovered elements, later to be identified as germanium, scandium, gallium, rhenium and technetium.

In this table Mendeleev separated the elements into their main and sub-groups, while he also confined his atomic weights to round numbers as he could not be sure of their accuracy. His predictions about the correct order for the platinum metals were fully confirmed a few years later by Karl Seubert (1851–1921), a student and later a colleague of Lothar Meyer's at Tübingen. Seubert also distrusted some of the old values and set about their re-determination, arriving at the following arrangement and so confirming Mendeleev's views (19):

| Ru | 101.4 | Rh | 102.7 | Pd | 106.35 |
| Os | 190.3 | Ir | 192.5 | Pt | 194.3 |

Some Spurious Platinum Metals

Several supposedly new elements of the platinum group were claimed before the publication of Mendeleev's Periodic Table, but the gaps he left, as already mentioned, led to further claims for the isolation of further members of the group. In 1877 Sergius Kern of the Obouchoff Steel Works in St. Petersburg wrote to *Chemical News* that he had "perceived the presence of a new metal of the platinum group which has been called by me davyum in honour of the great English chemist Sir Humphry Davy" (20). This claim was investigated in 1898 by Professor J. W. Mallet, an Irish chemist who had settled in America to become Professor of Chemistry at the University of Virginia, and who thought this might indeed be a member of the missing triad of platinum metals. While he was able to confirm Kern's experimental observations, he quickly showed that the new metal was merely

> "a mixture of iridium and rhodium with a little iron, and hence that we have not yet reason to believe in the existence of a third group of platinum metals" (21).

A further "discovery" was claimed by a French chemist, Antony Guyard in 1879:
> "Some years ago, about 1809, I discovered in some commercially fabricated platinum from Russian mineral a new member of the platinum group to which I give the name of Ouralium to commemorate its origin" (22).

The atomic weight was given as 187.25, its specific gravity as 20.25 and its ductility was said to be greater than that of platinum, but the experimental work was of a very low order and ouralium was again almost certainly a mixture of platinum with some iridium and rhodium. These two fallacious discoveries, together with several others, were reviewed in more detail by Dr. W. P. Griffith in 1968 (23).

The Modern Periodic Table

Mendeleev continued for the remainder of his life to take an active interest in his Periodic Law and used it as a base in his famous text-book, The Principles of Chemistry, first published in Russia in 1869 and in many later editions, with an English translation in 1891 and German and French versions a few years later. His chapter on the platinum metals opens with a statement of "the naturalness of the transition" from zirconium, niobium and molybdenum to silver, cadmium and iridium through ruthenium, rhodium and palladium, and similarly from tantalum and tungsten through osmium, iridium and platinum to gold and mercury. This is followed by an account of the chemistry of the platinum metals that would have been creditable to an author many years later, as for that matter would the whole of the book.

But it was not until the discovery of the electron by Sir J. J. Thomson in 1879 and then Moseley's work on the X-ray spectra of the elements that led to the concept of atomic number just before his death in 1915 in the European War that a sound theoretical basis could be established for the periodic system.

The Platinum Metals and their Neighbours
in the Periodic Table

	Group VIA	Group VIIA	Group VIII			Group IB
First long period	Cr_{24}	Mn_{25}	Fe_{26}	Co_{27}	Ni_{28}	Cu_{29}
Second long period	Mo_{42}	Tc_{43}	Ru_{44}	Rh_{45}	Pd_{46}	Ag_{47}
Third long period	W_{74}	Re_{75}	Os_{76}	Ir_{77}	Pt_{78}	Au_{79}

The part of the modern table in which the platinum metals occur is reproduced above with their atomic numbers, this including of course rhenium discovered in 1925 and predicted by Mendeleev as dri-manganese, and technetium, his eka-manganese, discovered only in 1937 in the bombardment of molybdenum by deuterons in a cyclotron. Just as Mendeleev emphasised, the greatest similarities are found in the vertical groups; there is a strong resemblance between ruthenium and osmium, between rhodium and iridium, and between palladium and platinum. At the same time there are obvious analogies in the horizontal series, between for example palladium and silver and between platinum and gold, while ruthenium and osmium more closely resemble technetium and rhenium, or in certain respects molybdenum and tungsten, than they do iron. Rhodium and iridium are more closely allied to cobalt than to any other metal, while platinum and palladium have close analogies with nickel.

In the two platinum metal triads the hardness and mechanical strength decrease from left to right and are greater in the second triad than in the first. Ruthenium and osmium, both close-packed hexagonal in crystal structure are brittle although they can be fabricated with difficulty at high temperatures, while palladium and platinum faced-centred cubic metals, are soft and readily workable in the cold. A review of these similarities in properties and of the relevant chemical properties of the group was contributed some years ago by the writers' colleague A. R. Powell (24).

References for Chapter 18

1 A. M. Ampère, *Ann. Chim.*, 1816, **1**, 295–308; **2**, 5–32

2 J. W. Döbereiner, *Ann. Phys. (Poggendorff)*, 1829, **15**, 301–307

3 J. H. Gladstone, *Phil. Mag.*, 1853, **5**, 313–320

4 E. Lenssen, *Ann. Chem. (Liebig)*, 1857, **103**, 121–131; **104**, 177–184

5 W. Odling, *Phil. Mag.*, 1857, **13**, 422–439; 480–497

6 H. Sainte-Claire Deville and H. Debray, *Ann. Chim.*, 1859, **56**, 385–389

7 C. Claus, *J. prakt. Chem.*, 1860, **79**, 28–59; **80**, 282–317; *Chem. News*, 1861, **3**, 194–195; 257–258

8 M. Faraday, A Lecture on Platinum, bound with The Chemical History of a Candle, London, 1861, 173–204; *Chem. News*, 1861, **3**, 136–141

9 W. Odling, A Manual of Chemistry, Part 1, London, 1861, 3

10 J. A. R. Newlands, *Chem. News*, 1863, **7**, 70–72; 1864, **10**, 59–60; 94–95; 1865, **12**. 83

11 J. A. R. Newlands, On the Discovery of the Periodic Law, London, 1884

12 W. Odling, *Q. J. Sci.*, 1864, **1**, 642–648

13 S. Cannizzaro, *Il Nuovo Cimento*, 1858, **7**, 321–366; English translation in Alembic Reprint 18; for an account of the Karlsruhe Congress see C. de Milt, *J. Chem. Ed.*, 1951, **28**, 421–425

14 J. L. Meyer, Die modernen Theorien der Chemie, Breslau, 1864

15 D. I. Mendeleev, *Zhur. Russ. Khim. Obshch.*, 1869, **1**, 60–77

16 D. I. Mendeleev, Proc. 2nd Meeting Scientists, 23 Aug, 1869, 62–71

17 J. L. Meyer, *Ann. Chem. (Liebig)*, 1870, Supp. VII, 354–364

18 D. I. Mendeleev, *Zhur. Russ. Khim. Obshch.*, 1871, **3**, 25–56; *Ann. Chem. (Liebig)*, 1871, Supp. VIII, 133–229

19 K. Seubert, *Ann. Chem. (Liebig)*, 1891, **261**, 272–279

20 S. Kern, *Chem. News*, 1877, **36**, 4; 114–115; 164

21 J. W. Mallet, *Am. Chem. J.*, 1898, **20**, 776

22 A. Guyard, *Moniteur Scientifique*, 1879, **9**, 795–797

23 W. P. Griffith, *Chemistry in Britain*, 1968, **4**, 430–434

24 A. R. Powell, *Platinum Metals Rev.*, 1960, **4**, 144–149

Henri Louis Le Chatelier
1850–1936

Born in Paris, Le Chatelier spent some time under Sainte-Claire Deville at the École Normale but his education was interrupted by the Franco/Prussian War. Later he studied at the École des Mines and became a mining engineer but in 1877 he returned there to teach chemistry, being appointed Professor ten years later. He was the first to employ a platinum against rhodium-platinum alloy thermocouple, so initiating a reliable means of determining high temperatures

19

Platinum in the Measurement of High Temperatures

"Much attention has lately been drawn to an alloy of pure platinum with 10 per cent of rhodium which has become important from the excellent service it has rendered in the determination of high temperatures."

EDWARD MATTHEY, 1892

Very soon after the unique properties of platinum – its very high melting point and its resistance to attack – were recognised its possibilities for the measurement of high temperatures were also grasped and led to an important use of the metal in many branches of industry. At first reliance was placed upon its

PYROMETRE DE PLATINE.

Le citoyen Guyton a présenté, à la séance de l'Institut du 26 floréal dernier, un instrument exécuté pour mesurer les degrés de la plus haute chaleur de nos fourneaux.

Il consiste en une verge ou lame de platine posée de champ dans une rainure pratiquée dans un tourteau d'argile réfractaire. Cette lame s'appuie à l'une de ses extrémités sur le massif qui termine la rainure; l'autre extrémité porte sur un levier coudé, dont la grande branche forme aiguille, sur un arc de cercle gradué. De sorte que le déplacement de cette aiguille marque l'allongement que la lame de métal prend par la chaleur.

The opening of the paper by Guyton de Morveau in the *Annales de Chimie* of 1803. This described a pyrometer he had invented, based upon the expansion of a platinum rod, to measure temperatures in furnaces and pottery kilns

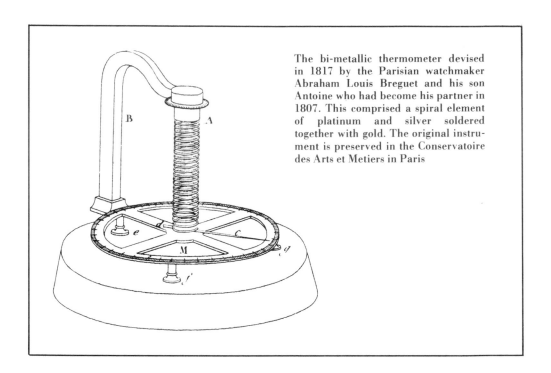

The bi-metallic thermometer devised in 1817 by the Parisian watchmaker Abraham Louis Breguet and his son Antoine who had become his partner in 1807. This comprised a spiral element of platinum and silver soldered together with gold. The original instrument is preserved in the Conservatoire des Arts et Metiers in Paris

coefficient of expansion, as in the very early pyrometer devised by Guyton de Morveau in 1803 and described on page 182. This relied merely upon the expansion of a platinum rod that operated a lever moving over a graduated scale, but the idea of using a bi-metallic device in which platinum was bonded to a metal of much lower coefficient of expansion was also adopted at an early stage. W. H. Wollaston apparently made use of this principle in 1807, when his note-books contain a reference to "Platina Thermometer for Mr. Tennant", one of two that he had caused to be made by Charles Malacrida, a London instrument maker. These were constructed from platinum bonded to copper by Charles Sylvester of Sheffield and formed the subject of a paper by Dr. J. A. Chaldecott (1).

A few years later the famous Parisian watchmaker Abraham Louis Breguet (1747–1823), together with his son Antoine, invented a bi-metallic thermometer in which the helical element was composed of strips of platinum and silver (2), while in 1821 Professor J. F. Daniell of King's College, London, reverting to a single rod of platinum, introduced a pyrometer in which the temperature was determined by the difference in expansion between it and an earthenware tube (3).

None of these instruments was capable of measuring really high temperatures, nor were they of great accuracy. A discovery was about to be made, however, that was to lead to one of the two reliable and accurate methods of temperature measurement that are still in extensive use in both manufacturing industry and scientific research.

348

The Discovery of Thermoelectricity

Shortly after the discovery by Ampère that a current flowing in a wire lying parallel with a magnetic needle had the power to deflect it, Thomas Johann Seebeck (1770–1831) found that when two different metals were joined in a closed circuit and the two junctions were kept at different temperatures an electric current would flow. Seebeck was a native of Reval in Estonia but left there at an early age to study medicine in Berlin and then moved to Jena where he was associated with Goethe, his patron Karl August the Duke of Weimar, Döbereiner and others of their circle. In 1818 he returned to Berlin, taking a post in the Academy there, and it was here that he began a long series of experiments on the magnetic character of the electric current. In August 1821 he announced his discovery of "thermomagnetism" as he called it to the Academy, describing the deflection of the magnetic needle arising from the difference in temperature of the metallic junctions, the variation of the effect with different metals and the increasing effect with rising temperatures (4). His discovery caused a great deal of interest among European scientists and his experiments were quickly repeated by many, including Michael Faraday. It did not occur to Seebeck to make use of his findings for the measurement of temperature, although very curiously he did employ them to check the purity of native platinum, finding that long exposure to a high temperature caused an alteration in the "thermomagnetic" action because of the oxidation of base metal impurities (5).

The first suggestion to make use of Seebeck's discovery as a means of measuring temperature came from Antoine Cesar Becquerel (1788–1878) in a paper read to the Académie Royal des Sciences in Paris on March 13th, 1826 (6). His investigations included observations of the needle deflection obtained with a number of combinations of metal wires when one junction was heated in a spirit lamp, and he deduced that, for certain of these combinations, the intensity of current developed was proportional to the rise in temperature. The most suitable combination, he decided, was a circuit consisting of platinum and palladium wires, and with this combination he obtained a straight line relationship between temperature and electromotive force up to 300°C and by extrapolation he was able to determine temperatures roughly up to 1350°C.

Becquerel further showed that the characteristics were independent of the diameter of the wire, and also that an impure platinum wire would give rise to a current if coupled with a pure platinum wire; he pointed out, in fact, the necessity for cleaning the platinum in nitric acid to avoid spurious effects due to contamination.

In 1836 Professor C. S. M. Pouillet (1790–1868), of Paris, also before the Académie Royal des Sciences, put forward his "magnetic pyrometer" and detailed its construction (7). This instrument, almost incredible by today's standards, comprised a platinum wire sealed into the breech of a gun, the wire passing up the barrel but prevented from touching the sides by a filling of magnesia or asbestos. The breech of the gun was then to be inserted into the hot zone.

In the course of his long and classic researches on heat Henri Regnault (1810–1878) made use of Pouillet's iron-platinum couple, but he found such irregularities that he emphatically condemned the whole idea of the thermo-electric method (8). Regnault's unhappy experiences were due partly to his use of iron as one element, and also to his failure to employ a high-resistance galvanometer. Later, in 1862, Edmond Becquerel (1820–1891) took up the study of his father's platinum-palladium thermocouple and used it as an intermediary with an air thermometer in determining the melting points of a number of substances (9). As a result of his researches he succeeded to some extent in rehabilitating the reputation of the thermocouple, and he derived an expression that was much too complex for the relationship between temperature and electromotive force.

The problem of devising an accurate relationship between these factors was further studied by Professor Peter Tait (1831–1901) of Edinburgh University (10). After a series of experiments with a number of combinations in an attempt to construct "thermoelectric diagrams" he concluded that the electromotive force was in general a parabolic function of the absolute temperature. He also reported that a very small amount of impurity, or even of permanent strain, is capable of considerably altering the line of a metal in the diagram. Professor Tait used

"platinum-iridium alloys containing respectively 5, 10 and 15 per cent of the latter metal. These were prepared for me from pure metals by Messrs. Johnson and Matthey".

This was the first use of iridium-platinum alloys coupled with pure platinum, but Tait's research did not lead to any serious application of this type of thermocouple.

The Work of Henri Le Chatelier

It is to Henri Louis Le Chatelier that we owe the successful practical use of the platinum thermocouple which for many years was in fact known by his name. Among his many activities Le Chatelier was engaged in the study of silicates and cements and needed a reliable method of measuring high temperatures. In 1886 he reported to the Académie des Sciences an investigation on the use of thermocouples for this purpose in which he had tried to verify the parabolic relationships found by Professor Tait between the electromotive force and the temperature of the hot junction, the cold junction being at 0°C. Using various metals and alloys against platinum, which he calibrated at the known melting points of lead, zinc, aluminium, silver, gold, copper and palladium, he obtained results that agreed with his calculations to within 20°C (11). He also came to the conclusion that of the various combinations he had used, platinum against 10 per cent rhodium-platinum gave the most consistent results.

Many years later, in the preface to a book published in 1912 in collaboration with G. K. Burgess he recalled this work:

350

"In 1885, when I attacked the problem of the measurement of high temperatures, it is fair to say there existed nothing definite available on this important question; we possessed only qualitative observations for temperatures above 500°C. Engaged at that time in industrial studies relative to the manufacture of cement, I sought a method which above all would be rapid and simple, and decided on the use of thermo-electric couples, intending to determine the order of magnitude of the sources of error noticed by Regnault. The readings of even a crude galvanometer might be very useful in technical work, provided the limitations of its accuracy were appreciated. I soon recognised that the errors attributed to this method could easily be eliminated by discarding in the construction of the couples certain metals, such as iron, nickel, and palladium, which give the rise to singular anomalies. Among the different metals and alloys studied, pure platinum and the alloy of platinum and rhodium which are still used today, gave the most satisfactory results." (12)

Le Chatelier devoted considerable time and effort to the development of the thermocouple pyrometer, and arranged for the instrument to be manufactured by Carpentier, the successor of the famous Ruhmkorff, at 20 Rue Delambre, Paris. The reputation of these instruments spread both rapidly and widely and they were adopted in a number of industries and laboratories. In 1890, for example, the great American metallurgist Professor Henry Marion Howe, wrote:

"Thanks to the labors of M. Le Chatelier, we have at last a pyrometer capable of measuring easily, accurately and rapidly extremely high temperatures, indeed, those approaching the melting point of platinum. And this is not an apparatus which each must construct for himself; it is for sale ready made. Indeed, it is so far simplified that it has actually entered into practical use for the control of high temperatures in steel works, glass works and gas works." (13)

The development of the Le Chatelier thermocouple also made the thermal analysis of steels practicable and this was immediately taken advantage of by Floris Osmond (1849–1912) of the Le Creusot works (14). In fact his later metallographic studies on carbon steels were carried out in Le Chatelier's laboratory.

The Contribution of Roberts-Austen

Engineers were becoming increasingly concerned at the lack of understanding of the properties of steels and other materials and the Institution of Mechanical Engineers therefore established its Alloys Research Committee. Their choice of an investigator was W. C. Roberts-Austen who combined the posts of Chemist to the Royal Mint and Professor of Metallurgy at the Royal School of Mines. In his first report (15) in 1890 to the Institution, Professor (later Sir William) Roberts-Austen wrote:

"In the present investigation it is necessary to measure much higher temperatures; and fortunately an accurate method is at hand. Early in 1889 I had occasion to employ the pyrometer devised by M. H. Le Chatelier, and was satisfied as to its being extremely trustworthy and convenient up to temperatures over 1000°C. or

1800°Fahr. The instrument in fact enabled me to confirm the fundamental observations of M. Osmond respecting the critical points of iron and steel, and to demonstrate the results in a lecture delivered before the members of the British Association in September 1889.''

Since 1875 Roberts-Austen had interested himself in the problems of liquidation or segregation of the constituents of alloys, and had been most painstaking in his measurement of temperatures using the laborious calorimetric methods then available. He therefore welcomed most readily the new type of instrument and proceeded to adapt it for the production of autographic records of the cooling and solidification of molten metals and alloys. His apparatus is illustrated on the facing page.

Problems of Homogeneity

Some doubt still remained, however, concerning the absolute reliability of the rhodium-platinum alloy, and Roberts-Austen referred to this:

> "It is asserted that even long wires of the platinum-rhodium alloy are homogeneous, and therefore do not give rise to subsidiary currents which would disturb the effect of the main current produced by heating the junction; but very careful experiments to determine whether this is the case have yet to be made.''

Sir William Chandler Roberts-Austen 1843–1902

Born William Chandler Roberts, he added the name of an uncle Austen in 1885. After studying at the Royal School of Mines he was appointed chemist to the Royal Mint in 1870 and later Professor of Metallurgy at the Royal School of Mines. The first investigator to the Alloys Research Committee of the Institution of Mechanical Engineers, Roberts-Austen was quick to appreciate the usefulness of Le Chatelier's thermocouple pyrometer in physical metallurgy and adapted it to produce autographic records of the cooling and solidification of molten metals and alloys

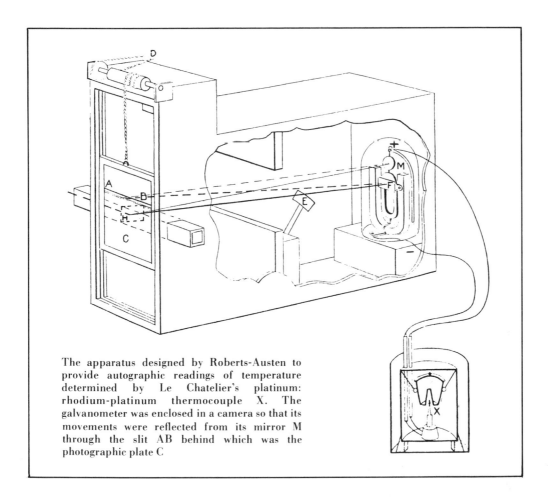

The apparatus designed by Roberts-Austen to provide autographic readings of temperature determined by Le Chatelier's platinum: rhodium-platinum thermocouple X. The galvanometer was enclosed in a camera so that its movements were reflected from its mirror M through the slit AB behind which was the photographic plate C

This uncertainty aroused the interest of Edward Matthey, who carried out a lengthy investigation on the liquidation of alloys of the platinum metals and reported his results in a paper to the Royal Society (16). On the rhodium-platinum alloys he had the following comment to make:

"Much attention has lately been drawn to an alloy of pure platinum, with 10 per cent of rhodium, which has become important from the excellent service it has rendered in the determination of high temperatures. The alloy of platinum with 10 per cent of rhodium is used with pure platinum as a thermocouple, and it is, therefore, interesting to be able to set at rest any doubt which might arise as to this alloy being uniform in composition when melted and drawn into wire."

Matthey prepared a melt of one and a half kilograms of 10 per cent rhodium-platinum, which he cast into a sphere of two inches diameter. The sphere was then sectioned, and samples were taken for analysis from a number of locations between the surface and the centre. The maximum difference between the centre

353

and the outside was found to be 0.06 per cent of platinum and 0.04 per cent of rhodium. He concluded:

> "This result proves that the alloy is not subject to liquation, and fully justifies the high opinion that H. Le Chatelier and Roberts-Austen have formed as to its suitability for thermometric measurements."

At much the same time, 1892, Edward Matthey was concerning himself with the extraction and refining of bismuth and he contributed a series of papers on this subject to the Royal Society. An extract from one of these papers (17), dealing with the temperature at which arsenic can be oxidised off from bismuth, reads at follows:

> "The work of Roberts-Austen has shown that a thermo-junction is practically the only form of pyrometer that can be used for delicate thermal investigations of this kind, but the question arose which particular thermo-junction should be adopted. Was it well to use the platinum-iridium one as advocated by Barus, or the platinum-rhodium one suggested by H. Le Chatelier? My previous work on the alloys of platinum and rhodium, lately published in the *Philosophical Transactions*, settled the question in favour of the rhodium-platinum thermo-junction, for I was satisfied that the alloy of platinum with 10 per cent of rhodium is as homogeneous as any known alloy could well be, and is therefore admirably adapted for use as a thermo-junction, pure platinum being the opposing metal."

Clearly the platinum thermocouple could not be used to measure temperatures higher than the melting point of the metal, and in the course of a lecture to the Royal Institution in 1892 Roberts-Austen referred to this:

> "Metals with higher fusion points than platinum are, however, available; thus iridium will only just melt in the flame produced by the combustion of pure and dry hydrogen and oxygen. By the kindness of Mr. Edward Matthey a thin rod of iridium has been prepared with much labour, and it can be used as a thermo-junction with a similar rod of iridium alloyed with 10 per cent of platinum. The junction may be readily welded in the electric arc, and by this means a temperature may be registered which careful laboratory experiments show to be close to 2000°C" (18).

The further development of platinum metal thermocouples and of their usefulness in industry will be reviewed later in this chapter as in the meantime an alternative method of temperature measurement was being developed although not for such an elevated range of temperatures.

The First Platinum Resistance Thermometer

The first suggestion for making use of the effect of temperature on the resistance of a metal for the determination of that temperature was due to Carl Wilhelm Siemens. He arrived penniless in England in 1843, just before his twentieth birthday, and after studying physics, chemistry and mathematics at Göttingen, and very soon established himself in the rapidly growing electric telegraph and submarine cable industry (19). During the autumn of 1860 the Siemens Company was charged by the British government to superintend the making and laying of

Sir William Siemens
1823–1883

Born Carl Wilhelm Siemens near Hanover, he came to England when not yet twenty and quickly made a reputation in the early days of the electrical industry. In 1860 he devised the first resistance thermometer, using copper wire, to check the temperature of coils of submarine cables, a step that saved an expensive cable from destruction by spontaneous overheating. Ten years later he introduced a platinum resistance thermometer for the measurement of high temperatures. He was the first President of the Society of Telegraph Engineers, later re-named the Institution of Electrical Engineers, and shortly before his death was knighted by Queen Victoria

Reproduced by courtesy of the BBC Hulton Picture Library

a cable between Rangoon and Singapore but the ship was delayed by storms and Siemens became concerned that the spontaneous generation of heat in the coils of cable lying in the hold could lead to its deterioration. He therefore devised an instrument based on the fact that the resistivity of a copper wire increases in a simple ratio with increasing temperature. In describing this new instrument in a letter to Professor John Tyndall at the Royal Institution he concluded:

"By substituting an open coil of platinum for the insulated copper coil this instrument would be found useful as a pyrometer" (20).

In 1862 Siemens, now naturalised as Charles William Siemens, was elected a Fellow of the Royal Society, a mark of the eminence he had achieved as a scientist, and although he continued to be actively engaged in the cable industry for the next few years he was eventually able to devote much of his energy to research.

In 1871 he was invited to give the Bakerian Lecture to the Royal Society and chose for his subject "On the Increase of Electrical Resistance in Conductors with Rise of Temperature and its Application to the Measure of Ordinary and Furnace Temperatures". This was unfortunately published only in abstract (21) but the manuscript is preserved in the archives of the Royal Society.

In his opening paragraph Siemens emphasised that researches on the effect

355

of temperature on resistance had been limited to the range of temperatures between the freezing and the boiling points of water and that platinum, the most suitable metal for extending the range, had been left out of consideration. In carrying out his own investigations he had used

"platinum wire of 0.021 inch diameter prepared by Johnson and Matthey by the old welding process, which gives a much more conductive and therefore purer wire than the more recent process by fusion in a De Ville furnace."

He continued by referring to the great utility of his first resistance thermometer in saving cables from destruction and then described his new instrument in which the platinum wire was wound in helical grooves on a cylinder of pipe clay contained in an iron tube

"for measuring with great accuracy the temperature at distant or inaccessible places including the interior of furnaces where metallurgical or other smelting operations are carried on."

Siemens also gave a lecture, "On Measuring Temperature by Electricity", to the Royal Institution in March 1872, in which he described his instrument as

"the result of occasional experimental research, spread over several years, and it aims at the accomplishment of a double purpose, that of measuring high temperatures, and of measuring with accuracy the temperatures of inaccessible or distant places" (22)

while earlier he had presented a paper to the Iron and Steel Institute at its meeting in Merthyr Tydfil in September 1870, in which he proposed the use of the new pyrometer for measuring the temperature in annealing ovens and of the hot blast supplied to blast furnaces and emphasised that he was not "seeking for any commercial reward, through the Patent Office or otherwise" (23).

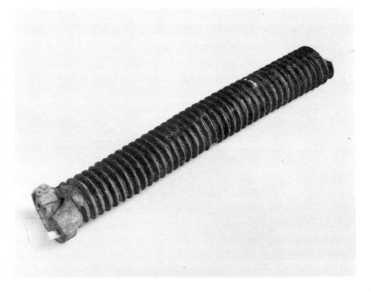

The first resistance thermometers, invented by Sir William Siemens in 1871, consisted of a coil of platinum wire wound on an insulating fire clay core and contained in an iron tube. The remaining portion of this instrument, which was in use in the Royal Arsenal at Woolwich in 1890, was examined by Callendar who found that the platinum wire had suffered a marked change in resistivity, mainly due to contamination by silica in the refractory body

Photograph by courtesy of the Science Museum

Siemens' lecture to the Royal Society naturally attracted the attention of the Council of the British Association, who promptly recommended at their meeting in Edinburgh in 1871 that a committee be formed, with power to add to its number, for "the purpose of testing the new pyrometer of Mr. Siemens". Unfortunately the committee, whose work was carried out by Professor George Carey Foster (1835–1919) in the Physics Department at University College, London, produced a most unfavourable report on the reliability of the instrument (24), although one member, Professor A. W. Williamson – the only chemist on the committee – considered that the observed deterioration of the platinum wire could have resulted from contamination by the reduction of silica from the fine-clay cylinder on which the coil was wound, a comment that was only too correct.

Callendar's Rehabilitation of the Resistance Thermometer

Siemens' resistance thermometer thus remained in disfavour for a number of years, sadly until after his death in 1883. Its rehabilitation was due to the foresight and the experimental skill of two remarkable men in the University of Cambridge. In 1884 J. J. Thomson was appointed, at the age of only 27, Professor of Physics and Director of the Cavendish Laboratory in succession to Lord Rayleigh, holding these posts, as is well known, with great distinction until his retirement in 1919. In the autumn of 1885 he received into his laboratory a new research student, H. L. Callendar, who had taken his degree in classics and mathematics, had never carried out any practical work in physics and had read scientific works only as a hobby. After a few weeks in the laboratory Thomson realised that Callendar had considerable gifts as a skilful experimenter and set about finding him a suitable research project that would give full play to his strong points and yet minimise his lack of experience. He decided that the most suitable work would be the accurate measurement of the resistance of platinum, its variation with temperature, and thus its use for the measurement of temperature. Many years later he wrote:

"Siemens had actually constructed a thermometer on this principle, but this was found to have grave defects which made accurate determinations of temperature impossible. The simplicity and convenience of using a piece of wire as a thermometer was so great that it seemed to me very desirable to make experiments to see if the failure of Siemens' instruments was inherent to the use of platinum as a measure of temperature, and not to a defect in the design of the instrument. Callendar took up this problem with great enthusiasm and showed that, if precuations are taken to keep the wire free from strain and contamination from vapours, it makes a thoroughly reliable and very convenient thermometer. This discovery, which put thermometry on an entirely new basis, increasing not only its accuracy at ordinary temperatures, but also extending this accuracy to temperatures far higher and far lower than those at which hitherto any measurements at all had been possible, was made with less than eight month's work" (25).

By taking care to avoid strain or contamination of the platinum wire Callendar established that its resistance was always the same at a given temperature – "at least this was the case with the specimens used in these experiments, obtained from the well-known firm of Johnson Matthey & Co". His work was carried out in most difficult conditions, on a window-sill in a passage between two rooms in the Cavendish Laboratory, but by sealing his platinum coil, wound on a piece of mica, inside the glass bulb of the air thermometer he was using as a standard he completely overcame the earlier troubles and developed a reliable formula relating change in resistance to temperature. In June 1886 he read a long paper to the Royal Society on his

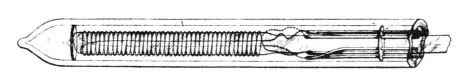

An early example of Callendar's improved resistance thermometer in which the platinum wire was wound on to a frame of mica and contained in a porcelain stem, so eliminating the risks of contamination

Ernest Howard Griffiths
1851–1932

findings (26), a contribution that earned him a Fellowship of Trinity College, and in the following year he filed his first patent on the resistance thermometer (27).

Unknown to Callendar however, there was another research about to start along similar lines, also in Cambridge. In 1888 those two founding fathers of physical metallurgy C. T. Heycock and F. H. Neville, studying the depression of the freezing points of metals by alloying additions and confined in their work to very low melting point solvents by the limitations of mercury thermometers, appealed for help to E. H. Griffiths, one of Neville's colleagues at Sidney Sussex College. Working in a crude wooden laboratory that had been built against the outside wall of the college grounds, Griffiths constructed a number of platinum resistance thermometers for Heycock and Neville and collaborated with them in calibrating these at a number of fixed points, including those of ice, steam, the boiling points of several organic compounds, and finally at the boiling point of sulphur, so making it possible to determine with accuracy the melting points of many metals and alloys.

In 1889 Griffiths became aware of Callendar's work and the two joined forces and succeeded in determining temperatures up to 1100°C (28). The accuracy and reliability of the device were now established, and Callendar and Griffiths approached Horace Darwin, the head of the Cambridge Instrument Company, who readily agreed to its manufacture, together with the necessary indicating equipment. These were quickly introduced into iron and steel making and into other industries where they have proved their usefulness over the years for both industrial temperature measurement and control and precise laboratory work (29). One early example was the use of the thermometer to determine steam temperatures, this having an important effect on the design of steam turbines.

Over the past three quarters of a century designs have improved to meet industrial needs, more manufacturers have engaged in its production, new techniques have been introduced, including the replacement of the coil of wire by thick films, and smaller and more robust thermometers have been made available. Production of platinum resistance thermometers for industrial use now runs to several millions a year (30), while the accuracy of the temperature-resistance characteristics is ensured by the continuing availability of the special high purity platinum upon which the functioning of the instrument is based.

Measuring the Temperature of Liquid Steel

To return now to the platinum thermocouple and its later applications, we find it in use in a wide variety of industries including glass manufacture, refractory making, the iron foundry and the nuclear energy industry, but almost certainly the most significant success it achieved was in the steel industry.

In the course of a discussion on modern methods of measuring temperature held by the Institution of Mechanical Engineers in 1913, opening with a comprehensive paper by R. S. Whipple of the Cambridge Scientific Instrument Company, Sir Robert Hadfield appealed for further work to be done in determining the temperature of molten metals:

"There was no doubt that the casting temperatures to which molten steel was heated were of greater importance than commonly imagined" (31).

Four years later a major discussion on pyrometers and pyrometry was organised by the Faraday Society and in a paper by Dr. W. H. Hatfield of the Brown Firth Research Laboratories he stated that he had used a similar instrument to determine the temperature of molten steel when poured into the ladle but added:

"Although the temperature at which steels are cast must have an influence upon their ultimate physical properties, no ready and really reliable method for measuring such temperatures from the works standpoint is available" (32).

It was not until twenty years later, however, that such a "ready and reliable method" was forthcoming. In 1937, after considerable co-operation with the steel industry, Dr. F. H. Schofield and his colleague A. Grace of the National

360

The measurement of the temperature of molten steel was for many years a much sought after means of controlling quality in the open hearth or other types of furnaces. The problem was eventually solved by Schofield and Grace at the National Physical Laboratory who devised a "quick-immersion" method employing a platinum:rhodium-platinum thermocouple led through a steel tube mounted on a trolley

Physical Laboratory put forward their quick-immersion method in which a platinum : rhodium-platinum thermocouple, sheathed in twin fine-clay insulators and led through a steel tube bent at a right angle at the hot junction and mounted on a trolley, could be employed to determine the temperature of molten steel in open hearth and electric arc furnaces and in Bessemer converters by dipping into the molten steel for about twenty seconds (33). This technique was rapidly adopted in the steel industry throughout the world and had a marked effect in improving quality. The difficulties that had to be surmounted were graphically described by two leading metallurgists, W. C. Heselwood and D. Manterfield of the United Steel Companies, who also played their part in the development of this procedure:

"Some twenty inches of molten steel at a temperature of about 1,620°C, covered by perhaps five inches of molten slag at 1660°C and contained in the hearth of a furnace typically fifty by fifteen by ten feet with refractory walls and roof at temperatures up to 1650°C, and through which roar at sixty miles an hour flame and gases at 1800°C; those are typical conditions within an open hearth steel making furnace and it is not surprising that the practical problem of measuring accurately the liquid steel temperatures remained unsolved for many years" (34).

361

These problems were of course solved, and in later years improvements and simplifications have been introduced into the "quick-immersion" method which continues to serve as a most valuable means of controlling the quality of steel. At the present time more than a hundred million readings a year are taken in the steel industry throughout the world, resulting in the production of more tons per hour than was formerly possible and longer furnace campaigns.

The International Temperature Scale

The accuracy and reliability of the enormous number of temperature measurements made every day in manufacturing industry with either the platinum resistance thermometer or with platinum thermocouples depend upon the acceptance of a recognised practical scale of temperature and its precise definition. Such a scale was first proposed by Callendar in a paper to the British Association in 1899 (35) and in 1903 the matter was placed in the care of the newly formed National Physical Laboratory at Teddington. After a great deal of study and discussions with the American Bureau of Standards, the Physikalische Technische Reichanstalt and the Bureau Internationale des Poids et Mesures, the work was interrupted by the first war but it was again taken up and the first International Practical Scale of Temperature was agreed upon in 1927. Several revisions have since taken place, but it is now firmly based upon the use of the platinum resistance thermometer from $-259.34°C$, the triple point of equilibrium between the solid, liquid and vapour phases of hydrogen, and $630.74°C$, the freezing point of antimony, and from that point on to the freezing point of gold at $1063°C$ by the platinum: 10 per cent rhodium-platinum thermocouple (36).

362

References for Chapter 19

1   J. A. Chaldecott, *Ann. Sci.*, 1971, **27**, 409–411; *Platinum Metals Rev.*, 1972, **16**, 57–58

2   A. L. B. and A. Breguet, *Ann. Chim.*, 1817, **5**, 312–315.

3   J. F. Daniell, *Quart. J. Sci.*, 1821, **11**, 309–320

4   T. J. Seebeck, Abhandlurgen der physikalische Klasse der Königlichen Akademie der Wissenschafter zu Berlin, 1822–23, 265–373

5   T. J. Seebeck, *J. tech. Ökon. Chem.*, 1828, **2**, 102–104

6   A. C. Becquerel, *Ann. Chim.*, 1827, **31**, 371–392

7   C. S. M. Pouillet, *Comptes rendus*, 1836, **3**, 782–790

8   H. V. Regnault, Relations des Experiences, Paris, 1847, **1**, 246

9   E. Becquerel, *Comptes rendus*, 1862, **55**, 826

10   P. G. Tait, *Trans. Roy. Soc. Edin.*, 1872–73, **27**, 125–140

11   H. L. Le Chatelier, *Comptes rendus*, 1886, **102**, 819–822

12   G. K. Burgess and H. Le Chatelier, The Measurement of High Temperatures, London, 1912, vi

13   H. M. Howe, *Eng. and Min. J.*, 1890, **50**, 426

14   F. Osmond, Methode Generale pour l'Analyse Micrographique des Aciers au Carbone, Paris, 1895

15   W. C. Roberts-Austen, 1st Report, Alloys Research Committee, *Inst. Mech. Eng.*, 1891

16   E. Matthey, *Phil. Trans. Roy. Soc.*, 1892, **183**, 629–645

17   E. Matthey, *Proc. Roy. Soc.*, 1893, **52**, 467–472

18   W. C. Roberts-Austen, *Proc. Roy. Inst.*, 1892, **13**, 502–518

19   L. B. Hunt, *Platinum Metals Rev.*, 1980, **24**, 104–112

20   C. W. Siemens, *Phil. Mag.*, 1861, **21**, 73

21   C. W. Siemens, *Proc. Roy. Soc.*, 1871, **19**, 443–445

22   C. W. Siemens, *Proc. Roy. Inst.*, 1872, **6**, 438–448

23   C. W. Siemens, *J. Iron Steel Inst.*, 1871, **1**, (i), 50–55

24   British Association Report, 1874, 242

25   J. J. Thomson, Recollections and Reflections, Cambridge, 1936, 132

26   H. L. Callendar, *Phil. Trans. Roy. Soc.*, 1887, **178**, 161–233

27   H. L. Callendar, British Patent 14,509 of 1887

28   H. L. Callendar and E. H. Griffiths, *Phil. Trans. Roy. Soc.*, 1891, **182**, 119–157

29   R. Price, *Platinum Metals Rev.*, 1959, **3**, 78–82

30   J. S. Johnston, in Temperature Measurement 1979, ed B. F. Billing and T. J. Quinn, Institute of Physics, London, 1975, 80

31   Sir Robert Hadfield, *Proc. Inst. Mech. Eng.*, 1913, (**3–4**), 766–768

32   W. H. Hatfield, *Trans. Faraday Soc.*, 1917–18, **13**, 289–294

33   F. H. Schofield, Iron Steel Inst., Special Report No. 16, 1937, 223–238

34   W. C. Heselwood and D. Manterfield, *Platinum Metals Rev.*, 1957, **1**, 110–118

35   H. L. Callendar, *Phil. Mag.*, 1899, **48**, 519; British Association Report, 1899, 242–243

36   C. R. Barber, *Platinum Metals Rev.*, 1969, **13**, 65–67; T. J. Quinn and T. R. D. Chandler, *ibid.*, 1972, **16**, 2–9

Joaquim Bishop
1806–1886

For some years assistant and instrument maker to Professor Robert
Hare at the University of Pennsylvania, Bishop set up in business in
Philadelphia in 1842 and put Hare's compound blowpipe to commercial
use in the melting of platinum. His activities in the manufacture of
platinum equipment expanded and his company later became part of
the Johnson Matthey group

20

Platinum Extraction and Fabrication on the American Continents

"The applications of platinum have steadily increased, and never so rapidly as in the last two decades. For many purposes no substitute for platinum has been found."

PROFESSOR JAMES LEWIS HOWE, 1900

The platinum industry, in so far as refining and manufacturing are concerned, took far longer to establish itself in the Americas than in Europe although, as related earlier, the original supplies of the native metal came from there, the principal source being New Granada, later securing its independence as the Republic of Colombia late in 1819. Shortly afterwards the new government decided to prohibit the export of platinum and to build up stocks with the idea of a platinum coinage in mind. In the following year, however, the Colombian Vice-President Francisco Antonio Zea (1766–1822) arrived in Europe with two objectives, to contract a loan of up to £5 million sterling and to seek the recognition by European powers of the newly established republic. In London Zea met Justus Erich Bollmann, by then a manufacturing chemist whose activities in the fabrication of platinum in the United States will be described a little later, who was instrumental in securing an advance of £20,000 towards a projected loan of £2 million that Zea eventually contracted in 1822. One half of this advance was contributed by Bollmann himself, and as a guarantee Zea, apparently unaware of the intention to mint a coinage, signed an agreement with him whereby 40,000 pounds of platinum (an impossibly large amount) supposedly held in the Colombian Mint in Santa Fé de Bogotá, would be released for sale in London, no doubt for use by Bollmann himself and by Wollaston, whose supplies had virtually ceased by this time.

Zea's authority to contract a loan was revoked in August 1821 by the Colombian governing council although news of this did not reach London until the October of the following year by which time Bollmann, who had visited Colombia in connection with the proposed shipment of platinum, had died in Jamaica on his way home, while Zea himself died only a month later.

The government declined to honour the agreement made between these two

After two centuries of hand operation by natives in the Chocó district of Colombia, mechanical equipment was brought in by the Compania Minera Chocó Pacifico in 1915. This is one of several dredges in operation in the main platinum-bearing area

men for the supply of platinum, no doubt because of their intention to mint a platinum coinage, authorised in September 1821 but never put in hand. These events, recently uncovered by Dr. J. A. Chaldecott and to be described by him in more detail in a forthcoming paper in *Platinum Metals Review*, were undoubtedly responsible for the statement by the unknown translator in his footnote to Baruel's paper in *The Philosophical Magazine* of 1822 referred to on page 190:

> "This valuable memoir derives particular interest from the large importation of the above ore, daily expected from South America, in consequence of the negotiation between M. Zea and some London merchants." (1).

This consignment never of course arrived, but its supposed existence has often been referred to, even by Sir Edward Thorpe, who wrote quite erroneously in his "Essays in Historical Chemistry":

> "Whole cargoes of the native metal are said to have lain unpurchased for years in London as it could not be turned to account." (2)

The possibility of minting a platinum coinage in Colombia arose again in 1829, prompted by the Russian moves in this direction, and Humboldt was consulted by Símon Bolívar, but with the resignation of the Liberator in 1830, and his death very soon afterwards, nothing more was heard of the idea.

366

Later Production in South America

Production of platinum in Colombia remained at a low level for many years, confined to the activities of native workers. In 1879, however, a small refining and assaying establishment was set up in Medellin by the Gutiérrez family. A younger member was sent over to Johnson Matthey in London to gain experience and the Fundicion Gutiérrez is still active at the present time.

In 1911 the Anglo-Colombian Development Company was formed jointly by Consolidated Gold Fields and Johnson Matthey to work the platinum deposits of the Chocó and dredging operations were begun, output beginning to increase by 1914. Later this company was taken over by the South American Gold and Platinum Company, formed in the United States with a subsidiary operating company, Compania Minera Chocó Pacifico. This organisation put a second dredge into service in 1920, a third in 1923 and a fourth in 1932 (3). Production of platinum thus increased considerably, reaching 30,000 ounces a year by 1920 but now representing a smaller part of world production.

The Fabrication of Platinum in the United States

Although as was said at the beginning of this chapter America was late in entering the platinum industry, there was one quite remarkable exception. In 1796 there arrived in New York the German physician and adventurer Justus Erich Bollmann already mentioned in his later dealing's with Zea. He had been imprisoned after two attempts to release Lafayette from his confinement by the Austrians and then expelled from the country. For his efforts on behalf of Lafayette he was given a warm welcome by George Washington and he established a merchanting business in Philadelphia. This proved to be unsuccessful, however, and he decided to take up practical chemistry, devoting his efforts to the production of malleable platinum and the development of its industrial uses.

No details of Bollmann's process have survived, but he could well have been aware of the publications of Richard Knight in 1880 and of Thomas Cock in 1807 referred to in Chapter 8. Cock had not attempted to remove the other platinum metals, although their discovery had been reported by Wollaston and Tennant between 1802 and 1804, but as a member of the American Philosophical Society since 1800 Bollmann could have learnt something from a paper read there in 1809, although not published until 1818, by Joseph Cloud of the Philadelphia Mint (4). This described the separation of palladium and rhodium from native platinum, and possibly he adopted this procedure. The first announcement of his success came in a note from Professor Thomas Cooper of Dickinson College in Pennsylvania, the editor of the *Emporium of Arts and Sciences*, in 1813 (5). This prompted a letter from Bollmann, part of which is reproduced here, in which he claimed to be the first to render platinum malleable in America "by means of a process which admits being executed on a large scale", going on to say that he had made sheets of thirteen inches square and that he

Justus Erich Bollmann
1769–1821

German physician, adventurer, merchant, economist, author and manufacturing chemist, Bollmann arrived in New York from England in 1796, after acquiring a knowledge of chemistry in Paris. He lived for most of the ensuing twenty years in Philadelphia where he became a member of the American Philosophical Society. He was the first to prepare malleable platinum on a commercial scale in the United States, and was responsible for providing the first platinum boiler used in America for concentrating the weak sulphuric acid produced by the lead chamber process

had in preparation platinum vessels that would hold from twenty to thirty gallons (6).

In a letter to the Governor of Pennsylvania, Simon Snyder, in 1814 (recently discovered by Dr. J. A. Chaldecott, to whom the present writer is indebted for these details) in the State Archives of Pennsylvania (7) Bollmann wrote:

"There has been a Boiler made here, of rolled Platina, for the Condensation of Oil of Vitriol, which holds 25 gallons and is the largest vessel of Platina, probably, in existence."

Part of Bollmann's letter of 16 June 1813 to Professor Thomas Cooper wherein he gives some information about his own production and use of malleable platinum. He also expressed the hope that his process would become beneficial to the arts, and to society. Cooper published the letter in August 1813 in his "Emporium of Arts and Sciences"

As the article stands it gives the impression, that I only do what Mr. Cloud did before, but better. The fact however is, that Mr. Cloud, whose merits I well know and fully acknowledge—does not seem to have intended more than to produce a cabinet piece of the greatest possible purity and specific gravity, and he has done so: but I have first rendered platina malleable in this country, by means of a process, which admits being executed on a *large* scale, and which, I hope, will become beneficial to the arts, and to society.—Pieces have been made of the weight of two pounds, and upwards. Sheets have been rolled of thirteen inches square, and vessels of platina are now making, and in preparation, which will hold from twenty to thirty gallons.

368

Dr. Chaldecott considers that this still was made for John Harrison of Philadelphia, the first successful large-scale manufacturer of sulphuric acid in the United States, who introduced the lead chamber process there around 1793. The still weighed 700 ounces, held 25 gallons, and was in service for some fifteen years (7).

Bollmann tried repeatedly to promote further applications for platinum, in glass manufacture, in the decoration of porcelain, for coins and medals and for standard weights and measures, but none of these came to anything. On a visit to England in 1814 he met Wollaston, from whom he would have learnt of the latter's success several years earlier in the production of malleable platinum and the making of large sulphuric acid boilers (7). In 1816 he left North America for England, never to return.

The Enterprise of Joaquim Bishop

After Bollmann's departure there appears to have been no refining or fabrication of platinum in the United States for some twenty-five years and the demand for laboratory apparatus and other uses was met by the European producers, each having its agents in New York and Philadelphia.

Then in 1842 an establishment came into existence that was to continue firmly until the present day. This was set up by Joaquim Bishop, the son of English parents who lived in Oporto in Portugal where he was born. The French wars disturbed the family, and in 1810 they emigrated to America, settling first in Baltimore and then in Philadelphia. In 1826 Joaquim became an apprentice to a manufacturing jeweller but left him to become a finisher in a brass foundry. His experience in metal working was combined with some scientific reading in his spare time and in 1832 he secured the position of assistant and instrument maker to Professor Hare in the University of Pennsylvania. Here he took part in building the apparatus for the later series of Hare's experiments in the melting of the platinum metals (described in Chapter 15) but he left the university in 1839 to become a "Machinist and Philosophical Instrument Maker".

In 1842 he was urged by his friends – most probably also by Hare – to take up the refining and fabrication of platinum, using the famous blow-pipe. Already in 1845 he was awarded a silver medal by the Franklin Institute for "skill and ingenuity in the manufacture of platinum scientific instruments". In 1858 he moved from Philadelphia to Radnor, and then again in 1865 to Sugartown where he bought a 43 acre estate and built a new home and workshop. There he set up a melting shop with equipment for the production of hydrogen and oxygen, a forge and rolling mills and a draw bench. The platinum was melted in 20-ounce ingots and he managed to secure and to treat native platinum from Russia as well as purchasing sheet from the agents of the European refiners. The staff numbered only two or three, with Bishop himself as skilled workman and salesman. The quality of both his metal and his workmanship were highly spoken of by contemporary users (8) and "from the beginning his several steps were crowned with remarkable success". (9)

369

The second platinum works established in 1865 by Joaquim Bishop at Sugartown, Chester County, Pennsylvania, with the founder on the left and his small staff. The plant included a melting shop with means for producing hydrogen and oxygen, a forge, one large and two small rolling mills, a wire drawing bench and a chemical laboratory

In 1876 he was awarded a bronze medal for an exhibit of platinum at the Philadelphia Centennial Exhibition, and he continued to be active in his business until two years before his death in 1886. His assistant and later partner Edwin Cox carried on the business until 1889, when Bishop's grandson and heir J. B. Matlack came of age. In 1903 the refinery and workshops were destroyed by fire, but the business continued, a new works being built at Malvern, and then in 1909 the firm was incorporated as J. Bishop and Co. Platinum Works. In this form it continued to expand, producing all the types of platinum apparatus needed and also taking up the manufacture of spinnerets for the American works established by Courtaulds in 1909 to manufacture artificial silk. Then in 1927 Johnson Matthey purchased a shareholding in the company, acquiring full control in 1933. Later the name was changed to Matthey Bishop Inc., and more recently to Johnson Matthey Inc. New and much greater facilities have of course been built in recent years to meet the ever increasing demand for the platinum metals, including a refining and chemical complex, a catalyst manufacturing plant and a new centre for metallurgical and fabricating operations.

Charles Engelhard

Bishop was not without competition in the platinum business. In 1875 Daniel W. Baker with his sons Charles and Cyrus set up in Newark, New Jersey, first as a manufacturing jeweller but later taking up the fabrication of platinum and by 1892 they had issued a catalogue of laboratory apparatus (10). Later another small competitor emerged, the Charles F. Croselmire Company, also in Newark. In the meantime there had arrived in New York in 1891 Charles Engelhard, the brother-in-law of Dr. Wilhelm Heraeus, to serve as the representative of the Hanau company. This was an event of great significance in the platinum industry and its future development in the United States. A man of great energy and vision, he embarked after some years in America on a policy of acquisition and expansion. In 1901 he acquired the Croselmire Company, a small manufacturer of platinum wire and sheet and on this basis formed the American Platinum Works, taking over the Baker firm in 1904.

When the extraction of platinum from Colombia, mentioned earlier in this chapter, yielded important quantities of metal during the first World War, a large refinery was erected by Baker and from then onwards the company played a leading part in the American industry and established branches in many parts

Charles Engelhard
1867–1950

A native of Hanau and brother-in-law of Dr. Wilhelm Heraeus, Engelhard was sent to New York in 1891 as agent for the Heraeus company. A man of considerable energy, he began to acquire several small companies engaged in the working of platinum and became President of Baker and Company, now known as Engelhard Industries. After the first world war he secured the supplies of platinum metals produced by the Mond Nickel Company and rose to a prominent position in the platinum industry of the world. The founder, Charles Engelhard, died in 1951 and was succeeded by his son Charles W. Engelhard (1917–1971) who in 1955 changed the style and title of his enterprise to Engelhard Industries.

Scientific Work on Platinum in the United States

From the earliest years of independence, and the high times of Benjamin Franklin, Philadelphia had been the cradle of scientific activity. The work of Robert Hare and his co-operation with Benjamin Silliman has been reviewed in Chapter 15, together with its continuation in France by Deville and Debray. Another early piece of research was carried out by Joseph Cloud (1770–1845), the chemist and refiner of the Philadelphia Mint from 1797 until 1836. In 1807 he received some small gold bars from Brazil that were only of a pale yellow colour and found that they contained a metal "that would resist the cupel and was soluble in the nitric and nitro-muriatic acids". In a paper to the American Philosophical Society he began:

> "Notwithstanding the numerous experiments that have been made by several eminent chemists, on a metallic substance, discovered by Dr. Wollaston, in combination with crude platinum and by him called palladium, there still remains much doubt with respect to the existence of such a simple substance" (11).

Cloud went on to prepare a specimen of this metal, finding that its properties did indeed agree with those described a few years earlier by Wollaston and confirming that palladium had been isolated from an entirely new source. Cloud's further paper on the separation of palladium and rhodium from native platinum has already been mentioned in connection with Bollmann's procedure (4).

But the leading research chemist in the United States for many years was Wolcott Gibbs, a native of New York who after graduating from Columbia College spent some time as assistant to Hare and then visited Europe to study in turn under Heinrich Rose, Liebig and Dumas. On his return he was appointed Professor of Chemistry at the newly established Free Academy which later became the City College of New York and then the City University there (12). Here he remained for fourteen years and in collaboration with Frederick Augustus Genth, formerly an assistant to Bunsen at Marburg, carried out a long series of researches on the platinum metals and their co-ordination compounds (13). Largely as a result of this work he was called to Harvard in 1863 as Professor of Natural Sciences, a post he retained until his retirement in 1887. Gibbs introduced into American science the German system of research as a means of chemical instruction (12), and one of his important achievements was the initiation of electrogravimetric analysis, employing a platinum crucible as the cathode connected to a Bunsen cell, the anode being a length of platinum wire immersed in the solution (14). In this way he made determinations of copper and nickel and pioneered a method of analysis that gave excellent service for very many years.

Gibbs also devoted his efforts to research on the separation of the platinum metals from each other and to their methods of analysis. For many years he edited the *American Journal of Science* established by Silliman, and he was one of the founders of the Academy of Sciences.

Yet another Philadelphian to interest himself in the platinum metals was Dr. William Henry Wahl (1848–1909). After studying at Heidelberg he was appointed secretary and editor to the Franklin Institute in 1871, a position he continued to hold until his death. He was especially concerned with electrodeposition, and in 1883 published a book on the subject (15) in which he dealt with both platinum and palladium plating and their electrolytes. In a later paper, "On the Electrodeposition of Platinum", read to the Franklin Institute in 1890 (16) he reviewed the whole problem of platinum plating, saying that while each of the earlier baths would yield satisfactory results for a time

"the peculiar difficulties met with in the practice of platinum plating render it impossible to maintain the chemical integrity of those electrolytes and in consequence thereof they soon become inefficient or inoperative by reason of contamination with the secondary products formed therein".

Wahl therefore experimented with anodes consisting of porous carbon impregnated with platinum to provide a means of dissolution in acidic

Wolcott Gibbs
1822–1908

Born in New York, Gibbs first attended Columbia College and then served as an assistant to Robert Hare at the University of Pennsylvania, later being appointed Professor of Chemistry at what is now the City University of New York. Here he introduced the German system of research as a means of instruction in chemistry while he carried out some of the earliest researches in the United States on the chemistry of the platinum metals. Later he was appointed Professor of Natural Sciences at Harvard

373

electrolytes, but his deposits were black and non-adherent. He then turned to alkaline baths containing oxalic acid and he was the first to employ an electrolyte free from chloride and to use "platinic hydrate" to maintain the metal content of his bath, so avoiding the build-up of undesirable compounds.

A major contribution, although of a rather different kind, to the history and the chemistry of the platinum metals began to be made in the late nineteenth century by James Lewis Howe, Professor of Chemistry at Washington and Lee University in Lexington, Virginia. His chemical researches were largely devoted to the compounds of ruthenium, a metal that he felt was the most interesting of the group and of which the least was known, including the confirmation and extension of the earlier work of Joly in France (17). But his most important contribution was his "Bibliography of the Metals of The Platinum Group", first published by the Smithsonian Institution in 1897 and followed by a number of further and more up-to-date editions. This was a most painstaking achievement, containing well over two thousand entries from some thirteen hundred authors,

Professor James Lewis Howe (1859–1955), was born at Newburyport in Massachusetts and studied under Wöhler at Göttingen from 1880 to 1882. In 1894 he was appointed Professor of Chemistry at Washington and Lee University at Lexington, Virginia, remaining there for almost fifty years. He became the leading authority on ruthenium and its compounds but his most valuable and best-known work was his "Bibliography of the Metals of the Platinum Group", first published in 1897 with a number of subsequent editions

SMITHSONIAN MISCELLANEOUS COLLECTIONS

1084

BIBLIOGRAPHY OF THE METALS OF
THE PLATINUM GROUP

PLATINUM, PALLADIUM,
IRIDIUM, RHODIUM, OSMIUM, RUTHENIUM

1748–1896

BY

JAS. LEWIS HOWE

CITY OF WASHINGTON
PUBLISHED BY THE SMITHSONIAN INSTITUTION
1897

**Thomas Alva Edison
1847–1931**

The great American inventor, born of poor parents and self-taught in chemistry and physics, was actively interested in the properties and usefulness of platinum. He purified the metal from discarded Grove cells and reduced it to wires of only 0.001 inch diameter for use in his early attempts to develop the incandescent lamp, while his observation that a platinum spiral heated in a vacuum gave a deposit of metal on the inside of the glass bulb led, in the hands of Sir Ambrose Fleming, to the development of the thermionic valve

extracted from a hundred scientific journals. Howe began by consulting the monograph compiled by Karl Klaus (page 250) which contained a fairly complete bibliography of the platinum metals up to 1861 but which included a number of errors that he had to rectify. His bibliography and his position as the leading American authority on the chemistry of the platinum metals led to his appointment in 1917 as chairman of a committee on platinum of the National Research Council.

Thomas Edison's Work with Platinum

The inventive genius of Thomas Alva Edison has already been referred to in connection with the incandescent electric lamp, but this was of course only one of his many achievements. His interest in platinum began as early as 1863 when, aged only sixteen, he was a night telegraph operator on the railway at Stratford Junction in Ontario. Finding that several old Grove batteries had been discarded, he obtained permission to remove the platinum electrodes and added them to his quite large stock of chemicals in the laboratory he had built in the cellar of his parent's house and where he spent most of the daylight hours. Later some of this platinum was almost certainly used, after being drawn to fine wire, in his initial experiments on the electric light.

In 1879 he presented a paper, "On the Phenomena of Heating Metals in Vacuo by Means of an Electric Current", to the meeting of the American Association for the Advancement of Science, held that year in Saratoga Springs, New York (18). This included an account of the behaviour of iridium-platinum alloy wires which, when brought to incandescence, gave rise to a deposit of platinum on the inner surface of the glass bulb.

In studying further this "Edison effect" (page 309) in 1884 he made lamp bulbs containing a thin plate of platinum placed between the limbs of the filament and found that if this plate was connected to the positive end a current would flow across the vacuum but that no current would flow if it were connected to the negative limb (19). Edison might therefore have discovered the grid, devised much later by his fellow American Lee de Forest (1873–1961) in 1907 (20) but unfortunately he was too far ahead of his time and the only use he could foresee was for an indicator to record variations in the potential of his lighting circuits.

A New Source of Platinum in Canada

In 1888 a group of gold miners working a small property in the Sudbury district of Ontario came upon a rich copper-nickel ore, a sample of which was sent to Francis Lewis Sperry, a chemist at the Canadian Copper Company. In making a fire assay for gold Sperry obtained a small white metallic bead which he passed along to his old professor Horace Lemuel Wells who, with his colleague Professor Samuel Penfield examined this further and identified the mineral as an arsenide of platinum to which they gave the name Sperrylite. Further samples were sent to Frank Wigglesworth Clarke, the chief chemist to the U.S. Geological Survey (an experienced operator with the platinum metals and the great worker on atomic weights) who reported that

"this is the first authentic instance of the occurrence of platinum in true metalliferous vein material, and it has therefore remarkable interest".

Clarke's analysis of the mineral showed only 0.025 ounce of platinum per ton, on which he commented:

"Whether it could be profitably extracted is an open question."

The copper-nickel deposit was worked with many difficulties by the Canadian Copper Company, together with the Orford Copper Company, separating the copper from the nickel but without recovering any of the platinum metals. In 1902 these two enterprises were merged into the International Nickel Company and more attention began to be paid to improved methods of producing nickel and to the recovery of the platinum metals, but for many years only a fraction of these was successfully extracted, the remainder going into the copper and nickel.

The first process for their recovery on a commercial scale stemmed from the activities of Ludwig Mond, a man of great energy and vision who after some years experience in chemical manufacture in Germany had settled at Widnes in England in 1867 and established himself in the chemical industry in partnership with John Brunner. One of the many problems he faced later in his career concerned the corrosion of nickel valves and the accumulation upon them of a

376

Ludwig Mond
1839–1909

Born in Cassel in the Grand Duchy of Hesse, Mond studied chemistry under Kolbe at Marburg and then with Bunsen at Heidelberg. After some years experience in the chemical industry in Germany and Holland he came to England in 1867 and began to build up his chemical enterprises in partnership with Brunner. Despite his many business problems he devoted some time to research and in 1889 devised a rudimentary fuel cell with platinum electrodes, while a year later he discovered the reaction between carbon monoxide and nickel that led to the carbonyl process for its refining. To make this process commercially successful required a source of mineral, and Mond's acquisition of a copper-nickel ore body in Canada led to the formation of the Mond Nickel Company and then to the extraction of the small proportion of the platinum metals from the residues of the carbonyl process

deposit of carbon, and this he set out to investigate in a laboratory he had set up in the stables of his private house in London, taking on a young Austrian chemist, Carl Langer who had studied under Victor Meyer at Zürich and then served as a research chemist at the Badische Anilin und Soda Fabrik at Ludwigshaven. In 1889 Mond and Langer, together with a young German chemist Friedrich Quinke and Mond's elder son Robert, discovered the formation of a gaseous compound of nickel and carbon monoxide when this gas was reacted with finely divided metallic nickel. This discovery was at once reported to the Chemical Society (23) but to Mond it suggested that here was a possible method of refining nickel, and he decided to build a large scale pilot plant for this purpose in the works of Henry Wiggin in Birmingham, Carl Langer undertaking its design and construction.

After some three years of experimentation Langer developed a reliable process and applied it to treat the copper-nickel sulphide matte that Wiggins were receiving from Canada, producing pure nickel and leaving a residue from which the platinum metals could be recovered.

If Mond was to enter the nickel business on a large scale he now needed supplies of suitable mineral, and having made this decision he engaged a German mining engineer, Dr. Bernard Mohr, to investigate possible sources,

To provide a source of mineral for his carbonyl process of refining nickel Ludwig Mond acquired a property in the Sudbury district of Ontario, opened a mine and erected a smelter. This shows the property, named the Victoria Mine, shortly after it commenced operations in 1900. Today the scene is very different, with a great complex of mines and metallurgical plants from which the platinum metals are eventually recovered

including those in Canada, and in 1899 Mond purchased two properties in the Sudbury area that he re-named the Victoria Mine and where he erected a smelter to produce a matte. A year later a nickel refinery was built at Clydach in South Wales, with Langer in charge, and this plant received the copper-nickel matte from Canada, producing pure nickel and copper sulphate. To link these two operations together the Mond Nickel Company was founded in 1901 with Carl Langer as managing director.

The presence of the platinum metals in the ore was confirmed, the refinery residues containing them began to accumulate, and considerable thought was given to the prospect of their recovery. Ludwig Mond had earlier interested himself in platinum and its properties, and one of the first inventions coming from his private laboratory stemmed from the original fuel cell devised by W. R. Grove in 1839 (described on page 211). Mond felt that his producer gas, rich in hydrogen, might be used to produce electricity. He and Langer therefore set about designing a "gas battery", using strips of platinum immersed in dilute sulphuric acid, then changing to porous diaphragms covered with fine platinum foil and coated with a thin film of platinum black. The investigation was reported in a paper to the Royal Society in 1889 (24), but unfortunately the life of this rudimentary fuel cell proved to be too short, and although he promised further studies the pressure upon Mond in his industrial empire prevented this. His interest in platinum continued, however, and in collaboration with Sir William Ramsay of University College he pursued a research on the occlusion of oxygen and hydrogen by platinum and palladium (25).

378

Carl Langer
1859–1935

Born in Moravia, now part of Czechoslovakia, Langer was brought up in Budapest and then studied chemistry under Victor Meyer at Zürich, obtaining his doctorate there in 1882. He then joined the Badische Anilin und Soda Fabrik as a research chemist, but Ludwig Mond invited him to London to become his personal assistant. After their discovery of the carbonyl process for refining nickel Langer designed and built the refinery at Clydach in South Wales and also investigated possible ore bodies in Canada. Later he supervised the development of a process for the extraction and refining of the platinum metals from the nickel residues

Thus the intention to recover the platinum metals was well in front of Ludwig's mind, and in 1902 samples of the residues were sent in to Johnson Matthey for analysis. This presented a number of difficulties because of their unusual and varying nature, but regular deliveries began a little later and the refining and preparation of pure platinum and palladium were undertaken on a contract basis by Johnson Matthey. Here again much difficulty was encountered, the residues containing only about 7 per cent of total platinum metals, but a process was duly worked out. This continued until 1919, by which time some 33,000 ounces of platinum and 53,000 of palladium had been produced, and then the Mond Nickel Company decided to undertake their own refining by a process that had been developed at Clydach by a young Swiss chemist, Dr. Christian Heberlein. A small-scale plant was set up in Southwark, across London Bridge, with Heberlein as chief chemist and Major Cuthbert Johnson (1882–1962), whose father had been associated with Ludwig Mond's early work in the alkali industry, as manager.

By 1924 the process had been improved, expansion became necessary, and a new refinery was erected at Acton to the west of London. Johnson was joined by Ralph Atkinson and later by Alan Raper, both former students of metallurgy under Professor Heycock at Cambridge. The complex series of smelting operations, electrolysis, dissolution of the platinum and palladium in aqua regia and

The first processing plant for the separation and purification of the platinum metals in the Acton refinery of the Mond Nickel Company. Originally built in 1924, major extensions were made in 1930 in order to treat the increasing quantities of platinum-containing residues from both the nickel carbonyl process and the electrolytic operations in Canada

then the separation of the insoluble metals rhodium, ruthenium and iridium, was described in a paper by Carl Langer and Cuthbert Johnson given to the Empire Mining and Metallurgical Congress held in Toronto in 1927 (26).

Then in 1929 the Mond Nickel Company, now led by Ludwig's son Alfred, created Lord Melchett in 1928, merged with the International Nickel Company, the primary reason for this being the development of a new copper-nickel deposit, named after its discovery the Frood mine, which was owned partly by each company. The mineral was richer in nickel than the other Canadian ores, this meaning of course a higher content of the platinum metals. Until now only a small proportion of these elements had been recovered by International Nickel, mainly from an electrolytic copper refining process that left them in the anode residues, but following the union with Mond all platinum metal refining was consolidated at Acton and the company became the world's largest producer, a position it continued to hold until the growth of the South African platinum mines that will be described in Chapter 23. Production rose to 300,000 ounces of total platinum metals a year and the marketing was entrusted to Charles Engelhard and his organisation, this again giving him dominance of the situation for a time.

In 1937 a more up-to-date and detailed account of refining operations at

380

Acton, and of the complex chemical engineering problems involved, was given by Johnson and Atkinson (27).

Platinum Mining in Alaska

In 1933 an entirely new source of platinum made its appearance in a quite unexpected quarter, at Goodnews Bay on the Bering Sea coast of Alaska. Alluvial platinum had been discovered there in 1926 by an Eskimo named Walter Smith while panning for gold. Hand mining began in the following year in the shallow gravels and continued until 1933, yielding some 3000 ounces of platinum that was purchased by Johnson Matthey for refining. Then in 1934 a dragline excavator and ancillary equipment were installed and the Goodnews Bay Mining Company was incorporated. To expand production a Yuba diesel-electric dredge was bought and output rose to 37,000 ounces in 1938 in the short mining season from May to November. In that year Johnson Matthey contracted to take the output on a long term basis, refining taking place in London until a new refinery was built at the Malvern plant of J. Bishop and Co. in

The only primary source of the platinum metals in the United States lies on the west coast of Alaska where the Goodnews Bay Mining Company began dredging operations in 1935. Before these can begin each season, lasting from May to November, some ten thousand tons of ice have to be removed from the working area and the dredge can then reach the platinum-bearing gravels lying fifty feet below water-level

381

When operations began at Goodnews Bay the only habitation was a small trading post in a log cabin. When a post office and a general store were established there a year later an official name had to be adopted for the settlement and "Platinum" was chosen

Pennsylvania in 1939. A detailed account of the development of the Goodnews Bay property, the only primary source of platinum in the United States, and also an important source of iridium, was given by their Vice-President, Charles Johnston in 1962 (28).

References for Chapter 20

1   E. Baruel, *Phil. Mag.*, 1822, **59**, 171–179
2   Sir Edward Thorpe, Essays in Historical Chemistry, London, 1911, 577
3   P. H. O'Neill, *Min. Eng.*, 1956, May, 496–500
4   J. Cloud, *Trans. Am. Phil. Soc.*, 1818, **1**, 161–165
5   T. Cooper, *Emporium Arts and Sciences*, 1813, **1**, 181
6   J. E. Bollmann, *ibid.*, **1**, 344–346
7   J. A. Chaldecott, *Platinum Metals Rev.*, 1981, **25**, 163–172
8   E. Child, The Tools of the Chemist, New York, 1940, 150
9   E. F. Smith, The Life of Robert Hare, Philadelphia, 1917, 5

10   Anon, *Chem. News*, 1894, **70**, 234

11   J. Cloud, *Trans. Amer. Phil. Soc.*, 1809, **6**, 407–411

12   G. Kauffman, *Platinum Metals Rev.*, 1972, **16**, 101–104

13   W. Gibbs, *Amer. J. Sci.*, 1861, **31**, 63–71; 1862, **34**, 341–356; 1864, **37**, 57–61; *Chem. News*, 1861, **3**, 130–131, 148–149; 1863, **7**, 61–63, 73–76, 97–98; 1864, **9**, 121–122

14   W. Gibbs, *Z. anal. Chem.*, 1864, **3**, 334–335

15   W. H. Wahl, Galvanoplastic Manipulations, Philadelphia, 1883, 354–364

16   W. H. Wahl, *J. Frankin Inst.*, 1890, **30**, 62–75

17   G. Kauffman, *J. Chem. Ed.*, 1968, **45**, 804–811; *Platinum Metals Rev.*, 1972, **16**, 140–144

18   T. A. Edison, *Proc. Amer. Assoc. Adv. Sci.*, 1880, 173–177

19   T. A. Edison, U.S. Patent 307,031 of 1884

20   L. de Forest, U.S. Patent 841,387 of 1907

21   H. L. Wells and S. L. Penfield, *Am. J. Sci.*, 1889, **37**, 67–73

22   F. W. Clarke and C. Catlett, *Am. J. Sci.*, 1889, **37**, 372–374

23   L. Mond, C. Langer and F. Quinke, *J. Chem. Soc.*, 1890, **57**, 749–753

24   L. Mond and C. Langer, *Proc. Roy. Soc.*, 1889, **46**, 296–304

25   L. Mond, W. Ramsay and J. Shields, *Proc. Roy. Soc.*, 1895, **58**, 242–243; 1897, **62**, 50–53

26   C. Langer and C. Johnson, *Trans. Inst. Canad. Min. Met. Soc.*, 1927, **30**, 903–909

27   C. Johnson and R. H. Atkinson, *Trans. Inst. Chem. Eng.*, 1937, **15**, 131–149

28   C. Johnston, *Platinum Metals Rev.*, 1962, **6**, 68–74; *South Afr. Min. Eng. J.*, 1962, June 15, 1297–1302

Friedrich Wilhelm Ostwald
1853–1932

Born in Riga, and for some years Professor of Chemistry there, Ostwald
accepted a similar chair in Leipzig in 1887 and established a leading
school of physical chemistry. His major achievements were in research
in catalysis and the development of the ammonia oxidation process for
the manufacture of nitric acid and thence of fertilisers. For this he was
awarded the Nobel Prize for Chemistry in 1909

21

The Growth of Industrial Catalysis with the Platinum Metals

"If one considers that the acceleration of reaction by catalytic means occurs without expenditure of energy or material, and is in this sense gratis, it is evident that the systematic use of catalysts may lead to the most far-reaching advances in technology."

WILHELM OSTWALD, 1901

The discovery of the great activity of platinum and palladium in the catalysis of chemical reactions and the early researches of the two Davys, Döbereiner, Faraday and others were reviewed in Chapter 12, together with the famous patent of Peregrine Phillips of Bristol in 1831 and the early practical applications of Frédéric Kuhlmann in the production of sulphuric and nitric acids in his chemical works in France in 1838.

For almost forty years no progress was made in the further application of catalysis in industry. The phenomenon was but little understood, while the chemical engineering techniques required to handle gases at high temperatures had not been developed. But as the dyestuffs industry grew the need for more concentrated sulphuric acid increased and two independent steps were taken in 1875. Then Dr. Rudolph Messel (1848–1920), who had come to London five years earlier after studying chemistry in Zürich, Heidelberg and Tübingen to join William Stevens Squire (1835–1906), later to found the firm of Spencer Chapman and Messel, devised a process of producing oleum by passing the vapour of ordinary sulphuric acid over platinised pumice at a red heat. The patent was filed in Squire's name only (1), and the process was put into operation, an account being given to the Chemical Society (2). Almost simultaneously a paper was published by Clemens Winkler (1838–1902), Professor of Chemistry at the Freiberg School of Mines, in which he proposed the use of platinised asbestos in what was virtually Peregrine Phillips method of employing sulphur dioxide and oxygen in stoichiometric proportions (3). Winkler did not patent his process, but used it in a chemical works in Freiberg of which he was a director.

Rudolf Theophil Josef Knietsch
1854–1906

A native of Oppeln in what is now Poland, Knietsch first became a mechanic and then studied chemistry in Berlin. In 1884 he joined the Badische Anilin und Soda Fabrik and carried out a long and successful investigation on the production of sulphuric acid by the oxidation of sulphur dioxide over a platinum catalyst. His study of varying conditions of temperature, the rate of flow of the reactants and the poisoning of the catalyst by arsenical fumes made possible the large scale production of acid by the contact process which then began to supersede the lead chamber method

Photograph by courtesy of Badische Anilin und Soda Fabrik

However, his method pointed the way for others until the researches of Rudolf Knietsch at the Badische Anilin und Soda Fabrik. In a lecture given to the Deutschen Chemischen Gesellschaft in 1901 (4) he reported an extensive series of investigations on the behaviour of platinum catalysts in varying conditions of temperature and showed clearly that the concept of using a stochiometric mixture of gases was fallacious. The contact process thus began to replace the lead chamber process (and so the days of the platinum boiler was also numbered) first in Germany and then in England and the United States. Very large quantities of platinum were consumed over a long period but during World War I the supply in Germany was interrupted and as a substitute vanadium pentoxide was used and began to be adopted in about 1926 by American acid manufacturers and later in England.

The Manufacture of Nitric Acid

While the use of a platinum catalyst for the production of sulphuric acid became one of the few applications of platinum to fall away, a very different state of affairs has characterised the production of nitric acid and here platinum is still in use in large quantities.

During the latter years of the nineteenth century discussion began to arise among men of science who were interested in the broader issues of their subject on what later became known as "The Nitrogen Problem". Typical of the expositions which now and then reached even the public press was the Presidential Address given by Sir William Crookes to the British Association for the Advancement of Science at its Bristol meeting in September 1898 (5). Crookes was concerned to show that at the prevailing rate of increase of population the world's supplies of wheat would soon prove insufficient, and that the land would not continue to produce the same yield year after year unless adequate quantities of nitrogenous manure were ploughed back. He appealed to the chemist to help remove the fear of famine by establishing a means of fixing atmospheric nitrogen, since the only available source – Chile saltpetre – might be exhausted in a comparatively short period of years.

This problem, of obtaining from the unlimited supplies of uncombined nitrogen in the atmosphere those compounds – principally ammonia and nitric acid – required for agricultural needs, was soon intensified by the realisation in a number of European countries that a precisely similar need for assured supplies of nitric acid existed in the manufacture of explosives, and that in the event of war the Chile nitrates might well prove to be inaccessible.

This is not to say that such thoughts inspired governmental action in any part of Europe; they were, in fact, confined to but a handful of scientists who could foresee their countries' long-term needs. One such man was Professor Wilhelm Pfeffer, (1845–1920), the famous botanist of the University of Bonn who in 1901 expressed his concern about the need for supplies of fixed nitrogen to his friend in the University of Leipzig, Professor Wilhelm Ostwald. At this time Ostwald had occupied the Chair of Chemistry at Leipzig for some fourteen years and had built up a school of physical chemistry, devoting much of his energy to investigating the effects of catalysts on chemical reactions. His response to Pfeffer's representations was immediate; it was obviously his duty as a chemist to play his part in making his country independent of Chile saltpetre, and in obtaining nitric acid from other sources.

Two possible lines of investigation presented themselves. Either free nitrogen and oxygen from the air could be combined, or ammonia, then readily available from the gas industry, could be oxidised to give nitric acid. As it seemed more simple to re-combine nitrogen which was already fixed than to fix free nitrogen, Ostwald decided to give his attention to the oxidation of ammonia.

The reaction was known, and Ostwald would have been well aware of the earlier work of Kuhlmann. It was clear to him that the theoretical basis of the ammonia oxidation reaction would have to be elucidated before it could be developed on a large scale, and experiments were begun by Dr. Eberhard Brauer, at that time Ostwald's private assistant. The first experiments were made using a clean glass tube only a few milimetres in diameter containing

The historic apparatus with which Ostwald and Brauer first studied the oxidation of ammonia over a platinum catalyst to produce nitric acid in the University of Leipzig in 1901. The investigation showed that the conversion was practicable and relatively simple but many problems had to be solved before a commercial process could be developed

platinised asbestos. Ammonia and air were passed over the catalyst in known quantities and with known velocities, and it was at once clear that the conversion to nitric acid was practicable and relatively simple to carry out, although some difficulties lay in the absorption of the reaction products. The historic apparatus used at this stage is shown above.

The first experiments using platinised asbestos gave only small yields and a platinum-lined tube proved little better. A new reaction tube was therefore made, consisting of a glass tube 2 mm in diameter in which was coiled a strip of platinum about 20 cm long. The whole tube was heated to redness, and the first experiment gave a converison of more than 50 per cent, while increasing the gas velocity gave a converison of 85 per cent.

Investigations were then carried out on the effects of variations in the ammonia: air ratio, in the time of contact and in the temperature of the catalyst. Thus were laid the foundations of a technical process for producing nitric acid from ammonia, but the translation from idea to practice presented many problems before the project was brought to fruition.

Ostwald filed patents for his procedure in 1902 (6) although his German patent was disallowed on account of Kuhlmann's earlier disclosures.

A small factory was made available to Ostwald and Brauer, and here a pilot

By 1904 a pilot scale ammonia oxidation plant comprising three reactors had been built in a small powder factory put at Ostwald's disposal by the Director of the German Explosives Combine. Porous platinum sheets were used as the catalysts, a yield of 75 per cent of nitric acid was obtained, and it was decided to erect a larger plant to produce 300 kilograms of acid a day

plant was developed. By 1904 the three converters illustrated above had been built and operated, and it was decided to erect a larger-scale plant at the Gewerkschaft des Steinkohlenbirgwerks Lothringen at Gerthe, near Bochum, to produce 300 kg per day of nitric acid.

This plant was brought into operation in May 1906 and fully proved the feasibility of the process. A larger-scale plant was then designed and built, and by the end of 1908 was producing some three tons of 53 per cent nitric acid per day.

The catalyst used at this time consisted of a roll of corrugated platinum strip about 2 cm wide and weighing about 50 g, heated initially by a hydrogen flame. The life of the catalyst was no more than a month or six weeks. The disadvantages of the process included the relatively large amount of platinum required per unit of acid produced, and the uncertainty of temperature control of the catalyst, but improvements were not long wanting.

The Platinum Gauze Catalyst

Professor Karl Kaiser, of the Technische Hochschule, Charlottenburg, attacked the problem, and filed patents in 1909 covering the pre-heating of the air to 300 or 400°C and the use of a layer, usually four in number, of platinum gauzes. He was the first to employ platinum in the form of gauze, and it is a tribute to his experimental skill that the precise form of gauze he settled on – wire 0.06 millimetre diameter woven to 1050 mesh per square centimetre – is still very largely employed. By 1912 Kaiser had a pilot plant in operation at Spandau, Berlin, but while this was inspected repeatedly by British, French and American industrialists, he failed to interest them in his process, although a plant was erected at Kharkov in Russia.

Further work was carried out by Nikodem Caro and Albert Frank at the Bayerische Stickstoffwerke. Several patents were filed during 1914, the process being based upon a single platinum gauze which was electrically heated. Progress was slow for a time, and numerous experimental plants failed, but the outbreak of war gave a much greater urge to the project and by 1916 the picture had changed radically. The Frank and Caro converter had by then been engineered by the Berlin-Anhaltische Maschinenbau A.G. (BAMAG), who had constructed more than thirty plants, first for the supply of nitric oxide to lead chamber sulphuric acid plants and later for nitric acid production. The single platinum gauze was subsequently replaced by multiple gauzes, and the electrical heating was discontinued. This type of plant supplied all the nitric acid required for explosives in Germany during the later years of the war. The converter had a diameter of 20 inches, the catalyst consisting of a layer of three platinum gauzes woven from 0.006 inch diameter wire of 80 mesh to the linear inch, operating at about 700°C.

A much greater catalyst life was obtained in this design of plant, extending to six months provided that conditions were uniform and that the gases were free from impurities that might have a poisoning effect.

The Synthesis of Ammonia

Shortly after Ostwald's development of the ammonia oxidation process the raw material began to become more readily available. The same considerations on the great importance of the fixation of nitrogen prompted Fritz Haber (1868–1934), then an assistant professor at the Karlsruhe Technische Hochschule, to investigate the catalytic formation of ammonia from its elements, nitrogen and hydrogen. This reaction had already been studied in 1881 by George Stillingfleet Johnson, a demonstrator in chemistry at King's College, London, who obtained ammonia in small quantity by passing the two gases over heated platinum sponge (9), and Ostwald had given some consideration to the process in 1904, but Haber established that a successful process depended upon the reaction being carried out under high pressure and at a high temperature. The investigation was taken over in 1909 by the Badische Anilin und Soda Fabrik who assigned Carl Bosch (1874–1940) to carry the project further. Haber had employed osmium as his catalyst (10) but the commercial success of the process required a metal that was both less expensive and available in greater quantity, and after some twenty thousand experiments by Alwin Mittasch (1869–1953), the head of catalyst research at BASF, a solution was finally arrived at with a mixture of iron and its oxides. Both Haber and Bosch were awarded Nobel Prizes for Chemistry, the latter commenting on the initial experiments at high pressures in the course of his address:

> "The two contact tubes, made by Mannesmann, had an operating life of eighty hours, then they burst. If we had filled them with osmium instead of the new catalyst the entire world stock of this precious metal, which we had by now bought, would have disappeared." (11)

The Production of Nitric Acid in America

At the beginning of the 1914 war the United States possessed no source of nitric acid other than Chile saltpetre, and it became distressingly evident that the nation was dependent upon a foreign country in this respect, while the production of nitric acid from this starting-point required large quantities of sulphuric acid already in short supply.

Cyanamide had been manufactured at Niagara Falls since 1909, and in 1916 the first American plant for the oxidation of ammonia produced from cyanamide was established by the American Cyanamid Company at Warners, New Jersey. The catalyst employed was a single platinum gauze, electrically heated. In the meantime, the ordnance department had decided to take action, and Dr. C. L. Parsons, of the Bureau of Mines, was asked to investigate European methods for nitrogen fixation. As a result the American Cyanamid Company was requested, in 1917, to form a subsidiary company, Air Nitrates Corporation, to act as agent for the United States Government for the construction and operation of a plant at Muscle Shoals, Alabama, to produce 110,000 tons a year of ammonium nitrate. This plant comprised some seven hundred catalyst units each containing a single rectangular platinum gauze woven from 0.003 inch diameter wire, 80 mesh, and heated electrically to 750°C. The total weight of platinum was a little over 300 oz, and the loading ratio about 1 kg per daily ton of ammonia.

Developments in Ammonia Oxidation in Great Britain

There had been little or no commercial interest in nitrogen fixation in Great Britain before the outbreak of war in 1914, and throughout the war period the supply of nitrogen products for munitions depended almost entirely on shipments of Chilean nitrate.

There were, however, a number of attempts to make nitric acid by the oxidation of ammonia, either from gas-liquor or cyanamide. An Ostwald plant was set up at Dagenham Dock by the Nitrogen Products Company in 1916–1917, but never achieved successful operation. The Gas Light and Coke Company developed a plant at Beckton using a pad of three or four flat platinum gauzes as catalyst, and attained an output of a ton of nitric acid per day.

A systematic investigation was undertaken, at the instigation of the Nitrogen

Sir Eric Rideal
1890–1974

Educated at Trinity College, Cambridge, and then at the University of Bonn, Rideal served in the Royal Engineers in World War I but was invalided out in 1916 and then joined the Munitions Invention Board together with J. R. Partington, J. A. Harker, H. C. Greenwood, E. B. Maxted and others with the object of establishing the ammonia oxidation process in England. Earlier, while on leave from France, he had studied this reaction in the Institute of Chemistry in London. The project, carried out at University College, London, led to the construction of a successful converter but on a very small scale. His distinguished career included much important research in catalysis

Products Committee, by J. R. Partington, E. K. (later Sir Eric) Rideal and others and was carried out in the laboratory of the Munitions Inventions Department (13). An effective design of converter was evolved, employing either an electrically heated pad with two gauzes or a thicker pad that was self-sustaining in temperature when reaction had been established. Somewhat similar converters were constructed by Brunner Mond & Company and by the United Alkali Company, both of whom turned to Johnson Matthey for advice on the production of the catalyst gauzes.

Although it came too late to be of service in the war, the decision taken in 1917 to erect a synthetic ammonia plant using the Haber-Bosch process led directly to the building of the Billingham plant by Synthetic Ammonia and Nitrates Ltd. (now Imperial Chemical Industries Ltd.). The ammonia plant first came into operation in December 1923 and the nitric acid plant – the first successful large-scale plant in this country – during 1927. An account of the early years of this development has been given by A. W. Holmes (14).

Although the process remains unchanged in principle – and even in some details such as the mesh sizes of the gauze pads – the size and complexity of the plant units has been tremendously increased.

The platinum gauze catalyst, supplied by Johnson Matthey, used in the researches carried out for the Munitions Invention Department in 1916. Measuring only six inches by four inches, it comprised two gauzes mounted in an aluminium frame with silver leads for the heating current

After the war the work of the Munitions Invention Department was taken over by Brunner Mond and Company (later to become part of Imperial Chemical Industries) and this atmospheric pressure ammonia oxidation plant was installed at Billingham in 1927. The platinum gauzes were twenty inches in diameter, by contrast with those now employed running up to five metres in diameter

From the 50 grams of corrugated foil in an Ostwald unit, the weight of platinum in a single converter has steadily increased until it may now reach from 20 to 30 kilograms, while the diameter of the rhodium-platinum gauzes, introduced in 1928 by E. I. Du Pont as an improvement on the pure platinum formerly used (15), can be as great as five metres.

The Manufacture of Hydrogen Cyanide

Another process, developed some years later by Leonid Andrussow, like Ostwald a native of Riga, at the I.G. Farbenindustrie plant in Mannheim, also makes use of woven gauzes of rhodium-platinum alloy to convert methane, ammonia and air to hydrogen cyanide (16). This is required in enormous quantities for the manufacture of acrylic resins such as polymethyl methacrylate, known in Britain as Perspex, in America as Lucite and in Germany as Plexiglas, and adiponitrile, an intermediate in the production of Nylon (17). Operating temperatures in the process are appreciably higher than in ammonia oxidation plants, ranging up to 1200°C, so that the mechanical strength of rhodium-platinum at high temperatures and its resistance to oxidation play an important role in addition to its catalytic activity.

Catalysis in the Organic Chemical Industry

The wider adoption of catalytic reactions with the platinum metals in the manufacture of organic chemicals, eventually to achieve immense significance in the pharmaceutical, dyestuffs, plastics and synthetic fibre industries, occurred much later than was the case with inorganic products. A great deal of the basic research had been carried out by the beginning of the twentieth century and even before, but the transition into commercial applications was slow in development.

As early as 1874 Professor Prosper de Wilde of the University of Brussels discovered that acetylene could be hydrogenated to ethylene and then to ethane over a platinum catalyst (18) while in 1894 Professor Paul Sabatier and his assistant the Abbé Jean Baptiste Senderens (1856–1936) at the University of Toulouse published the first of their very numerous papers on catalysis (19). Sabatier had been intrigued by Ludwig Mond's discovery in 1890 of the reaction between nickel and carbon monoxide (page 377) and in 1902 he and Senderens reduced carbon monoxide to methane over a nickel catalyst (20). By 1911 Sabatier had reported at length on the many hydrogenation and dehydrogenation reactions that could be carried out in the laboratory and he became the leading authority on catalysis of his time although he made no attempt to introduce any industrial processes.

Finely Divided Platinum and Palladium

Much of the early work involved the use of very finely divided metals, generally in a colloidal state. Carl Ludwig Paal (1860–1935) Professor of Chemistry in the University of Erlangen and later in Leipzig, made a long series of studies on the

Paul Sabatier
1854–1941

Born at Careassonne, after studying at
the École Normale in Paris Sabatier
became an assistant to Marcelin Berthelot
at the College de France. In 1882 he
moved to Toulouse, being appointed
Professor of Chemistry in 1884 and
remaining there for the remainder of
his life despite an offer of a chair at the
Sorbonne in succession to Moissan. He
was a pioneer in the field of catalysis
and was awarded the Nobel Prize in
chemistry for this work in 1912. His
many papers on the subject were summarised, together with the work of
others, in his book, "La Catalyse en
Chimie Organique", published in 1913

preparation of colloidal platinum and palladium and of their effectiveness on
catalytic reactions (21) while Aladar Skita (1876–1953), Professor of Chemistry
at Karlsruhe, pursued similar investigations on the hydrogenation of aldehydes
and ketones with colloidal platinum and palladium (22) and the two
collaborated in 1909 in filing a patent for causing these reactions (23). But the
use of colloidal preparations was not a practical proposition outside the
laboratory because of the difficulty of separating them from the reaction
products and attention turned to the so-called "blacks", a finely divided form of
the metal containing an uncertain amount of oxygen. Platinum black had been
discovered by Döbereiner in 1833 (page 222), although the product described by
Zeise in 1827 (page 264) was possibly of the same nature, but a reliable method
for its preparation was first devised by Oscar Loew (1844–1941), a plant
physiologist in Munich, in 1890 (24). His method was improved by Richard
Willstätter (1872–1942) in 1912 while he was for a period Director of the Kaiser
Wilhelm Institute after a long series of investigations on the hydrogenation of
aromatics (25). Professor Gustave Vavon of the University of Nancy also carried
out a massive research on the hydrogenation of aldehydes and ketones in the
presence of platinum black, describing these in his doctoral thesis to the

Roger Adams
1889–1971

A graduate of Harvard, Adams spent some time studying under Professor Richard Willstätter at the Kaiser Wilhelm Institute in Berlin and in 1916 he was appointed Professor of Chemistry at the University of Illinois where, apart from intervals of government service during two world wars, he remained until his retirement. Under Willstätter he had been engaged in the preparation of platinum black for use as a catalyst and on his return from the first war in 1919 he successfully developed a procedure for its production in a state of high activity and reliability. This useful catalyst still bears his name and is widely used, particularly in the pharmaceutical industry

University of Paris in 1914 (26), while Vladimir Ipatieff (1867–1952) in St. Petersburg, another prolific worker in catalysis, after a series of investigations with nickel, studied a number of catalytic reductions with palladium black in 1912 (27).

At about this time Nicolai Dmitrievich Zelinsky (1861–1953) also began his long series of researches, converting cyclohexane into benzene with both platinum and palladium blacks as catalysts and continuing these investigations for many years (28).

During this early period, however, the platinum blacks often showed a low or a varying activity and it was not until 1919, when the problem was tackled by Professor Roger Adams who had spent some time under Willstätter at the Kaiser Wilhelm Institute, that a product of uniform activity was obtained consistently. Searching for an active catalyst for organic reductions, Adams and his students developed a successful procedure for what is still known as Adams' Platinum Oxide Catalyst (29). An account of their work with comments by Professor Adams himself may be found in *Platinum Metals Review* (30). This catalyst was at first prepared by individual workers in their laboratories, but before long it came into use in the pharmaceutical industry and the demand increased. Scaling up was undertaken by platinum refiners in the United States, while in England Johnson Matthey collaborated with May and Baker to develop a process for its preparation in relatively large batches for use in a variety of liquid phase hydrogenation reactions (31).

Supported Platinum and Palladium Catalysts

These early forms of finely divided platinum and palladium catalysts were, however, largely superseded by supported catalysts, more especially of palladium in the first place, to make more effective use of the metal and to enable a wider range of reaction conditions to be met. Among a great many materials used as supports, including alumina, asbestos and silica gel, the most generally useful has been activated charcoal, and palladium-on-charcoal catalysts have played an important part in low pressure liquid phase hydrogenation reactions in the pharmaceutical industry to produce vitamins, cortisone and dihydrostreptomycin among other products. Their usefulness, and also that of platinum-on-charcoal in one establishment, Merck of New Jersey, has been described by W. H. Jones (32).

The Growth of Commercial Processes

Slowly processes based upon catalysis began to come into industrial use for the production in large quantities of chemicals that were otherwise difficult or impossible to produce, although not at first with platinum metals catalysts. The first major liquid phase processes were for the conversion of animal and vegetable oils into edible fats, generally with finely divided nickel catalysts.

In gas phase reactions the first recorded processes, as mentioned earlier, was devised by Sabatier and Sanderens in 1902 for the production of methane from carbon monoxide and hydrogen, also over a nickel catalyst (20) while, following up this work in 1923, a major step forward was made by Franz Fischer (1877–1935) and Hans Tropsch (1889–1935) at the Kaiser Wilhelm Institut für Kohlenforschung at Mülheim in the Ruhr, in developing their well known synthesis of liquid hydrocarbons by the gasification of coal and by reacting the hydrogen and carbon monoxide produced in the presence of a catalyst, first of cobalt and later of iron (33).

But the great stimulus to the use of the platinum metals as catalysts came when petroleum began to replace coal tar as the major source of organic chemicals, and with the realisation that the platinum metals, although more expensive initially, often displayed greater activity and made it possible to carry out commercially important reactions at appreciably lower temperatures and pressures than those necessary with base metal catalysts. Greater product selectivity could be achieved, while the platinum metals could readily be recovered and recycled, making their use much more commercially attractive.

The Production of High Octane Fuels and Aromatic Chemicals

As long ago as 1894 Francis Clifford Phillips (1850–1920), Professor of Chemistry at Western University in Allegheny, Pennsylvania – another researcher well in advance of industrial exploitation – studied the nature and constituents of the natural gas and petroleum found in his native state and carried out a long series of investigations on the oxidation of hydrocarbons over finely divided platinum, palladium, iridium, rhodium and osmium supported on asbestos (34).

Before World War II the catalytic reforming of petroleum to increase the octane rating of petrol was introduced in Britain, the United States and Germany, using a molybdenum on alumina catalyst, but this was found to be uneconomical and was superseded by a process developed by Universal Oil Products and known as "Platforming" (35). This was devised from the great expertise on catalysis built up in the later thirties under the leadership of Ipatieff and Tropsch, both of whom had by then joined U.O.P., and by one of Ipatieff's first students at Northwestern University, Vladimir Haensel. The process involved the reforming of crude naphthas to aromatic hydrocarbons, particularly benzene, toluene and the xylenes, over a platinum on alumina

Vladimir Haensel

Born in Germany, Haensel received his early training in Moscow and then under the leading catalytic expert Professor Vladimir Ipatieff at Northwestern University in Elvaston, Illinois. He joined Universal Oil Products in 1937, working with Ipatieff who divided his time between teaching and directing research there for many years. His major contribution was the development of a platinum on alumina catalyst that made it possible to produce not only high octane petrol but also a range of aromatic hydrocarbons from crude petroleum. The process, known as Platforming, has been adopted on a world wide basis. In 1964 he was appointed Vice-President and Director of Research at U.O.P., while in 1974 he was awarded the National Medal of Science by the U.S. Government for "his outstanding research in the catalytic reforming of hydrocarbons which has greatly enhanced the economic value of our petroleum natural resources".

398

One of the early Platforming units commissioned by British Petroleum for the production of both high-octane petrol and a range of aromatic chemicals. The process, licensed from Universal Oil Products, employs a platinum-on-alumina catalyst and great numbers of plants of this type were erected in all parts of the world

catalyst, and apart from yielding the high octane petrol needed for modern automobile engines opened the way to the tremendous growth in the production of synthetic fibres, plastics, synthetic rubbers, insecticides and many other chemical products.

Catalyst requirements were met by Universal Oil Products for many users, but in 1953 catalyst manufacturing facilities were set up in the United Kingdom by Universal-Matthey Products, a subsidiary company of Universal Oil Products and Johnson Matthey, in order to meet the growing demand from European licensees of the Platforming process, while a few years later a similar plant was established in Cologne.

The initiative taken by Universal Oil Products was quickly followed by others in the petroleum industry in developing broadly similar reforming processes employing platinum catalysts (36). In fact the response was enormous, and by the mid-1950s plants were being built in many countries of the world (37). Platinum reforming became one of the most versatile procedures available

399

to the oil industry – as well as a major user of platinum – and has continued to provide a wide range of intermediates for the chemical industries. The benzene produced has many uses, including the manufacture of styrene and polystyrene, of cyclohexane for the production of Nylon (first discovered by W. H. Carothers whose doctoral thesis under Professor Roger Adams dealt with the catalytic hydrogenation of aldehydes with platinum black (29)), as well as of phenol for phenolic resins, dichlorobenzene for dystuffs and maleic anhydride for polyester resins. The toluene produced finds extensive use as a solvent for nitrocellulose lacquers, while ortho-xylene yields phthalic anhydride for plasticisers, dyes and pigments and para-xylene is used to produce terephthalic acid for polyester fibres.

Thus the many types of synthetic materials that provide our fuel, our clothing and the many other items made from plastics depend for their production upon large-scale industrial processes in which the vital part is played by platinum.

References for Chapter 21

1   W. S. Squire, British Patent 3278 of 1875

2   R. Messel and W. S. Squire, *Chemical News*, 1876, **33**, 177

3   C. Winkler, *Poly. J. (Dingler)*, 1875, **218**, 128–139

4   R. Knietsche, *Ber. Deutsh. Chem. Gesellschaft*, 1901, **34**, 4069–4115; *J. Soc. Chem. Ind.*, 1902, **21**, 172–173

5   Sir William Crookes, British Association Report, 1898, 3–38

6   W. Ostwald, British Patents 698 and 8300 of 1902; *Chem. Zeitung*, 1903, **27**, 457–458

7   K. Kaiser, German Patent 271,517; British Patent 20,325 of 1910; U.S. Patent 987,375 of 1911

8   A. Frank and N. Caro, German Patents 286991, 304269, 303822

9   G. S. Johnson, *J. Chem. Soc.*, 1881, **39**, 128–133

10 F. Haber, *Z. Elektrochem.*, 1910, **16**, 244–246; Badische Anilin und Soda Fabrik, German Patent 223408 of 1910

11 E. Farber, Nobel Prize Winners in Chemistry, New York, 1953, 126

12 E. J. Pranke, *Chem. and Met. Eng.*, 1918, **19**, 395–396

13 Ministry of Munitions, Munitions Invention Dept., H.M.S.O. London, 1919; J. R. Partington and L. H. Parker, The Nitrogen Industry, London, 1923

14 A. W. Holmes, *Platinum Metals Rev.*, 1959, **3**, 2–8

15 E. I. Du Pont de Nemours, British Patent 306382 of 1928

16 L. Andrussow, German patent 549055 of 1932; *Z. angewand. Chem.*, 1935, **48**, 593–595

17 J. M. Pirie, *Platinum Metals Rev.*, 1958, **2**, 7–11

18 P. de Wilde, *Berichte*, 1874, **7**, 352–357

19   P. Sabatier and J. B. Senderens, *Comptes rendus*, 1897, **124**, 616–618; 1358–1361

20   P. Sabatier and J. B. Senderens, *Comptes rendus*, 1902, **134**, 514–516; 689–691

21   C. L. Paal, *Berichte*, 1907, **40**, 2201–2200; 1908, **41**, 805–817; 2273–2282

22   A. Skita, *Z. angewand. Chem.*, 1913, **26**, (i), 601–602

23   A. Skita and C. Paal, German Patent 230724 of 1909

24   O. Loew, *Berichte*, 1890, **23**, 289–290

25   R. Willstätter and D. Hatt, *Berichte*, 1912, **45**, 1164–1481

26   G. Vavon, *Ann. Chim.*, 1914, **1**, 144–200

27   V. N. Ipatieff, *Berichte*, 1912, **45**, 3218–3226

28   N. D. Zelinsky, *J. Russian Phys. Chem. Soc.*, 1912, **44**, 274–275

29   V. Voorhees and R. Adams, *J. Amer. Chem. Soc.*, 1922, **44**, 1397–1405; W. H. Carothers and R. Adams, *J. Amer. Chem. Soc.*, 1923, **45**, 1071–1086

30   L. B. Hunt, *Platinum Metals Rev.*, 1962, **6**, 150–152

31   D. H. O. John, *Chem. and Ind.*, 1944, **43**, 256

32   W. H. Jones, *Platinum Metals Rev.*, 1958, **2**, 86–89

33   F. Fischer and H. Tropsch, *Brennstoff Chem.*, 1923, **4**, 276–285; 1924, **5**, 201–208

34   F. C. Phillips, *Am. Chem. J.* 1894, **16**, 163–187; 255–277; 340–365; 406–429

35   V. Haensel, U.S. Patent 2,479,109 of 1949

36   S. W. Curry, *Platinum Metals Rev.*, 1957, **1**, 38–43; H. Connor, *Platinum Metals Rev.*, 1961, **5**, 9–12

37   B. M. Glover, *Platinum Metals Rev.*, 1962, **6**, 86–91

Nikolai Semenovich Kurnakov
1860–1941

One of the principal founders of the modern platinum industry in the
U.S.S.R., Kurnakov was first a student and then in 1893 Professor of
Inorganic Chemistry in the Mining Institute in St. Petersburg. His work
on the complex compounds of the platinum metals materially assisted
refining methods and on the death of Chugaev in 1922 he was appointed
Director of the Platinum Institute

Photograph by courtesy of the late Academician
I. I. Chernyaev and Professor George Kauffman

22

Production of the Platinum Metals in Soviet Russia

"Entering again the world platinum market after the Revolution of 1917, our country not only had to re-establish the former output of platinum but also to re-organise its extraction on the basis of modern technology."

O. E. ZVYAGINTSEV, 1927

For many years before the outbreak of the first world war the platinum industry in Russia was dominated by the European refiners. The two large producers, Count Demidov and Count Schuvalov, were supplying them with large quantities of native metal from the Urals, while a small refinery and workshop had been set up in 1875 in St. Petersburg jointly by Johnson Matthey and Desmoutis Quennessen to meet the very limited Russian needs for platinum laboratory apparatus. This control of the industry by foreigners was much resented by both the intelligentsia and the politicians, and in 1910 a conference was arranged by the Ministry of Trade and Industry, presided over by Professor Kurnakov, to consider the question of the refining of platinum in Russia. Recommendations were made, but no action was taken, and then in 1913 the Ministry imposed a duty of 30 per cent on exported platinum and a year later, on the outbreak of war, prohibited its export altogether. In the same year the setting up of a refinery was authorised and this was built at Ekaterinburg (now Sverdlosk), the centre of the Ural mining industry. The Quennessen company was to take on the financing, Heraeus to provide a suitable engineer, and Johnson Matthey to design the buildings and plant, to provide the methods of refining and to prepare and ship the equipment. The first local collaborator was the largely uncommitted mining concern, the Nikolai-Pavdinsky Mining Company, which was to find the site and to build the refinery. Many delays occurred, however, and it was not until after the outbreak of war in August 1914 that building began. Operations began on a small scale in 1915 with N. N. Baraboshkin, a former student of Kurnakov's at the Mining Institute of St. Petersburg, as manager, and by the end of 1917, when all Russian mineral

403

One of the American built dredges in operation in 1923 to extract native platinum from the alluvial deposits in the river beds of the Urals. The operating season extended only from May until the end of November. During the next five years several larger dredges were installed, the crude platinum being transferred to a refinery built at Sverdlosk

resources were nationalised, output had increased to some 30,000 ounces a year. The refinery was then taken over by the new government under the terms of a state monopoly for the production, refining and marketing of platinum. The general disorder during the first years of the revolution greatly reduced production, however, which fell to only around 6,000 ounces in 1922 (1).

A Research Institute for the Platinum Metals

During the war the shortage of platinum for the contact process for the manufacture of sulphuric acid had caused great concern – the large Tenteleev organisation in St. Petersburg had been producing acid by this process since 1900 – and in 1915 a commission was set up for the study of all Russian natural resources, with a section devoted to platinum under Kurnakov. An appeal was made to Lev Aleksandrovich Chugaev, Professor of Chemistry at the University of St. Petersburg, and as well as urging the creation of a state monopoly he proposed the establishment of a research institute for investigations on all the platinum metals, their methods of refining and analysis, their alloys and their co-ordination compounds. After the revolution this scheme came to fruition and in 1918 Chugaev was appointed director of the newly formed Institute for the Study of Platinum and other Noble Metals (2). This organisation at once set out to provide the refinery with improved methods, as well as conducting many

Lev Aleksandrovich Chugaev
1873–1922

Born in Moscow where he studied at the University, in 1908 Chugaev became Professor of Inorganic Chemistry in the University of St. Petersburg, succeeding Mendeleev, and remained there until his early death at only forty-nine. Apart from his researches on platinum complexes he proposed the establishment of an institute for comprehensive research on the platinum metals, a project that came to fruition in 1918 when he was appointed the first Director of the Platinum Institute of the Academy of Sciences in Petrograd, now Leningrad

researches on the chemistry of the platinum metals, particularly on their co-ordination compounds. Their results were published in a journal established in 1920, the *Izvestia* of the Platinum Institute, the first and for three decades the only journal devoted exclusively to the platinum metals (3). Unfortunately Chugaev died in 1922 at the early age of 49 and he was then succeeded by Kurnakov who continued to stimulate the refinery management at Sverdlovsk and to contribute to the study of the co-ordination compounds of the platinum metals. In 1934 the Platinum Institute, the Institute of Physicochemical Analysis and the General Chemistry Laboratory of the Academy of Sciences were combined into the Institute of General Chemistry in Moscow, with Kurnakov as Director. On his death in 1941 this was re-named the N.S. Kurnakov Institute in his honour. An appreciation of his life and work has been compiled by Professor Kauffman and will be published shortly (4).

Chugaev's proposal for a nationalised industry had, however, already been adopted with the formation late in 1921 of a commercial trust, The State Association of Platinum Miners of the Urals, abbreviated to "Uralplatin". The new organisation found that nearly all the old dredges used to excavate the

405

Il'ya Il'ich Chernyaev
1893–1966

A former student of Chugaev's, in 1918 Chernyaev joined the Institute for the Study of Platinum of the Academy of Sciences in Petrograd and remained there until his death, succeeding Kurnakov as Director in 1941. There he carried out important work on the refining of the platinum metals and their preparation in a state of high purity. He was also co-editor of the *Izvestia* of the Platinum Institute from 1947 until 1955 when it ceased publication

alluvial platinum from the river beds had ceased to operate for lack of spare parts, and new American dredges were purchased, beginning in 1925 and continuing for several years. They could operate only from May to December as they were forced to stop working when the accumulations of ice became too great (5).

In 1921 a representative of Johnson Matthey, Mr. A. B. Coussmaker, then the company's mining engineer who had considerable experience in the platinum fields in the Urals, secured an interview with Maxim Litvinov with a view to obtaining supplies of native metal for refining. Litvinov, later the Soviet Ambassador in London and then Foreign Minister in Moscow, emphasised, however, that his government was well aware of the extent to which the world was dependent upon Russian supplies of platinum and that they firmly intended to do the refining themselves.

Refining and separation of the platinum metals progressed in the Ekaterinburg plant under the guidance of Baraboshkin and with the close co-operation of

To publish the results of research in the newly founded Institute for the Study of Platinum and other Noble Metals Chugaev founded a journal, the *Izvestia*, first published in 1920. This included a massive series of papers by Chugaev himself and by his colleagues on the chemistry and metallurgy of all the platinum metals. This shows the cover of the volume for 1936, with the title given in French, and then edited by N. S. Kurnakov and O. E. Zvyagintsev. The journal ceased publication in 1955

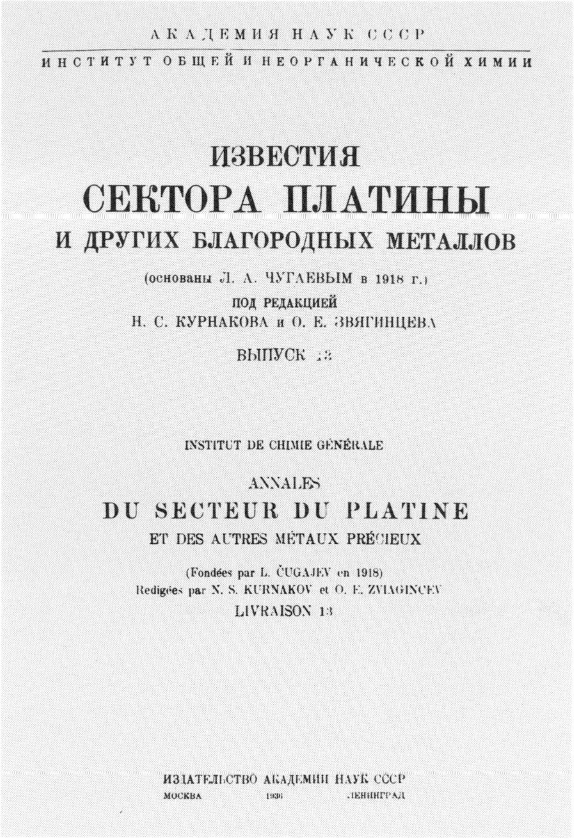

АКАДЕМИЯ НАУК СССР
ИНСТИТУТ ОБЩЕЙ И НЕОРГАНИЧЕСКОЙ ХИМИИ

ИЗВЕСТИЯ
СЕКТОРА ПЛАТИНЫ
И ДРУГИХ БЛАГОРОДНЫХ МЕТАЛЛОВ

(основаны Л. А. ЧУГАЕВЫМ в 1918 г.)

ПОД РЕДАКЦИЕЙ
Н. С. КУРНАКОВА и О. Е. ЗВЯГИНЦЕВА

ВЫПУСК 13

INSTITUT DE CHIMIE GÉNÉRALE

ANNALES
DU SECTEUR DU PLATINE
ET DES AUTRES MÉTAUX PRÉCIEUX

(Fondées par L. ČUGAJEV en 1918)
Rédigées par N. S. KURNAKOV et O. E. ZVIAGINCEV

LIVRAISON 13

ИЗДАТЕЛЬСТВО АКАДЕМИИ НАУК СССР
МОСКВА   1936   ЛЕНИНГРАД

Kurnakov and his staff in improving the processes; high purity platinum, palladium, iridium and rhodium were successfully produced, and a state factory in Moscow carried out the fabrication work.

By 1924 negotiations with the Russians yielded an agreement for Johnson Matthey to take the whole of their output of refined metal for the following year, estimated at 70,000 ounces. This was shared with the other European platinum companies, and further quantities were made available on the same basis in 1926 and 1927, but then a marketing company was set up in Berlin by the Russian government for the direct disposal of their metal.

The Discovery of Platinum in Siberia

Extraction of the alluvial platinum in the Urals continued, and a detailed survey of the mineral resources there was undertaken by a leading Swiss geologist, Professor Louis Duparc, with a Russian assistant, Marguerite Tikanovitch (6), but it was realised that a search must be made for the primary

After the discovery of a large nickel-copper ore body in north-western Siberia the extraction of the platinum metals from this source began to increase. In the early period before World War II the mineral was shipped to Murmansk and then to a refinery at Monchegorsk where the nickel was refined electrolytically, the platinum metals accumulating in the anode residues and then being separated and refined individually

rock from which these water-borne deposits emanated. In 1919 while prospecting for coal, Russian geologists discovered a large nickel and copper ore body in the far north-west of Siberia and in 1924, after some samples had been sent to Leningrad, this was found to contain the platinum metals. Some years passed before any further action was taken and then in 1935 the Noril'sk Mining and Metallurgical Combine was established to exploit this large body of mineral. First of all a railway had to be built to link the future mining operations with the port of Dudinka on the River Yenisei, open only for four months of the year. By 1938 the railway had been completed despite the most harsh Arctic conditions, and extraction operations could begin. Initially the ore was shipped from Dudinka to Murmansk and then to Monchegorsk where a nickel refinery had been opened in 1938, but in 1940 Noril'sk had its own smelter and refinery and became a vital source of nickel and copper during the second world war. The ore is first treated along similar lines to those in use at Sudbury, smelted to a matte and the final refining of copper and nickel carried out by electrolytic methods. From these two last operations the anode residues containing the platinum metals are flown to a refinery built at Krasnoyarsk, the capital city of the region nearly a thousand miles to the south on the Trans-Siberian Railway. Here the individual platinum metals are separated and refined. By far the major part of the Russian output of platinum metals now comes from this source.

408

In more recent years a large smelting complex has been built at Noril'sk to refine both nickel and copper. The final anode residues are then treated at a platinum refinery at Krasnoyarsk, nearly a thousand miles to the south

The ore-body at Noril'sk is, unlike the rich platinum mineral from the Urals, much richer in palladium than in platinum, while the fact that these metals emerge only as by-products from the extraction of nickel and copper means that production is governed by the demand for these two base metals. Output figures for Russian platinum are most difficult to establish, but production has increased many times since the opening of the Noril'sk operations in 1940 and for a time constituted the greater proportion of world supplies until it was surpassed by the increasing output from the South African mines, to be reviewed in the next chapter.

References for Chapter 22

1   E. K. Fritsman, *Ann. Inst. Platine*, 1927, **5**, 23–74
2   G. B. Kauffman, *Platinum Metals Rev.*, 1973, **17**, 144–148
3   G. B. Kauffman, *Platinum Metals Rev.*, 1974, **18**, 142–148
4   G. B. Kauffman, *Platinum Metals Rev.*, forthcoming
5   J. B. Bubb, *Eng. and Min. J.*, 1928, **126**, 284–286
6   L. Duparc and M. N. Tikanovitch, Le Platine et les Gites platiniferes de l'Oural et du Monde, Geneva, 1920

Arthur Blakeney Coussmaker
1885–1974

After gaining valuable experience in the platinum mining area of the Urals prior to the 1914–1918 war Coussmaker was appointed a Director of Johnson Matthey and determined to seek a major new source of native platinum to expand their refining and fabricating activities. The discovery of the Merensky Reef in 1924 made it possible for him to achieve this objective

From a portrait by David Jaeger in the
possession of Johnson Matthey

23

The Discovery of the World's Greatest Platinum Resources

"This alluvial platinum must come from somewhere in the surrounding basic rocks, and if only we can find it in payable quantities in the main source, the mother reef, then we would have discovered something far bigger and far more important than the Russian and American deposits"

HANS MERENSKY, 1924

The cessation of supplies of crude platinum from Russia faced Johnson Matthey with a serious problem and a challenge. This was taken up by A. B. Coussmaker, the company's mining engineer who had been appointed a director in 1925 and who now resolved to search tirelessly for new sources of platinum and to restore the pre-eminent position in the industry that had been built up by George Matthey and John Sellon. Before the first world war he had made a series of prospecting missions in Canada, Australia and New Zealand without success, but an entirely new era was now about to open.

Rumours of platinum being found in South Africa had begun to circulate in 1923 and had alerted a number of prospectors. Then in June 1924 a small bottle of greyish-white concentrates arrived by post in the Johannesburg office of the consulting geologist Dr. Hans Merensky. This had been sent by H.C. Dunne whose brother-in-law Andries Lombaard, a farmer with some experience of panning for gold, had found what he thought was evidence of platinum in one of the streams on his farm at Maandagshoek to the north of Lydenburg in the Transvaal. Analysis quickly confirmed the presence of platinum, as well as of rhodium and iridium, and Merensky immediately went to investigate. Together with Lombaard and two of his wife's cousins, Schalk and Willem Schoeman, he examined a large stretch of country around the stream and within a few weeks, with great geological insight and deductive reasoning, had located the basic mineral in a reef running parallel to the mountain range in a northerly direction and again in a southerly direction, extending in all some sixty miles. Merensky's first thought was to christen it "The Lombaard Reef", but his colleagues over-ruled him and it was named "The Merensky Reef".

Finance was provided by a small syndicate, soon to become Lydenburg Platinum Limited, but this company was later taken over by a more substantial concern controlled by Consolidated Gold Fields who began mining operations.

Hans Merensky explored still further and traced the reef at Potgietersrust to the north-west, and then much further afield at Rustenburg, some sixty miles west of Pretoria, the indications in the latter area pointing to richer deposits. This caused the emphasis to move away from Lydenburg and Potgietersrust to Rustenburg and so to the most regular and most valuable part of the platinum-bearing reef.

The Merensky Reef forms a layer in the Bushveld Igneous Complex, an irregular oval or saucer-shaped area of some 15,000 square miles in the Central Transvaal. The reef varies in depth, from outcrops at the surface down to about 3,000 feet, but averages only three feet in thickness. A detailed account of its mineralogy was given in 1929 by Dr. Percy A. Wagner, for many years the mining geologist to the South African government, who not only dedicated his book on the platinum deposits and mines of South Africa to Merensky but wrote in his preface:

"The story of the opening up of these deposits – which transcend in magnitude and importance anything that had hitherto been dreamt of in the way of platinum

The beginning of the great discovery of the platinum metals in the Transvaal. Dr. Hans Merensky, standing on the left, is explaining the features of the deposit to a party of visitors, including the then Prime Minister of South Africa, General Hertzog, in 1925

The scene on the same part of the Merensky Reef thirty years later, the surface plant of Rustenburg Platinum Mines giving some indication of the great activity underground. The shallower parts of the mine, which now stretches for over thirty miles, are worked from inclined haulages, while the deeper areas are opened up from vertical shafts ranging in depth from 500 to 3,000 feet

occurrences – has often been told, but the writer feels that sufficient credit has never been given to Dr. Hans Merensky for the part that he played in this epic of mineral exploration" (1).

The geological features of the Bushveld Igneous Complex and its resources of platinum were later reviewed by C.A. Cousins (2).

The discovery of the Merensky Reef was quickly followed by a boom in platinum mining and a great many small companies were floated, generally by those whose experience had been confined to the mining of gold with its unlimited market and fixed price and who gave little or no thought to the means of extraction from the complex minerals.

In June 1925 A.B. Coussmaker, keenly interested in this new source of platinum, went out to South Africa and with E.C. Deering from the Research Department as analyst, made contact with the owners of the several properties, sampling the mineral from outcrops throughout the reef and bringing to the survey his outstanding knowledge of platinum from its geology to its marketing. He was fully aware of the serious problem of devising methods for the extraction of the platinum metals and on his return to London in August he assigned the task to the Research Department, headed by A.R. Powell, who was assisted

413

Hans Merensky
1871–1952

The son of a German medical missionary to the Transvaal, Merensky returned to study geology at the Technical High School in Breslau and then at the University of Berlin. Going back to South Africa in 1904 he set up as a consulting geologist and mining engineer but he was interned for five years during the 1914–1918 war. The following years were difficult ones, but in 1924, hearing of the discovery of alluvial platinum in the Transvaal, he immediately prospected for the basic rock and quickly discovered the extensive reef that was named after him, so initiating the development of a great platinum industry

a little later by Deering, now back from South Africa with considerable knowledge of the deposits. The minerals were of a type not previously worked in quantity, and the platinum metals were associated with the sulphides of nickel, copper and iron, but intensive work over the next two years yielded a successful and economical process for treating the flotation concentrates by smelting in blast furnaces to a nickel-copper matte and treating this to obtain a residue carrying the platinum metals that could then be refined by the normal methods employed for alluvial platinum. A patent covering this process was filed on May 1, 1928 (3).

In the meantime large samples of concentrates had been sent by the mining companies to Krupp Grusonwerk in Germany, to various American refiners, to the Rand Mines laboratory in Johannesburg and to the Chemical and Metallurgical Corporation in England, a company associated with Consolidated Gold Fields having a hydro-metallurgical plant treating lead and zinc ores at Runcorn in Cheshire.

Nothing was heard of any methods of treatment tried in America. A process devised by Krupps, smelting to a matte and then leaching with sulphuric acid, gave an extraction of 78 per cent of the platinum metals, the Rand Mines

414

developed a chlorination process that gave good results on a small scale but later proved to be impracticable on larger scale trials (4), while the Chemical and Metallurgical Corporation also relied upon dissolution in hydrochloric acid and chlorine followed by precipitation of the platinum metals with zinc (5), but by early 1930 they had decided to abandon the project (6).

The only workable process remained that devised by Powell and Deering, and Johnson Matthey secured the appointment of refiners of the whole of the output of the only two remaining mining companies, all of the smaller concerns having failed or abandoned their operations. Those still active were more substantial concerns, Waterval Platinum, controlled by Consolidated Gold Fields, and Eerstegeluk Platinum, owned by Potgietersrust Platinum and controlled by Johannesburg Consolidated Investment Trust, both operating adjacent properties on a small scale at Rustenburg with small gravity concentration and flotation plants. In 1928 Coussmaker again set out for South Africa and visited these two companies. From this and his earlier survey he came to the firm conclusion that this section of the reef was the most promising and could well become an important source of platinum. He discussed the position fully with the two companies and stressed the great advantages to be derived from their

Alan Richard Powell
1894–1975

Joining Johnson Matthey in 1918 to establish a research department, Powell remained to manage a growing activity for thirty-six years. In 1926 he initiated work on a process for the extraction of the platinum metals from the newly discovered and complex source of mineral in South Africa and with his colleagues successfully developed a method of treatment that led to Johnson Matthey becoming the sole refiners to Rustenburg Platinum Mines. His many contributions in the refining, purification and analysis of the platinum metals were recognised in 1953 by his election as a Fellow of the Royal Society

415

Ernest Charles Deering

Educated at King's College, London, and joining Johnson Matthey in 1918, Deering was sent out to South Africa in 1925 to set up a laboratory in Johannesburg for the analyses of samples from the many platinum properties that were being established. Returning to London a year later he joined Powell in the Research Department in developing the process for extracting the platinum metals, and also the nickel and copper, from Rustenburg mineral and then put the process into production in a smelting works he designed and built at Brimsdown on the outskirts of London. He remained in charge of this plant until his appointment as a director in 1949, becoming Chairman in 1964 until his retirement in 1966

amalgamation. This rather difficult proposition was made a little easier as both Consolidated Gold Fields and Johannesburg Consolidated Investment Trust had been substantial shareholders in Johnson Mathey since 1918 when, on John Sellon's death, his holdings had been offered to them in equal proportions, while for many years, until the building of the Rand Refinery in 1922, the gold from both companies' mines had been refined by Johnson Matthey.

Negotiations went ahead for some time, with the co-operation of Lord Brabourne (1863–1933) a director of both Consolidated Gold Fields and Johnson Matthey, Dr. James Gunson Lawn (1868–1952), a director and the consulting engineer to Johannesburg Consolidated, and John Alexander Agnew (1872–1939), the distinguished mining engineer who was then the chairman of Waterval Platinum and also a director of Consolidated Gold Fields. Initially co-operation was confined to joint investigation of the problems of extraction by the consulting metallurgists, and when the Third Empire Mining and Metallurgical Congress met in Johannesburg in April 1930 a paper on "The Metallurgy of Transvaal Platinum Ores" was presented by F. Wartenweiler of Johannesburg Consolidated and A. King of Consolidated Gold Fields. This recorded their early work on flotation but they emphasised that "the extraction of platinum to

Operations at Rustenburg Platinum Mines include the haulage of the mineral by trains of 15-ton trucks to the main hoisting shafts. To produce one ounce of platinum ten tons of ore have to be brought to the surface, crushed, milled and treated by flotation and smelting processes before refining can begin

marketable form presented most difficult metallurgical problems and needed a vast amount of research and experimental work" (7).

Finally in 1931 the merger was achieved with the formation of Rustenburg Platinum Mines, the shares being held as to 52.5 per cent by Potgietersrust Platinum and 47.5 per cent by the Waterval Platinum Mining Company, with Johnson Matthey continuing to be responsible for the refining and marketing of all the platinum metals produced. Coussmaker was on the way to seeing his great resolve come to fruition.

Unfortunately the Great Depression was now causing growing problems in finding a market for the increasing output from Rustenburg and operations there were reluctantly discontinued in April 1932 while serious thought was given by the parent companies to their complete withdrawal from the mining of platinum in South Africa. Coussmaker's faith in the future and in the development of new industrial uses, as well as in the great potential of Rustenburg as a producer, remained unshaken and again he urged the desirability of keeping the mines open on a maintenance basis, prepared to resume operations when a sufficiently attractive market seemed assured (8). This policy was accepted, and by August 1933 a courageous decision was made to re-open the Rustenburg mine.

417

After the development of a process for the extraction of the platinum metals from the Rustenburg Platinum Mines this smelting plant was built by Johnson Matthey at Brimsdown near London. Here the matte produced from the flotation concentrates was smelted to yield crude nickel and copper which were then refined electrolytically, leaving the platinum metals in the anode residues to be separated and refined in the wet process refinery

For the first time platinum and its allied metals were being mined as primary products, with nickel and copper as by-products, by contrast with the mines in Canada and Siberia, essentially sources of nickel and copper with the platinum metals as by-products. This difference was to become of great importance in enabling Rustenburg to respond more quickly and effectively to increasing demand and to become the world's largest producer of platinum.

Meanwhile progress was being made in the extraction and refining of the platinum metals, but as the depth of the mine increased it was found that the ore was gradually changing from a weathered and oxidised form to an unaltered sulphide and that the Powell-Deering process needed to be modified accordingly. In 1929 a smelting plant had been designed and built by Deering at Brimsdown to the north of London, capable of expansion to meet increasing demand, and here further experimental work had to be carried out. Many of the operations were novel and new problems and difficulties were frequently encountered but successfully overcome.

The early years following the re-opening of the mine in 1933 were not without their difficulties. Production gradually increased, but demand was small and in the year 1938 the through-put of platinum in the Johnson Matthey refineries

In the earlier years the flotation concentrates were smelted to a matte in a series of blast furnaces, but later submerged arc electric furnaces were installed at Rustenburg to cope with increasing demand. Each has six four-feet diameter consumable electrodes in line

amounted to only 25,000 ounces. During the war this reached 40,000 ounces a year, but thereafter continual expansion began in order to meet the steadily increasing demand, this stemming largely from the adoption of the melting of optical glass in platinum equipment, from the increasing output of fibre glass and from the extensive use of platinum thermocouples in the steel industry. As the scale of operations increased it was found desirable to carry out certain of the earlier operations in South Africa and a smelting plant was erected at Rustenburg to deal with the flotation concentrates and to yield a matte suitable for treatment at Brimsdown. Later, to cope with the increasing output from the mines, arrangements were made to duplicate the primary stage of the Brimsdown process and a company known as Matte Smelters was formed in South Africa as a joint subsidiary of Rustenburg and Johnson Matthey.

In 1948 the Union Platinum Mining Company, also operating on the Merensky Reef some 60 miles further north, was acquired by Rustenburg and its property now constitutes the Union Section of Rustenburg Platinum Mines. Two years later the quantities of matte being received at the Johnson Matthey refinery at Brimsdown had increased to the extent that a completely new plant was added there. A detailed account of the complex and lengthy cycle of

419

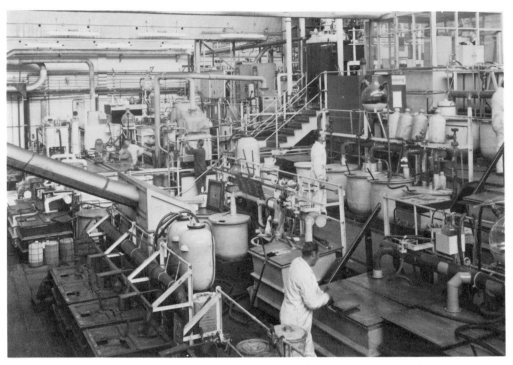

The final section of the Johnson Matthey platinum refinery, the largest of its kind in the world. This plant, built in 1956, replaced an older and smaller refinery and handled the precipitation, re-dissolving and re-precipitation of platinum and palladium, while the insoluble metals rhodium, ruthenium, iridium, and osmium were treated in another part of the refinery

operations required to effect the complete recovery of all six metals in a state of high purity was given by two of Deering's colleagues, A.F.S. Gouldsmith and B. Wilson (9).

Annual production had now reached 70,000 ounces, but then demand from the petroleum industry for the reforming of crude naphthas to both high octane petrol and a range of aromatics, described on page 398, began to take effect and by 1955 the annual production of platinum had risen to 200,000 ounces. A more extensive expansion programme at the mine was then set in train together with a major expansion of the Johnson Matthey refineries, and output was increased to 850,000 ounces, reaching a million ounces a year by 1973. Output of the other five platinum metals increased of course in the same proportion.

To anticipate this history a little, annual output of platinum from Rustenburg has now risen to approximately one-and-a-quarter million ounces, with more than 40,000 workers employed. This achievement has been brought about by a series of capital expenditure programmes aimed at matching supply with demand, by bringing new mining areas into operation, by the mechanisation of mining methods and by the development of improved methods of extraction. Rustenburg Platinum Mines is now the largest underground mining operation in the world, extending over thirty-three miles in length, and constituting the world's largest source of platinum and its allied metals.

A further development in 1972 was the formation of Matthey Rustenburg Refiners, jointly owned by Rustenburg Platinum Mines and Johnson Matthey, which now undertakes the whole of the extraction and refining operations in three plants. That at Rustenburg treats the matte and separates the base metals from the platinum metals and also refines the nickel and copper, leaving concentrates containing around 50 per cent of total platinum metals. The two plants at Royston in England and at Wadeville near Johannesburg then separate and refine the individual metals to the high degree of purity required for their many applications by means of a number of selective precipitation techniques from solutions of the mixed metals. During the past few years, however, research by Johnson Matthey and development work by Matthey Rustenburg Refiners have shown that improved recoveries may be obtained by a solvent extraction process and a new refinery is being built at Royston to operate this process. The new technique reduces both the number of refining stages and the time required, improves the yield and enables increased automation to be utilised in the refinery (10).

Thus together Rustenburg Platinum Mines and Johnson Matthey ensure to industry throughout the world adequate and continuing supplies of the platinum metals. The known reserves in the Merensky Reef run to over 300 million ounces of platinum alone, while below this is another reef containing as much platinum again.

References for Chapter 23

1   P.A. Wagner, Platinum Deposits and Mines of South Africa, London, 1929; reprinted Cape Town, 1973

2   C.A. Cousins, *Platinum Metals Rev.*, 1959, **3**, 94–99

3   A.R. Powell and E.C. Deering, British Patent 316,063 of 1928

4   R.A. Cooper and F.W. Watson, *J. Chem. Met. Min. Soc. South Africa*, 1929, **29**, 220–230

5   S.C. Smith, British Patent 289,220 of 1928

6   *Chem. Trade J.*, 1930, **86**, 357 and 405

7   F. Wartenweiler and A. King, *Proc. Third Empire Min. and Met. Congress,* 1930, Part III, 331–356

8   A.B. Coussmaker, Personal communication, 1960

9   A.F.S. Gouldsmith and B. Wilson, *Platinum Metals Rev.,* 1963, **7**, 136–143

10   M.J. Cleare, P. Charlesworth and D.J. Bryson, *J. Chem. Tech. Biotechnol.,* 1979, **29**, 210–214; P. Charlesworth, *Platinum Metals Rev.,* 1981, **25**, 106–112

24

The Story Continues . . .

*"Wollaston was a man of great imagination,
but I doubt if even he could have foreseen the
immense importance of the platinum metals in
modern industry"*

H.R.H. THE DUKE OF EDINBURGH, 1953

The story of platinum and its allied metals, their discovery, mining, extraction, chemistry, fabrication and uses, has now been carried along to around the
nineteen-fifties and to a stage when the literature on the subject has grown to
enormous proportions and is readily available to those interested. But the continuing technological progress in which these metals find an increasing number
of important applications requires a brief final chapter. The last thirty years
have in fact seen more new developments in the uses of these metals than
occurred during the whole of their earlier history and there is every indication
that research now being carried out will lead to further new industrial applications. It was in fact to ensure that information was readily available on the
researches being carried out on the properties of platinum and its allied metals
and on their further potential applications that Johnson Matthey, supported by
Rustenburg Platinum Mines, undertook in 1957 the publication of the quarterly
journal *Platinum Metals Review*. In this way a link is provided between researchers
and the many industrial scientists and technologists who have the responsibility
of finding practical solutions to the material problems of modern technology.

Part of the increasing volume of production has certainly been absorbed by
the growth of applications that were established some time ago. These include
electrical contacts for an enormous range of relays, sensitive switches and
measuring instruments in which a low and stable contact resistance must be
maintained, and either the carrying of a current without overheating or the
closing of the circuit after a long period of idleness – or both – are vital to their
operation and reliability. For some applications platinum has been used for
many years, but alloys containing iridium, ruthenium or nickel are more often
selected for their greater hardness. Palladium is used extensively in telephone
relays, but again alloys of palladium with either silver or copper are more frequently employed in many other types of relay and control equipment.

Great numbers of relays manufactured over the past fifty years and more have been fitted with platinum or palladium contacts to ensure their reliability. These, made by Bell Telephone Laboratories, have palladium contacts and have an expectation of life of at least two hundred million operations

Another well-established application that has grown considerably is in the measurement of high temperatures. The platinum resistance thermometer has found many more uses while the addition to the wire-wound form of devices based upon thick film technology has still further extended their range of usefulness from industrial control to domestic appliances, medical applications and food processing. The platinum:rhodium-platinum thermocouple, still very much in demand, is obviously not capable of measuring temperatures approaching the melting point of platinum, and to meet the need for higher determinations other combinations have been developed such as 20 per cent rhodium-platinum: 40 per cent rhodium-platinum for use up to almost 1900°C, and iridium: 40 per cent iridium-rhodium for measuring temperatures up to 2000°C.

Again, the world-wide increase in the need for nitrate fertilisers to enhance our food supplies, particularly in the developing countries, has led to an enormous increase in the use of rhodium-platinum catalyst gauzes for the oxidation of ammonia to nitric acid. At the present time some 60 million tons of acid are produced annually throughout the world by this process.

The glass manufacturing industry, which began to use platinum melting equipment in 1934 for the production of lamp bulbs, has also greatly increased its demand for platinum in the melting of improved optical glasses for spectacles, camera lenses and television tubes, while the manufacture of fibre glass for insulation and for a growing range of fibre-reinforced products is requiring increasing quantities of rhodium-platinum alloys for the bushings through which the glass is extruded at very high temperatures. A further development from this, of more recent origin, is the concept of "fibre optics", glass fibres to replace coper wires in telecommunication systems, their production depending upon the use of special high purity platinum equipment.

For fifty years or more optical glass has been melted in platinum crucibles in order to avoid contamination although more recently continuous melting in platinum equipment has been adopted

The electrodeposition of platinum, first developed by Alfred Smee working in the Bank of England in 1840 (page 211), developed relatively slowly, and though platinum plating found a number of applications it was eventually replaced by the advent of rhodium plating. Described first by Professor Colin Fink of Columbia University in 1933, the electrodeposition of rhodium offered considerably greater hardness, higher reflectivity and greater resistance to wear combined with reasonable ease of deposition. During World War II it played an important part as a sliding contact surface in the radio-frequency equipment then being developed so intensively. More recently the electrodeposition of palladium has established itself as a replacement for the more expensive gold deposits used in electronic equipment, while ruthenium plating is also receiving more consideration in the same field.

Platinum plating has returned, however, in a new form in recent years by the replacement of aqueous electrolytes by a bath of molten cyanides. By this process much thicker deposits of platinum having greater ductility and freedom from porosity are now being applied to some of the refractory metals such as molybdenum and tungsten for use at high temperatures. One such example is the use of platinum clad molybdenum wire in quartz-halogen lamps.

Growing Uses based upon Old Discoveries

But in recent years there has developed an increasing demand for the platinum metals for applications that did not formerly exist and were probably quite unforeseen thirty years ago, although in a few cases the scientific seeds had

424

An economical method of preventing the corrosion of ships' hulls and steel structures exposed to marine conditions is provided by cathodic protection systems in which a platinum-clad anode is connected to the steel and a small direct current is passed, ensuring freedom from attack. The nuclear-powered United States submarine "Seawolf" was one of the first vessels to be fitted with a platinum cathodic protection system in 1956 and since then many ocean-going vessels and a number of off-shore oil rigs have been similarly protected

been sown very many years before. One such example is the technique of cathodic protection of steel structures and sea-going vessels.

As long ago as 1824 Sir Humphry Davy was consulted by the British Admiralty who were concerned about

"the rapid decay of the copper sheeting of His Majesty's ships of war and the uncertainty of the time of its duration".

Davy proposed the attachment of a small piece of zinc to nullify electrochemical action on the copper sheathing, while he also investigated an impressed current system, but reliable batteries had not then been developed. It was not until 1956 that the United States Navy seriously began to experiment with platinum-clad titanium anodes for the protection of their ships and submarines. A great deal of development work had to be undertaken by those concerned in the design and construction of cathodic protection systems, and in more recent years niobium has replaced titanium as the substrate metal, but today this invaluable means of avoiding corrosion is an established technique on many types of ships and steel structures, including off-shore oil rigs – themselves undreamt of thirty years ago.

Much has been heard of the fuel cell as a source of power combining high

The high catalytic activity of platinum, together with its great resistance to oxidation, account for one of its major uses, in the form of woven rhodium-platinum alloy gauzes, in the oxidation of ammonia to nitric acid, a reaction taking place at around 850°C. This shows a new pack of gauzes being installed

Among the most severe conditions in which platinum is employed are those in the manufacture of fibre glass. Molten glass at temperatures up to 1400°C flows rapidly through a great number of precisely dimensioned orifices in a rhodium-platinum alloy trough or bushing, these orifices having to retain their exact size and alignment over a long period of use

The fuel cell, in which electricity is generated by the reaction between either hydrogen or a hydrocarbon with oxygen at electrodes coated with a catalyst, has been the subject of extensive development in recent years. Working in collaboration with United Technologies Corporation in the United States, Johnson Matthey developed platinum-based catalysts for this purpose and large 4.5 megawatt units are now being installed in both New York and Tokyo. This shows a 1 megawatt pilot plant that has been operating for some years, the six fuel cell stacks in the foreground containing the platinum catalyst

A complex platinum compound, *cis*-diamminedichloroplatinum II, first discovered as a potential means of treating some types of cancer by Professor Barnett Rosenberg, has received the approval of the government authorities of the United Kingdom and the United States for use in chemotherapy. It is now being manufactured by Johnson Matthey Incorporated in the United States in equipment specially designed for clinical conditions

thermal efficiency with very low risk of environmental pollution. After many difficulties and extensive research projects, commercial megawatt generating systems are now in the course of installation in the United States and Japan, these relying upon a coating of platinum dispersed on carbon applied to the electrodes to activate the electrochemical reaction that converts hydrogen, or a hydrocarbon, and oxygen into electrical energy. The origin of the fuel cell goes back, however, as far as 1842, when the London scientist, W. R. Grove, published a paper on "A Gaseous Voltaic Battery" in which he described the first practical fuel cell constructed from platinum foil coated with spongy platinum and with dilute sulphuric acid as the electrolyte (see page 210).

The great advances in microelectronics have also relied to some extent upon the platinum metals in the form of a range of metallising preparations based upon platinum and palladium, or their alloys with gold or silver, to provide thick conductive films on ceramic substrates, while ruthenium is used similarly for resistors. These compositions consist of the powdered metal or alloy mixed with finely ground particles of glass and suspended in an organic medium for application by screen printing on to selected areas and firing. They owe their original concept to the famous German chemist, Martin Klaproth who, working with the Berlin Porcelain works in 1788, first succeeded in decorating porcelain with such a preparation as described on pages 125 to 127. Naturally, rather more scientific methods are employed nowadays in their compounding and application.

An important application of palladium that also had its origins over a century ago is in the supply to industry of pure hydrogen. The selective diffusion of hydrogen through a palladium membrane was discovered by Thomas Graham in 1866, as described on page 266, but unfortunately it was later found that serious dimensional changes occurred when palladium is alternately heated and cooled in hydrogen. The discovery in the 1950s that an alloy of palladium with silver – in a narrow range of proportions – not only gave a higher rate of

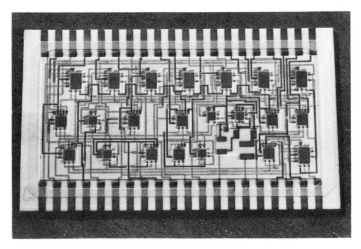

A range of metallising preparations including platinum and platinum-gold, platinum-silver, palladium-gold and palladium-silver alloys is employed for the production of thick film conductors in multi-layer hybrid circuits such as this, fabricated by E.R.A. Technology Ltd

428

transfer but remained dimensionally stable, made it possible to manufacture diffusion units for the production of high purity hydrogen for a number of industrial purposes, while in more recent years generators have been made available for yielding hydrogen by the cracking of a methanol-steam mixture followed by separation of the high purity hydrogen from the carbon monoxide through silver-palladium alloy membranes.

Modern Industrial Applications

Still further uses of the platinum metals stem from much more recent research and development.

One such major application of platinum lies in the field of pollution control or abatement. From the manufacture of nitric acid and the elimination of noxious tail gases with a platinum catalyst to many of the more everyday industrial processes such as plastics manufacture, printing, wire enamelling, paint drying and many others, platinum catalytic combustion systems can now eliminate many of the toxic gaseous waste products that were formerly a cause of unpleasantness, irritation, or even disease.

A more extensive use in recent years is in the control of automobile exhaust gases. Platinum metal catalysts systems have been successfully developed to eliminate from the atmosphere the three principal forms of noxious emissions from petrol engines, carbon monoxide, nitrogen oxides and hydrocarbons, while similar catalysts for the control of both gaseous and particulate emissions from diesel-engined vehicles are in the course of development.

Again in the field of automobile engineering, research being undertaken by Ricardo Consulting Engineers jointly with Johnson Matthey shows promise in the development of an engine with a platinum catalyst instead of spark ignition, this both reducing dangerous emissions and giving more economical running.

Improvements in the performance of gas turbine blades have also been the subject of a number of joint studies leading to the development of a platinum aluminide diffusion coating to increase their corrosion resistance and durability at high operating temperatures in aircraft engines. Further research by Johnson Matthey has shown that small additions of platinum group metals as alloying elements in new nickel-based superalloys can appreciably enhance their resistance to oxidation and corrosion while retaining their original mechanical properties.

In an entirely different field, the use of a complex compound known as *cis*-dichlorodiammineplatinum(II) in cancer therapy has resulted from outstanding research and development following the discovery by Professor Barnett Rosenberg in 1967 of this method of treatment. This original compound, known commercially as Cisplatin or Neoplatin, is now well established, but further research is leading to an improved range of anti-tumour drugs that are now undergoing clinical trials.

429

During the last few years effective and durable platinum catalyst systems have been introduced by several platinum manufacturing companies to meet the regulations covering the emission of exhaust gases from automobiles and well over 50 million cars have now been fitted with these devices. Further research by Johnson Matthey has yielded a lead-tolerant catalyst system that has shown its durability for 50,000 kilometres road usage. Here a Morris Marina is undergoing laboratory tests in which the exhaust emissions are being measured

The production of hydrogen for a number of industrial processes is now well established by a process based upon the catalytic cracking of a methanol-steam mixture and then the separation of high purity hydrogen by diffusion through silver-palladium alloy membranes. This typical installation, manufactured by Johnson Matthey, comprises the generator and the diffusion unit inside the red cylinder, with the control panels on the left

Some of the important economic advantages obtainable by the use of platinum group metal homogeneous catalysts are well illustrated by the process developed jointly by Union Carbide, Davy McKee and Johnson Matthey for the hydroformylation of propylene to butyraldehyde, required among other uses in the production of flexible polyvinyl chloride. This plant, one of a number already working, is operated by Chemische Werke Hüls in West Germany and has a designed capacity of 250,000 tons a year, employing a complex rhodium compound as the catalyst

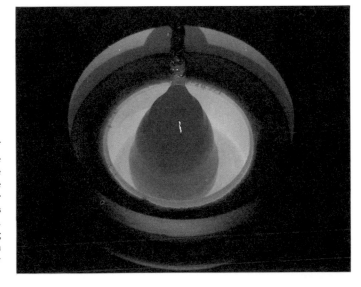

One modern application of iridium is as a crucible material for the growth of the large and perfect single crystals of semi-conductor materials required for lasers and other electronic devices. Here a single crystal is being pulled from an iridium crucible at approximately 1600°C

Heterogeneous and Homogeneous Catalysis

The great advances made in the employment of the platinum metals in heterogeneous catalysis have been described in Chapter 21. The many types of chemical products now produced by the reforming of crude petroleum, economically and in bulk, by such catalytic reactions include high-octane petrol and a range of compounds such as benzene, toluene and the xylenes, all the latter serving as vital intermediates for the manufacture of synthetic fibres, plastics, dyestuffs and other products required in modern ways of life. But alongside these reactions an essentially different form of catalysis has been developing more recently.

The chemistry of the organometallic compounds of all six platinum metals has been the subject of investigations by very many academic workers during the past ten years or so, the principal industrial consequence being the increasing use of homogeneous catalysis in which the catalyst is soluble in the reactant. This has advantages in terms of high activity per unit weight of metal, high selectivity and long life of the catalyst, with obvious savings in operational costs, while in some cases this is the only possible route to the desired product. Higher yields and operation at lower pressures and temperatures are also made possible by comparison with heterogeneous catalytic reactions. Processes recently introduced include the conversion of ethylene to vinyl acetate, of propylene to acetone, and of methanol to acetic acid, all employing palladium compounds as catalysts, while one other large scale process is the hydroformylation reaction first discovered by Professor Sir Geoffrey Wilkinson at Imperial College and then developed jointly by Union Carbide, Johnson Matthey and Davy McKee which converts propylene into aldehydes such as n-butyraldehyde using a soluble rhodium co-ordination compound. In this last process an output of some two million tons of the product, an important intermediate in the manufacture of polyvinyl chloride, involves the use of less than one ton of rhodium.

Metallurgical Developments

The metallurgy of the platinum metals and their alloys has, of course, also developed, with studies of their constitutional diagrams and the provision of new or improved alloys. For example, a series of platinum alloys having a wide range of resistivities combined with low temperature coefficients of resistance has been made available for use in precision potentiometers and transducers in industrial control equipment, while a cobalt-platinum alloy provides an exceptionally powerful permanent magnet that can be used in complex shapes or small sizes that would be impracticable with the conventional permanent magnet materials.

But the most outstanding achievement during recent years has been the successful production of dispersion strengthened platinum and certain of its alloys. These materials contain a uniform distribution of extremely fine refractory precipitate dispersed throughout the mass, and the mechanically worked material develops a highly fibrous recrystallation structure on annealing

432

and is unusually stable. In this condition it is many times more resistant to creep failure at elevated temperatures and thus when under stress at 1400°C dispersion strengthened pure platinum can endure for at least twice as long as an alloy of platinum with 40 per cent rhodium, previously regarded as the strongest commercially available high-temperature alloy. Such materials can be used with advantage for the construction of equipment required to operate in air at very high temperatures; for example, dispersion strengthened rhodium-platinum and platinum have both found widespread use in the production of optical and fibre glass where their high strength and resistance to contamination have resulted in considerable process improvements.

Thus the story of platinum and its allied metals continues, with new applications continually developing based upon their remarkably useful combinations of properties. From their past history it is evident that still further uses will be found in the course of time and that they will continue to make significant contributions to the material needs of our increasingly complex and rapidly changing world.

Name Index

Page numbers in bold type indicate portraits

437

Subject Index

448